D1420609

ULTRA HIGH DILUTION
PHYSIOLOGY AND PHYSICS

Ultra High Dilution
Physiology and Physics

Edited by

P.C. ENDLER
Ludwig Boltzmann Institut für Homoeopathie,
Graz, Austria

and

J. SCHULTE
National Superconducting Laboratory,
Michigan State University, U.S.A.

With contributions by:

Georgos Anagnostatos, David Auerbach, Madeleine Bastide, Alexander Berezin, Jacques Benveniste, Jean Cambar, Massimo Citro, Peter Fisher, Emilio del Giudice, Christian Endler, Max Haidvogl, Franz Moser, Menachem Oberbaum, Fritz Popp, Bernard Poitevin, Waltraud Pongratz, Marco Righetti, Jürgen Schulte, Cyril Smith, Harald Walach, Roeland van Wijk and colleagues

KLUWER ACADEMIC PUBLISHERS
DORDRECHT / BOSTON / LONDON

A C.I.P. Catalogue record for this book is available from the Library of Congress.

ISBN 0-7923-2676-8

Published by Kluwer Academic Publishers,
P.O. Box 17, 3300 AA Dordrecht, The Netherlands.

Kluwer Academic Publishers incorporates
the publishing programmes of
D. Reidel, Martinus Nijhoff, Dr W. Junk and MTP Press.

Sold and distributed in the U.S.A. and Canada
by Kluwer Academic Publishers,
101 Philip Drive, Norwell, MA 02061, U.S.A.

In all other countries, sold and distributed
by Kluwer Academic Publishers Group,
P.O. Box 322, 3300 AH Dordrecht, The Netherlands.

Arranged and printed with the help of
Bundesministerium für Wissenschaft und Forschung, Vienna, Austria
Land Steiermark, Austria
Stadt Graz, Austria
Ludwig Boltzmann-Institut für Homoeopathie, Austria
Realised in the perspective of the
"Homeomet" European Union Research Programme.

Printed on acid-free paper

Typesetting: pro graphia H. Zahradnik/C.E./W.P, Graz, Austria

02-0496-70 ts
Reprinted 1996

Printed in the Netherlands

CONTENTS

PREFATORY WORD

The scientific investigation of ultra high dilutions has been triggered by recognition of the therapeutic efficacy of homoeopathy. Strictly speaking, homoeopathy refers to the therapeutic use of substances to treat syndromes in patients similar to those which they can cause in healthy probands. But homoeopathic medicines are often used in extremely high dilutions - including ultra high (or ultramolecular) dilutions at which, according to Avogadro's Law, it is extremely unlikely that any molecule of the starting substance persists.

Its use of ultra high dilutions appeared, and to many scientists still appears, to make homoeopathy a scientific absurdity. But this is a two-edged sword: if, despite their apparent absurdity, ultra high dilutions do indeed have effects, the apparent absurdity reflects a deficiency in scientific understanding, not homoeopathy.

This book marks a significant stage in a remarkable and tortuous scientific metamorphosis. The roots of homoeopathy are empirical: the practice of 'proving' potential medicines on healthy volunteers is fundamental to homoeopathy and was the first systematic study of drug action in the history of medicine. The early homoeopaths were denounced by their medical contemporaries for 'empiricism'. The concepts of homoeopathy, particularly ultra high dilutions, did not conform to the mechanistic science of the late 19th and early 20th centuries. Avogadro proposed his law in 1811, but it was not generally accepted until 1848. From this time on, the metaphysical school of homoeopathy gradually became dominant, and homoeopathy withdrew from the scientific arena.

The increasingly metaphysical tone in turn has led to the marginalisation and decline of homoeopathy, a decline which seemed terminal by the third quarter of the 20th century. But the last 15 years have seen a remarkable reversal of fortune. Driven by public demand more interested in practical results than scientific theory and by growing evidence from controlled clinical trials, homoeopathy has staged a strong resurgence. But the 'empirical' label still sticks: we now have good clinical evidence that ultra high dilutions have therapeutic effects, but we do not know how or why.

This is the first multi-disciplinary work to address the how and why of the actions of ultra high dilutions. It is clear that information storage in ultra high dilutions will be central to any understanding of their action. Part 1 of the book (Physiology) reports a number of methods with the potential to become a standard bioassay. Part 2 of the book (Physics) focuses on theoretical models of information storage. Part 3 is dedicated to biophysical joint-ventures and Part 4 to the historical and medical background of ultra-high dilution research. Most beneficial to the science of ultra-high dilutions would be an assay which can be readily reproduced in any suitably equipped laboratory, an assay that also allows experimental investigation of the physical hypotheses. This book points in this direction.

The investigation of ultra-high dilutions is one of the most exciting and dynamic growing edges of science. It is generating genuinely new scientific concepts, and may have consequences extending far beyond medicine and biology. Ultra High Dilution - Physiology and Physics makes an important contribution to this nascent science, bringing together many of the most important workers in the field. It is a fertile source of references and concepts, and deserves wide attention in the scientific community.

Peter Fisher

Royal London Homoeopathic Hospital
May 1993

PREFACE

The idea of editing this book was born in the winter of 1988/ 1989. Christian Endler was organizing the workshop 'Wasser und Information' (water and information) in Austria [1], and Jürgen Schulte was working on a publication of his results on atomic cluster stabilities and long-range electromagnetic interaction in atomic clusters. It was Franz Moser from the Technical University of Graz who brought these two together. After a talk that Moser had given in Bremen, Schulte explained to him his ideas about clusters and long range interaction, and his concern about reliable theories and experiments in research on ultra high dilutions (UHD) and homoeopathy. He was suggested to be a speaker at the Austrian workshop. Reviewing the contributions of this workshop and the current literature on UHD and homoeopathy, especially the PhD thesis by Giesela King [2] and the excellent survey by Marco Righetti [3], we decided to work on a book in order to critically encourage more scientists to work and publish in this field with a high scientific standard. What we had in mind was a useful contribution to the goal to lift research on UHD and homoeopathy to an internationally acceptable scientific standard, to encourage international scientists to work in this area and to establish UHD and homoeopathy in academic science.

Delayed by our individual academic careers in our specific fields, and delayed by lack of funds it took us about four years to finish this book. This was, of course, a time of development of research in the field, and our publication project, mostly focusing on the actual situation and including our own research on a non-violent bio-assay (model on development of amphibia [4]) seemed increasingly worthwile. After looking at the final version of this book, we must say that we have not reached our ideal goal that we aimed at in 1989. We wish to emphasize that comments from the readers are highly appreciated by the editors. However, we think we came close to the goal we aimed at, and we hope that the reader will enjoy this book and be stimulated and encouraged to review her or his own research, and will recommend this book to friends and colleagues. The book can be taken as a reference for research on UHD. Although it is not meant to be a complete review, it may well have a stimulating effect in the adventure of research, as it marks representative corner-stones on the way of our understanding of UHDs in terms of physiology and physics, including electromagnetic interactions, and it might stimulate research in a variety of other fields in science.

Our research and new ideas presented in this book would not have been possible without previous work in more conventional fields of science [5], those who brought the issues of UHDs and homoeopathy to our interest [6], recent valuable publications [7], financial support [8] and the support of those who encouraged us to venture on the work of bringing physiology and physics closer [9]; we would like to thank our colleagues [10], friends and families for their support, including some wonderful piano music, and for keeping us on the track. We are especially grateful to the contributing authors, the lecturers [11] and the publisher for their excellent collaboration and patience.

For the final finishing we are grateful to the valuable comments of the referees Robert G. Jahn (Aerospace Science, Princeton Engineering Anomalies Research, USA), Bryan D. Josephson (Cavendish Laboratory, Cambridge, UK), Madeleine Bastide (Fac. de Pharmacie,Univ. de Montpellier, France), further independent referees who have reviewed the manuscript to this book, as well as to all the people who cannot be mentioned here because of the lack of space.

This book is dedicated to our readers and teachers, our parents
and specially to Ernst and Yvonne Clar.

Graz (Styria, Austria) and East Lansing (Michigan, USA) 1993

Christian Endler   Jürgen Schulte

ANNOTATIONS AND REFERENCES

1. This workshop has been organized under the auspices of the I.S.M.F. and the Physiological Institute of the University of Graz, and has been published in a book distributed by Haug Verlag, Heidelberg.

2. King G. Experimental Investigations for the Purpose of Scientific Proving of the Efficacy of Homoeopathic Preparations – A Literature Review about Publications from English-speaking Countries. Thesis, Tierärztliche Hochschule Hannover, FRG, 1988.

3. Righetti M. Forschung in der Homöopathie - Wissenschaftliche Grundlagen, Problematik und Ergebnisse. Göttingen 1988: Burgdorf.

4. Endler, Pongratz et al. (Conference American Association Advancement Science, Boston 1993; J. Veterinary & Human Toxicology, in press).

5. Both research on longevity (Endler) as well as on applied system theory (Schulte) revealed para-doxical, non-linear effects of certain stimuli. Also, the application of low doses or high dilutions of substances to living systems provides many examples for such paradoxical effects.

6. R.F. Lexner; P. Andersch; F. Dellmour (Homöopathie, Facultas Universitätsverlag, Wien); M. Dorcsi (Handbuch der Homöopathie, Orac Verlag, München); R. Livingstone (Homoeopathy, Evergreen Medicine. Poole: Asher Asher Press); R. Schwarz; G. Vithoulkas (Medizin der Zukunft, Homöopathie. Kassel: G. Wenderoth Verlag).

7. Albrecht H, Franz G (eds.). Naturheilverfahren, Zum Stand der Forschung. Berlin: Springer 1990.
   Bastide M. Signal and Images. Paris: Alpha Bleue 1991.
   Bellavite P, Signorini A. Fondamenti teorici e sperimentali della medicina omeopatica. Palermo: Nuova Ipsa Editore 1992.
   Bischof M, Rohner F. In: ZDN und FFB (eds.): Dokumentation der besonderen Therapie-richtungen und natürlichen Heilweisen in Europa. Essen: Verlag für Ganzheitsmedizin 1992.
   Doutremepuich C. Ultra Low Doses. London: Taylor and Francis 1991.
   Harisch G, Kretschmer M. Jenseits vom Milligramm. Berlin: Springer 1990.
   Kokoschinegg P. Wasserstrukturen. In (1).
   Leopold H. Masse, Energie, Information. In (1).
   Majerus M. Kritische Begutachtung der wissenschaftlichen Beweisführung in der homöopathischen Grundlagenforschung (francophone literature).Thesis, Tierärztliche Hochschule Hannover, FRG, 1990.
   Melchart D, Wagner H. Naturheilverfahren. Grundlagen einer autoregulativen Medizin. Stuttgart: Schattauer 1993.
   Popp FA (ed.). Electromagnetic Bio-Information. Vienna: Urban & Schwarzenberg 1989
   Projektträgerschaft Forschung im Dienste der Gesundheit. Unkonventionelle Medizinische Richtungen. Bestandsaufnahme zur Forschungssituation. Bremerhaven: Wissenschaftsverlag NW 1992.
   Resch G, Gutmann V. Wissenschaftliche Grundlagen der Homöopathie (Scientific Bases of Homoeopathy). Berg /Starnberger See: O Verlag 1986 (1987).
   Trincher K. Die Anomalien des Wassers und der Temperaturbereich des Lebens. In (1).
   Weingärtner O. Homöopathische Potenzen. Berlin: Springer 1992.
   ZDN (ed.). Homeopathy in Focus. Essen: Verlag für Ganzheitsmedizin 1990.

8. University of Graz, Ludwig Boltzmann Society, Steiermärkische Landesregierung, HomInt.

9. M. Haidvogl, G. Karmapa, D. Lama and others.

10. V. Antonchenko, P. Bellavite, Z. Bentwich, J. Benveniste, C. Bornoroni, J. Cambar, R.W. Davey, A. Delinick, F. Dellmour, M. Dicke, P. Dorfman, M. Dorsci, C. Doutremepuich, G. Fachbach, K. Flyborg, J. Gnaiger, S. Grievetz, F. Gross, G. Gutenberger, K. Hagmüller, M. Haidvogl, G. Heinrich, H. Heran, R. Hill, M. Hoffmann, W. Hohenau, J. Hornung, H. Hubacek, Th. Kartnig, G. Kastberger, Th. Kenner, C. Kern, V. Klima, P. König, K. Kratky, W. Kropp, E. Lehner, R. Lexner, K. Linde, W. Ludwig, M. Moser, F. Muhry, N. Mystafaef, S. Novic, M. Oberbaum, B. Paletta, K. Peithner, W. Pongratz, F.A. Popp, R. and M. Reilly -Taylor, G. Resch, B. Rubik, A. Scott-Morley, F. Senekowitsch, A. Stacher, C.W. Smith, E. Stabenteiner, D. Ullman, F. Varga, R. van Wijk, G. Vithoulkas, H. Wagner, F.A.C. Wiegant, F. Wieland, H. Zahradnik and many others.

11. E. Lamonte, G.Krenn and the senior editor's sister M. Koch-Endler. The senior editor wants to point out that this book has been written in a language which is not his mother tongue. This is why he asks the English-speaking readers not to be too strict with their judgement.

CONTRIBUTORS

THE AUTHORS

Anagnostatos G.S.
Institute of Nuclear Physics
National Center for Scientific Research
Aghia Paraskevi
GR - 15310 Attiki

Andersch-Hartner Peter
Burenstraße 49
8020 Graz

Auerbach David
Max-Planck-Institute of Fluid Dynamics
Bunsenstr. 10
D - 3400 Göttingen

Benveniste Jacques
INSERM U200
Université Paris-Sud
32 Rue.des Carnets
F - 92140 Clamart

Berezin A.A.
Department of Engineering Physics
McMaster University
CAN -L8S 4M1 Ontario, Hamilton

Cambar Jean
Groupe d'Etude de Physiologie et
Physiopathologie Renale
Faculté de Pharmacie
Université de Bordeaux
F - Bordeaux

Citro Massimo
I.R.M.M./Via Cibrario 33
I - 10143 Torino

Del Giudice Emilio
Department of Nuclear Physics INFN
University of Milano
Via Celoria 16
I - 20133 Milano

Haidvogl Max
Ludwig Boltzmann Institut
für Homöopathie
Dürerg. 4
A - 8010 Graz

Endler P.C.
Ludwig Boltzmann Institut
für Homöopathie
Dürerg. 4
A - 8010 Graz

Gehrer Michael
Forschungsstelle für
niederenergetische Bio-Information
Wittenbauerstr. 137
A - 8042 Graz

Hilgers Helge
Zoologisches Institut
der Universität Wien
Althanstr. 14
1010 Wien

Moser Franz
Institut für Verfahrenstechnik
Technische Universität Graz
Inffeldg. 25
A - 8010 Graz

Poitevin Bernard
Laboratoires Boiron
20, rue de la Libération
F - 69110 Ste. Foy-les-Lyon

Oberbaum Menachem
Ruth Ben Ari Institute for Clinical
Immunology
Kaplan Hospital
Israel - Rehovot

Pongratz Waltraud
Ludwig Boltzmann Institut für
Homöopathie
Dürerg. 4
A - 8010 Graz

Popp F.-A.
International Institute of Biophysics
Technology-Center
Opelstr.10
D - 67661 Kaiserslautern 25

Righetti Marco
Sonntagsteig 3
CH - 8006 Zürich

Schulte Jürgen
Department of Theoretical Physics
University of Michigan
East Lansing
USA - MI - 48824 1321

Scott-Morley A.J.
Poole Acupuncture Clinic
103 North Road, Parkstone
GB - BH 14 0LT Poole, Dorset

Smith C.W.
Department of Electric and Electronic
Engineering
University of Salford
GB -827221 Salford

Waltl Karl
Ludwig Boltzmann Institut für
Homöopathie
Dürerg. 4
A - 8010 Graz

Van Wijk Roeland
Department of Molecular Cell Biology
University of Utrecht
Padualaan 8
NL - 3584 CH Utrecht

Wiegant F.A.C.
Department of Molecular Cell Biology
University of Utrecht
Padualaan 8
NL - 3584 CH Utrecht

THE EDITORS

Endler P.C., affiliated to the Ludwig Boltzmann Institut für Homöopathie, Graz, Austria; senior scientist at the Research Site for Low Energy Bio-Information, Graz, Austria; guest lecturer at the Centers for Complementary Health Science of the University of Exeter, Great Britain, and the University of Urbino, Italy; member of the Groupe Int. Rech. Doses Infinitesimales and the European Committee Res. Homoeopathy.

Schulte Jürgen, National Superconducting Cyclotron Laboratory, Theory, Michigan State University, USA; affiliated to the Research Site for Low Energy Bio-Information, Graz, Austria.

THE AUTHOR OF THE PREFATORY WORD

Fisher Peter, Royal London Homoeopathic Hospital, The Faculty of Homoeopathy. Editor of the British Homoeopathic Journal, London.

The final versions of all CONTRIBUTIONS have been accepted in 1993.

INTRODUCTION

Research on ultra-high dilutions, and the interaction of ultra-high dilutions and living systems, has reached a level of quality and popularity that it is about to be taken seriously by current orthodox sciences. By the term ultra-high dilutions (UHDs), within the context of this book, we mean standardized aqueous or aqueous-alcoholic solutions where a substance has been diluted through a special dilution process in such a way that the concentration ratio of solute to solvent becomes of the order of Avogadro's number, or even far below it. The characteristics and effects of UHDs seem to contradict the principles of pharmacology. Thus, a considerable number of researchers have become interested in this subject. At sufficiently low concentration no pharmacological effect of the original substance is expected. Therefore, researchers came up with the idea that the substance might transfer information to the solution, and through an appropriate process the information might be stored or imprinted. This new idea caused a breakthrough in research on UHDs, and has extended the pharmacological concepts. The idea of information transfer to the solution by means of the supply of kinetic energy through a process of subsequent dilution has been adopted from the so-called method of "homoeopathy" known in medical therapy. Today, homoeopathy, as a science, may be placed somewhere between the sciences of biochemistry, biophysics, and physiology.

The objective of research on UHDs is to investigate whether characteristics and effects of such dilutions can be quantified. It is especially interesting to study the possible mechanism of the information transfer from the original substance to the solution - i.e. of the interaction between the molecular mother substance and the solvent - (A), the mechanism of long term information storage in UHDs - i.e. of molecule-specific information in the solvent - (B), the physiological basis of the sensitivity of living systems towards UHDs (C) and the mechanism of the interaction of the UHD and the organism, including the effects in the living organism caused by this interaction (D).

Although the experimental basis can still be questioned in most cases, researchers can start from remarkable fundamental observations in order to explain the topics (A) - (D). For some topics in question, it seems to be evident that the information transferred is stored in the specific properties of the dilution substance by means of long-range electromagnetic fields. Also, the interaction of the UHD and the organism seems to be based on the interaction of long-range electromagnetic fields. Some recent experiments and theories supporting this assumption will be presented and discussed in this book.

The special composition of this book has been formed by the intensive and close collaboration of a biologist (P.C. Endler, experimental biology) and a physicist (J. Schulte, theoretical physics), and the extreme interdisciplinarity of this subject. We prefer to introduce the reader into the principles of ultra-high dilutions and homoeopathy by using data collected in the biological laboratory as well as by presenting elaborated theories.

Part 1 (Physiology), includes an initial chapter on dose-dependent effects in low and very low doses and then emphasizes some exemplary promising recent physiological experiments using UHDs in detail. In Part 2 (Physics) we present a critical dictionary of experiments in physics, followed by some physical theories and models on UHD physics, which are currently most elaborated and fundamental. Furthermore, in Part 3 biophysical contributions and papers on preliminary interdisciplinary joint ventures are presented. Part 4 brings in contributions on the historical and medical context of UHD research. The volume further includes a prospect on Preliminary Elements of a Theory on UHDs and a general SUMMARY (p. 263f).

This book has been written to encourage high standard research on complementary medicine, as well as to serve as additional standard reference on current most promising experiments and theories in research on UHDs and homoeopathy, respectively, and to encourage interdisciplinary research. A.o., we have tried to show that guidepost research on living systems can also well be performed on non-invasive, non-violent biological models.

Great attention has been paid to keeping the scientific contributions of experiments and theories as understandable as possible, thus the authors have been asked to be very specific on important parts of their contributions. Summaries of the contributions as well as titles with regard to the whole book have been introduced by the editors in some cases. We hope that our didactic approach is appreciated by the advanced reader, too.

The topics covered here are not solely contributing and interesting to the research on UHDs (homoeopathy) and related therapy methods using structured solutions or electromagnetic fields, respectively, but also to a variety of different sciences such as ultra low dose and contamination science. Research on UHDs is about to become a model for interdisciplinary research. Also, it might serve as an example of related physical and physiological processes.

"Dear colleague, can you believe these strange results?"

PART 1: PHYSIOLOGY

"In complex dynamical systems, nonlinearities and feedback loops often lead to a surprising behaviour. The short-term effect of a stimulus may be reversed by the feedback loops. Thus, the relation between 'cause' and 'effect' is much more complicated than usually imagined. In the first place, it is important to get an appreciable effect altogether, which is the case if there is some kind of resonance between the stimulus and the system. It depends on details of the system (and the interaction) whether it reacts in the expected manner or just the other way round."

K. Kratky, Meeting University Urbino, 1992

HORMESIS: DOSE-DEPENDENT REVERSE EFFECTS OF LOW AND VERY LOW DOSES

M. Oberbaum and J. Cambar

SUMMARY

In various fields of biology, dose-dependent reverse effects of low and very low doses of substances have been described, and the concept of 'hormesis' has been established. In a simple form, it says that in many cases a stimulatory effect occurs in biological systems after exposure to a low concentration of an otherwise toxic agent. In this contribution, respective examples are given.

ZUSAMMENFASSUNG

In verschiedenen Gebieten der Biologie wurden dosisabhängige Umkehreffekte niedriger und sehr niedriger Substanzdosen beschrieben, und es wurde der Begriff "Hormesis" geprägt. Einfach ausgedrückt besagt dieser, daß in vielen Fällen in biologischen Systemen nach Einwirkung einer niedrigen Konzentration einer ansonsten toxischen Substanz eine stimulierende Wirkung auftritt. Im vorliegenden Beitrag werden entsprechende Beispiele diskutiert.

INTRODUCTION

[By the Law of Similarity, we mean that, when a substance in a high dose is able to induce defined symptoms in a healthy living system, it is also able under certain circumstances to cure these symptoms when applied in a low or very low dose, see Glossary.] Even the most conservative scientist cannot criticize the "Law of Similarity" in homoeopathy. With a little bit of good will, it is possible to find the use of this law in conventional medicine, even if not called by name.

For example, mercury salts, as part of their toxicologic picture, cause oliguria and anuria; many of the classical diuretic drugs are mercury derivatives (e.g. mersalyl and chlormerodrin). The same holds true for the diuretics of the sulfonamide family (chlorthalidone, mefrusid, and chlorthiazide). Another example is digitalis, which in toxic quantities causes a picture of arrhythmic tachycardia; one of its main indications is the treatment of atrial fibrillation.

The "Law of Similarity" can also be seen in conventional pharmacology in the example of emetin. At a high dosage it causes a picture of diarrhea and ulcers of the colon, resembling amebiasis; in conventional medicine it is used as a drug against amebiasis (1).

There is also a field in conventional science, far away at the fringe of it, which uses a certain aspect of the "Law of Similarity". We refer to HORMESIS. Even though it is a part of conventional science, it has earned ridicule, scorn and suspicion: Schulz, one of the founders of this field has been labelled "The Greifswald homoeopath" (3), and that was not meant as a compliment. Even though many papers showing hormetic effects have been published in the last century, the category "hormesis" has been included in the Biological Abstracts only in the last decade. Other retrieval systems such as Index Medicus, Current Contents, Excerpta Medica and others, do not include either this term or its synonyms (Arndt- Schulz law, Hueppe's Rule, Hormoligosis).

The definition of hormesis has been slightly altered in the approximately 100 years of its existence, especially in the last decades since our understanding of the meaning of this strange phenomenon has been mainly extended concerning longevity hormesis.

5

P.C. Endler and J. Schulte (eds.), Ultra High Dilution, 5–18.
© 1994 Kluwer Academic Publishers. Printed in the Netherlands.

Hormesis is, according to Neafsey (2), "a term which has been applied to a variety of stimulatory responses to low doses of otherwise toxic substances which improve health and enhance longevity. Improved fecundity, hatching, viability, growth, weight gain, development, reproductive life span and wound healing in exposed animals" are regarded as hormetic effects, as are "P-450 enzyme induction, decreased tumor incidence, increased resistance to infection and disease, radiation resistance and longevity enhancement."

MAIN PART

The "birth" of the "Arndt-Schultz Law"

In 1877 Hugo Schulz published a paper in which he claimed that the effect of a stimulus on a living cell is indirect and proportional to its intensity or quantity (4). At about the same time, the psychiatrist Rudolf Arndt developed his "Basic Law of Biology" which states that "weak stimuli slightly accelerate the vital activity, middle strong stimuli raise it, strong ones suppress it and very strong ones halt it" (5).

After Schultz had met Arndt, it came to him like "a sudden enlightenment" that he had the experimental proof of Arndt's basic postulate. He spent the next several years on further proof of this law (6).

In 1888, Schulz published the results of an experiment that examined the activity of various kinds of yeast poison such as iodine, bromine, mercuric chloride, arsenious acid, chromic acid, salicylic acid and formic acid in subtoxic concentrations. He showed that almost all of these materials have a stimulatory effect on yeast metabolism when given in low concentrations. As an index he used CO_2 production. According to his observation, the yeast increased CO_2 production by 103-106 times (5).

On the basis of his own studies and those of Arndt, Schulz formulated what became later known as the "Arndt-Schulz Law": "Every stimulus on a living cell elicits an activity, which is inversely proportional to the intensity of the stimulus" (5).

Regarding the practical application to this law, Schulz opined that if the law were correct, "it must be possible to use a sufficiently small quantity of a correctly chosen, organ-specific drug in order to raise the depressed vital energies to an extent that the physiological norm will be restored completely or as much as possible" (4).

Schulz and Arndt both asserted that the stimulation-depression phenomenon was a fundamental characteristic of the excitability of 'protoplasm' (6).

A short time later, Hueppe described similar phenomena in bacteria (7).

At the beginning, the work of Arndt and Schulz earned reactions that ranged from total rejection to reservations concerning the generality of the law. A.J. Clark (cited from 6), for example, claimed that polyphasic responses to drugs are often seen in complex systems, and diphasic actions on tissues are not unusual. Since phasic phenomena are also observed in preparations of purified enzymes, there is no need to attribute "mysterious" properties to living tissues. Clark also showed that there are many cases where high and low doses of drugs produce a single graded response. For example, acetylcholine has an inhibitory effect on frog heart and a contractile effect on the fibers of the rectus abdominis. The positive effect of a low concentration of cyanide on organisms or isolated organs is attributed to the inactivation of inhibitory trace metals. Stimulation of growth can be explained by assuming that partial destruction of a population (e.g., yeast) provide food sources to the survivors.

In the 1920s there was a renaissance of interest in hormesis, and many scientists published papers in this field, including a repetition of Schulz's work. They examined the effects of many toxic substances given in very low concentrations (8-23), and from their findings one can conclude that hormesis is most probably a universal phenomenon. This is true even though most of the reported works of that time do not fulfill the requirements of modern science concerning sample size and rigorous statistics. Those studies examined a large range of toxic substances or elements in very low concentrations, including heavy metals (8), pesticides (9), antibiotics (10-12), trace elements (13), chlorinated hydrocarbons (14), and physical agents like heat, cold, oxygen, and X-rays (15-18). The experiments were performed on bacteria (10-13), yeast (19), fungi (20), protozoa (21), algae (22), plants (23), insects (17), fish (16), and mammals (15). Among the studies conducted in order to repeat the work of Schulz, that of Holz et al. (24) deserves special consideration. The scientists showed that high and low concentrations of "yeast poison" lead to the mobilization of glycogen from the yeast cell. Contrary to the high concentrations which suppress the metabolism of glucose, low concentrations do not affect the metabolism of glucose. According to Holz et al., the lack of suppressive activity on glucose metabolism of "yeast poison" at low concentrations is responsible for the "stimulation" of CO_2 consumption and therefore they cannot confirm Schulz's interpretation (cited from 25).

In 1943 Southam and Ehrlich (20) reported the growth-enhancing effect of the toxic extract of western red cedar (Thuja plicata) if given in low doses to plant pathogenic fungi. At a high concentration, the extract reveals a fungistatic or fungicidal activity. Southam and Ehrlich used concentrations that are 10-100 fold lower than the growth-suppressive concentrations. The authors explained the growth enhancing effect as an initial response followed by progressive desensitization to subinhibitory concentrations of a toxic constituent of the extract. They termed this phenomenon h o r m e s i s (from Greek: to enhance) and defined it as a "stimulatory effect of subinhibitory concentrations of any toxic substance on any organism."

In the 1950s Wilder (26) proposed a new formulation of the ArndtSchulz law. He claimed that, besides the intensity, "also the direction of a response of a body function to any agent depends to a large degree on the initial level of that function at the start of the experiment. The higher this 'initial level', the smaller is the response to function raising, the greater is the response to function-depressing agents. At more extreme initial levels there is a progressive tendency to 'no response' and to 'paradoxic reactions', i.e., a reversal of the usual direction of response." By the same token, the lower the "initial level" the greater the response to 'function raising' agents and the less the response to "function depressing" ones.

Wilder noted that his rule contradicts the common sense opinion in medicine: "It is necessary to emphasize that most investigators do not take the initial levels into account at all. If they do, they usually operate with the tacit assumption that the opposite of our law is true: e.g., that a hypertensive individual will respond to adrenalin with a higher rise in pressure than one that is normotensive."

Thus, while the Arndt- Schulz law was formulated in terms of pharmacological dose, Wilder expressed his rule in terms of the varying sensitivity of different organisms to a given dose level:

what is a "small dose" for one individual may be "medium" for another and a "large" dose for a third one, yielding different effects in each individual. In all cases, however, the phenomenon of interest is the "reversal of the usual direction of response", whether it is a function of change in dose size or of altered sensitivity of the organism (cited from 27).

In 1956, Luckey (28) published a paper reporting that germ-free birds grow more rapidly when minute amounts of antibiotics are added to their feed. Since the birds were germ-free, the enhancing influence of the antibiotics could not be attributed to changes in the bacterial flora. In Luckey's opinion, the antibiotics had a stimulatory effect on the birds. He called this phenomenon "hormoligosis" (hormo- enhance, oligo- small amount), that is, a substance which is stimulatory in small quantities, harmful in higher quantities (therapeutic), and toxic in high quantities. Townsend and Luckey (29) published a list of 100 substances known to be capable of causing an inhibition at high concentrations and stimulation at low concentration in 1960. As references for their work, they used three standard text books in pharmacology, namely, Goodman & Gilman, Sollman, and Drill. They suggested to add to their list all substances known to be stimulatory, since at sufficiently high concentration, they become toxic. Townsend and Luckey distinguished four basic forms of dose-response curves (Figure 1), namely:

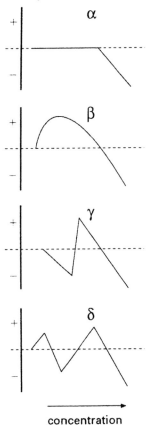

concentration

A curve: The best known pattern. It describes a situation of no effect of a toxic substance from low to subtoxic concentrations. Above a borderline concentration, it suppresses progressively. This curve describes the behaviour of substances or drugs that do not induce a stimulatory response. Substances described by this pattern do not appear in the list of Townsend and Luckey.

B curve: The most frequently observed pattern that describes the behaviour of most of the drugs examined by Townsend and Luckey. It shows one peak of a typical stimulation at subtoxic concentration, and progressive inhibition as the concentration reaches a toxic level. According to Stebbing (30), any substance for which there are precise and comprehensive data, has the B- curve pattern. As more concentrations are examined, the form of the curve becomes similar to that of the B curve (38).

C and D curves: These curves are seen with low concentrations, but data on them are very rare and therefore their validity remains in question.

Fig. 1: The four basic types of dose-response curves: A-curve describing the simple augmentation of toxicity, when increasing the dose concentration. B-curve, describing an initial stimulation followed by toxicity augmentation (hormesis). C-curve, where an initial depression is followed by a stimulation and an uncompensated toxicity, when increasing the dose (rare). D-curve, where a stimulation is followed by an inhibition, which in turn is followed by a stimulation, ending with an uncompensated toxicity (rare). (Modified from ref. 30.)

Luckey continued to develop the hormesis concept. According to his definition, "any agent be it physical, chemical or biological, may be expected to be stimulatory when it is given in much smaller doses than that found to be harmful. The recipient organism will respond more readily if it exists in suboptimum conditions of nurture and nature" (ref. 31 p. 88).

All substances that exhibit a stimulating effect are called "hormetica" by Luckey , the stimulating effect of low concentrations he calls "hormoligosis," and the phenomenon as such "hormesis" (31). According to Luckey, vitamins and trace elements also belong to the hormetica (32). In Luckey's view every toxicant has a concentration at which it does not have any effect (zero equivalent point) (Fig. 2), but this point varies with diet, environment, condition and animal species or strain. Like Schulz, Luckey defines hormesis as a biological law.

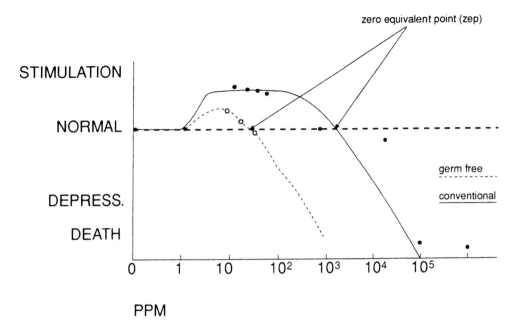

Fig. 2: Dose-response curve of oxytetracycline as a hormetic substance. The oxytetracycline was continuously administered to chicks for 4 weeks. The growth response of "classic chicks" is represented by the solid line, and that of the germfree chicks is represented by the broken line. The difference between the two groups suggests that microorganisms could have a particular influence on hormetic effects. (Modified from ref. 31.)

Despite the questions concerning many elements of Luckey's concept, the term "hormesis" has been accepted for phenomena in various areas in which high concentrations suppress and low ones stimulate. For example: growth and development, death, life expectancy, hatching success, reproductive life span, behavioral parameter, P-450 enzyme induction, fecundity, cancer reduction, disease resistance, viability, respiration, radiation resistance, wound healing, resistance to infection and longevity enhancement (30,33-36) (cited from 6).

From the work that has been done to date, it seems that there are no taxonomic limitations to growth hormesis as well as no limitations concerning the substances that cause this phenomenon. Hormesis occurs with the same pattern in different organisms that are exposed to low concentrations of different toxic substances (5, 7, 19, 36-38). For this reason, the control mechanisms of this phenomenon must be at low organizational steps, but are not lost at high organizational steps (30,39).

Sagen (40-41) is of the opinion that the stimulating, hormetic effect of a toxic substance does not exclude a toxic effect on other parameters (Fig. 3). Hormesis and toxicity are understood as two parallel existing phenomena. The overwhelming of one of the two depends on the dose, environmental conditions, age and class of organism.

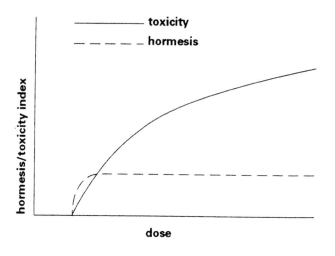

Fig. 3: Hormesis and toxicity as parallel occurring phenomena. The overwhelming action of one on the other depends here on dose (in addition to environmental conditions, age and class of the organism) (Modified from ref. 41.)

In the 1980s, an empirical mathematical formulation was proposed to account for and predict this phenomenon (42-44):

$$E(D) = a0 - a_1 D + a_2 D_2; \quad ai > 0$$

where E(D) is effect (e.g. of cancer) at dose D, and ai values are constants. At low doses, E(D) is negative and therefore the beneficial effect overshadows the harmful one. Conversely, at high doses the harmful effect predominates (Fig. 4) (cited from 6).

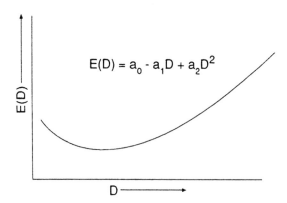

$$E(D) = a_0 - a_1 D + a_2 D^2$$

Fig. 4: Linear quadratic model of hormesis and toxicity. E(D) versus D where E(D) is effect (e.g. of cancer) at dose D. At low doses, E(D) is negative and the hormetic, beneficial effect overshadows the harmful one. At high doses the harmful, toxic effect predominates. (Modified from ref. 43.)

Possible mode of action

The explanation of the phenomenon of hormesis still needs clarification, especially since biochemical mechanisms are unknown. Liu et al. (43) suggest that the immune system might be involved in radiation hormesis, stimulating the defence mechanism. Congdon (44) thinks that a rebound regenerative hyperplasia in the early phase of chronic exposure could explain radiation longevity. Feinendegen et al. (45) see the inhibition of thymidine kinase by acute radiation as a possible explanation of hormetic effects. That is, thymidine kinase can be inhibited by increased concentration of intracellular free radical, liberated by low dose radiation. This inhibition of thymidine kinase activity could slow down the DNA synthesis, giving more time for DNA repair.

Some authors suggest that there are cases in which chelation or complexation is involved. Jones (46), for example, claims that copper enhances the growth of bacteria by displacing micronutrients, which then allows growth. The same opinion is held by Davis and Hidu (47).

Nickel and Finlay (23) assume that antibiotics indirectly enhance the growth of plants by binding toxic metabolites as complexes or by increasing the permeability of the cell wall, thereby leading to an increase in the uptake of water; this uptake increases growth. Also Dunstal et al. (48) suggest an increase in the permeability of the cell wall and, thereby enhanced availability of the conditions necessary for growth.

Fabricant (49) was able to show a compensatory cell proliferation response in mice exposed to low-dose continuous irradiation. These mice exhibited an increase in cell-birth rate, a decrease in cell-loss in some tissues, and shortening of the stem-cell cycle. The increase in cell-birth rate was due to regenerative proliferation of the surviving cells, migration of cells from unirradiated areas of the tissue which causes an increase in the size of proliferative compartments, an increase in stem- cell compartment output, and an increase in transit time through the proliferative compartment.

Kondo (50) suggested that radiation-induced hormesis might be a result of 'altruistic cell suicide'. The elimination of the injured cell benefits the whole organism if the cell suicide stimulates the proliferation of healthy cells. The author calls this event 'cell replacement repair' in order to distinguish it from DNA repair, which he usually regards as being incomplete.

Kondo thinks that in the James and Makinodan study (51), suicide of some radiosensitive pre-T cells stimulated proliferation and activation of precursor T cells, thus resulting in augmented effector T cell production in normal mice. With a larger number of radio-sensitive cells in the immunologically-depressed mice, the hormetic effect was exaggerated. Kondo suggested that continuous up-regulated immunity would augment antitumor responses which could result in longevity hormesis, if spontaneously occurring tumors respond to the cytotoxicity of killer cells.

Another possibility in order to explain hormetic mechanisms is by assuming that the life-enhancing effect is caused by stress protein synthesis. This is suggested by Naefsey (2) relating to the work of Haidrick et al. (52). They showed that the antioxidant activity of 2-mercaptoethanol retarded free radical damage in spleen lymphocytes in exposed mice. As 2-mercaptoethanol is also known as an inducer of glucose-regulated stress protein in mammalian cells (53), it is possible that there is a connection between the enhanced longevity resulting from 2-mercaptoethanol administration and the production of stress protein.

The most extensive work has been done by Stebbing (25,38,39,5456). In his view the only element common to all the phenomena that show hormetic effects is the dose-response curve (39). In his first experiment on the colonial hydroid Campanularia flexulosa, Stebbing showed that a low concentration of copper ions enhances growth, whereas a high concentration suppresses it.

His assumption was that there is a low concentration of copper ions in the sea water, and therefore their supplementation in the experiment created an optimal environment by raising the copper concentration, and growth was thereby enhanced.

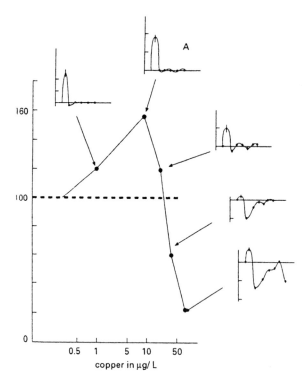

This assumption was rejected when it became clear that a similar phenomenon was observed also with other toxicants (cadmium ions, two synthetic biocides and a low concentration of salts) (38). In the experiments described below, Stebbing et al. (54) showed that the growth rate after exposure to a toxic agent oscillates with an amplitude which decreases until equilibrium is reached (Fig. 5-A). In another system - Rhodotorula rubra - Stebbing was able to repeat these observations (56) and to obtain a similar curve (Fig. 5-B).

Fig. 5: Curve A - Mean size of the colonies of Laomedea flexuosa after exposure to different doses of copper. Curve B - Mean size of cultures of Rhodotorula rubra after exposure to different concentrations of cadmium for 8 hours. The small curves describe the oscillative perturbation of the growth rate after the exposure to the toxicants (Modified from ref. 51.)

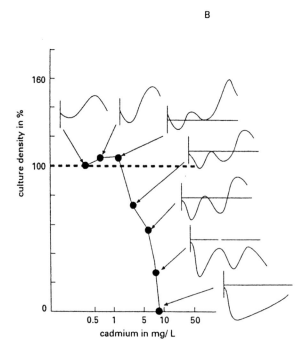

In his opinion, this is an answer of the control system to an input which disturbs the equilibrium. Stebbing (39) mentions that other scientists also found similar oscillatoric perturbation of the growth rate following exposure to toxic substances. He drew the conclusion that growth hormesis may be "the consequence of regulatory overcorrection by biosynthetic control mechanisms to low levels of inhibitory challenge resulting in growth that is greater than normal" (30). In other words, Stebbings believes that the growth increase is not in itself an adaptation in order to reduce the growth inhibition. It is, rather, a consequence of the adaptive behaviour of rate-sensitive control mechanisms. As hormesis can be seen as the consequence of regulatory overcorrection, so is inhibition the consequence of regulatory undercorrection. At high concentrations of toxic agents, the ability to contraregulate becomes "exhausted" rapidly and leads to a fast, non-compensable decrease in the growth rate.

Hormesis can be considered to be a simple cybernetic system according to Stebbing (30). The oscillatory behaviour of the system depends on exactly how the receptor, the integration system and the effector react to the specific properties of the perturbation. Stebbing also considers that the amplitude of the initial overcorrection is not alone responsible for the strength of the hormetic effect. For example, the strongest hormetic effect in Fig. 5-B is not exerted by a concentration of $1m$ Cu^{++}/l, at which the initial overcorrection is the strongest, but rather by a concentration of $5m$ Cu^{++}/l.

Longevity hormesis

Until the 1960s, hormesis research concentrated on the aspect of growth hormesis. Sacher and Trucco (57) collected more and more evidence that low doses of toxicants increase and improve survival in animals. The improvement in survival was attributed to induced increase in the informational content of systems which is equivalent to a decrease in the fluctuation variance of the system. They were, in fact, the first to connect longevity enhancement with the Arndt - Schulz Law.

During the next decade, more and more evidence concerning the longevity-prolonging effect of low doses of toxicants accumulated. Much effort was expended on differentiating the hormetic effects from other life-prolonging effects, such as reduction of caloric intake (2,59-61). In 1977 Sacher published "Life Table Modification and Life Prolongation" (58). This book chapter, which was described by Boxenbaum et al. (6) as "his classic work", contained some restrictions to the classical definition of hormesis. He distinguishes "proper action" from real hormetic action.

"Proper action" is the reduction of accumulation of ageing lesion by a specific biochemical mechanism through a direct protection against age-dependent deterioration, e.g. vitamins, steroids, other hormones and essential substances. On the contrary, a hormetic effect is a non-specific effect, being induced by a variety of chemically different organic and inorganic substances, radiation and other stimuli (2).

According to Sacher, a hormetic response is a function of the "state" of an organism, rather than of the stimulant. Perhaps the most important point raised by Sacher is that hormesis does not increase the potential longevity. Rather it enables the biological system to approach its potential longevity. Therefore, a hormetic effect can be observed only in depressed or ill individuals and not in healthy ones. The hormetic effect is not a "real-life prolongation" but rather the nullification of some deleterious effects from the environment (2).

Hormesis in homoeopathic research

Over 100 toxicology-orientated papers, reporting attempts to prove the biological activities of homoeopathic remedies, have been published in recent years and can be regarded as experiments to examine hormetic effects. In these, the influence of serially diluted toxic substances on biological systems intoxicated with the same or a similar toxic substance, were measured to evaluate the very elimination kinetics or the survival of the systems.

The most frequently used toxicants were heavy metals. Their effects were examined on cells (63), animals (64-66), isolated organs (67,68) and plants (69,70). Only a few experiments had been performed with high dilutions, that is, dilutions above the Avogadro number. Most of these works lack the required scientific qualities to be published in conventional literature, and indeed, only a few of them could find access to conventional scientific papers (71,72).

DISCUSSION

The mechanisms underlying hormesis are still not known, but according to the observations amassed to date, especially in the field of longevity hormesis, four characteristics are exhibited by any possible mechanism (2):

1. The hormetic effect occurs after exposure to a low concentration of an otherwise noxious agent.

2. The hormetic effect is reversible. It does not accumulate after termination of exposure to the causal agent.

3. The hormetic effect can be induced by a variety of chemically unrelated substances.

4. The mechanism has to be evolutionarily conserved.

A relatively large number of papers has been published on hormesis. Those cited in this chapter represent just a small collection of them. Hormetic phenomena are also very often observed in clinical and experimental research, but generally are ignored as 'artefacts'. This results most probably from the contradiction to the basic paradigm which suggests that the stronger the stimulus, the stronger is the effect; and the weaker the stimulus, the weaker is its effect. An enhancing effect of low doses of a noxious agent is not recognized according to this paradigm. A toxic agent can reduce its toxic activity only when the quantity is reduced, but it cannot enhance or have a positive effect.

This is most probably the reason why conventional science pays little attention to this field. Only in a few scientific indexes the term 'hormesis' can be found. Publications suggesting that carcinogens can also have a 'cancer protective' effect (73) are published without arousing any reaction.

A further problem is that there exists no proof of a possible mode of action of this phenomenon. Most of the research work has been conducted to show that the phenomenon exists and that it is a general rule. Very few works have been performed to understand the underlying mechanisms.

Hormesis is not homoeopathy. The "classic" hormetic systems act at concentrations that are much higher than those used in homoeopathy. Therefore, the problem of explaining how an 'ultra-high dilution' that does not contain even a single molecule yet which still possesses biological activity, is avoided. Contrary to homoeopathy, it does not need any special technique to prepare the 'active agent'. However, one cannot overlook some similarities between homoeopathy and hormesis, beyond those of the basic laws:

There is a lack of evidence that homoeopathy has prophylactic properties. Homoeopathic treatment requires symptoms. This means a disease, a suppressed system. As Sacher showed (58), this is a basic requirement of a hormetic system, a system which is depressed or ill. Longevity hormesis is, in his view, not a "real life prolongation" but rather the nullification of some deleterious effects on the environment (2).

Another similarity is the rationale in diluting and potentiating homoeopathic remedies: A substance which is toxic, or suppressive, is diluted. By dilution its noxious properties disappear and its enhancing properties emerge. The greater the dilution is, the stronger is the enhancing activity of the previous toxic substance.

As mentioned above, it is obvious that hormesis and homoeopathy are not synonymous and not special cases of each other. Nevertheless it is almost impossible to overlook the similarities between them, especially in those areas that contradict the conventional paradigms.

This may suggest that there is some connection between homoeopathy and hormesis, or a common basis.

That common basis might be an information process. It has been shown that hormesis may be explained by biochemical modification related to every toxic effect in many cases: No common biochemical mechanism exists, but each one is exactly the right mechanism which enables the biological system to adapt through counteracting a particular toxic effect. The adaptability of the living system could mean that it is able to recognize the aggression and to receive it as an information. This would explain the variability of the defences observed in the various hormetic models. In the same way a homoeopathic remedy represents the informative image of the pathology of the subject and helps the organism to react. This hypothesis was formulated by Lagache (74,75) and Bastide and Lagache (76,77). For example, toxic concentrations of a heavy metal kill cells or animals. When the concentration is decreased, a point is reached where toxicity ends. This "toxic end-point concentration" is followed by atoxic concentrations unable to kill the living system but able to inform the living system about the properties of the toxicant (Fig. 6).

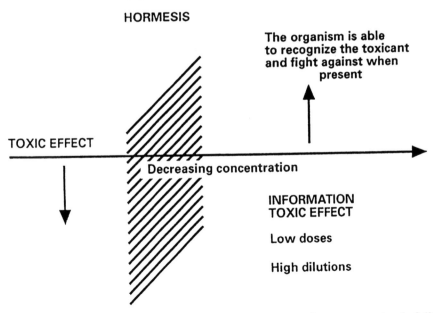

Fig. 6: Hormesis as an information process. The toxic end-point concentration is followed by atoxic concentrations unable to kill the living system but able to inform the living system about the properties of the toxicant. (Modified from Bastide M, Lagache A. Essential Elements I. Atelier Alpha Bleue. Paris p. 18.)

The living system is now prepared to recognize the toxicant and to fight against it by developing the antagonistic mechanism that is adapted to the toxicant. The same information is obtained with high dilutions of toxicants as shown by Larue et al.(70) and Cal et al. (71). This new paradigm, designed as 'paradigm of signifiers' presents hormesis as a general application of a process of information.

ANNOTATION

Thanks are especially due to D. Bransky, S. Shoshan, Y. Shoenfeld and E. Shoe.
The remark in [¹ was added by the editor.

REFERENCES

1. Jouanny J. The essentials of homoeopathic Therapeutics. 14-15. Laboratoires Boiron, Lyon-France: 1985.
2. Neafsey PJ. Longevity Hormesis. A review. Mechanism of Ageing and Development. 1990; 51, 1-31.
3. Boxenbaum H, Neafsey P, Fournier D. Hormesis, Gomperz function and risk assessment. Drug Met Rev 1988; 19,2: 195-229.
4. Schulz H. Ueber die Theorie der Arzneimittelwirkung. Virchows Archiv 1877; 108: 423-434.
5. Martius F. Das Arndt- Schulz Grundgesetz. Muench Med Wschr 1923; 70, 31: 1005-1006.
6. Schulz H. Ueber Hefegifte. Pflueger Archiv fuer Physiologie 1988; XLII: 517-541.
7. Hueppe F. The Principles of Bacteriology Chicago: Open Court 1896.
8. Stare FJ. Arsenic as growth promotor. Nutr Rev 1956; 14, 206-209.
9 . Chapman RK, Allen TC. Stimulation and suppression of some vegetable plants by DDT. J. Econ Ent 1948; 41, 616-623.
10. Finklestone- Sayliss, Sheff MD, Paine CG, Patrick LB. The bacteriostatic action of p- amino- benzene-sulphonamide upon haemolytic streptococci. The Lancet October 1937; 792-795.
11. Randall WA, Price CW, Welch H. Demonstration of hormesis (increase in fatality rate) by penicillin. Am J Public Health 1947; 37, 421-425.
12. Miller MW, Green CA, Kitchen H. Biphasic action of penicillin and other sulphonamide similarity. Nature 1945; 155, 210-211.
13. Hochkiss M. Studies on salt action. VI. The stimulating and inhibitive effect of certain cations upon bacterial growth. J Bacteriol 1923; 8, 141-162.
14. Hammett FS, Reiman SP. Effects of methylcholanthrene on developmental growth of obelia geniculata. Am J Cancer 1935; 25, 807-808.
15. Binger CAL, Faulkner JM, Moor RL. Oxygen poisoning in mammals. J Exper Med 1927;45, 849-864.
16. Faulkner JM, Birger CAL. Oxygen poisoning in cold blooded animals. J Exper Med 1927; 45, 865-871.
17. Davey WP. Prolongation of life of Tribolium confusum apparently due to small doses of X- Rays. J Exp Zool 1919; 28, 447-458.
18. Taylor NB, Weld CB. A study of the action of irradiated ergosterol and of its relationship to parathyroid function. J Cand Me. Assoc 1931; 25,20-34.
19. Branham SE. The effects of certain chemical compounds upon course of gas production by Baker's yeast. J Bakteriol 1929; 18, 247-264.
20. Southam CM, Ehrlich J. Effects of western red cedar heartwood on certain wood- decaying fungi in culture. Phytopathology 1948; 33, 515-524.
21. Mottram JC. An increase in the rate of growth of paramecium subjected to the blastogenic hydrocarbon 3:4 benzpyren. Nature 1939; 144, 154.
22. Dunstan WM, Atkinson LP, Natoli J. Stimulation and inhibition of phytoplankton growth by low molecular weight hydrocarbons. Mar Biol 1975; 31, 305-310.
23. Nickel LG, Finlay AC. Growth modifiers: antibiotics and their effects on plant growth. Agr Food Chem 1954; 2, 178-182.
24. Holz P, Exner M, Schuman HJ. Zellreiz und Zellschaedigung: Untersuchungen ueber die "Reizgaerung" und "Reizatmung" der Hefezelle. Naunyn- Schmiedeberg Archiv fuer exptl. Pathol und Pharmakol 1948; 205, 243-275.
25. Linde K. Dosisabhaengige Umkehreffekte - eine differenzierende Literaturbetrachtung. Thesis, Ludwig- Maximilian University, Munich 1991.
26. Wilder J. The law of initial value in neurology and psychiatry: Facts and problems. J Nerv Mental Dis 1957; 125, 73-86.
27. Coulter HL. Homoeopathic Science and Modern Medicine, the Physics of Healing with Microdoses. Ritchmond: North Atlantic Books 1981.

28. Luckey TD. In: Proceeding of the first international conference on the use of antibiotics in agriculture, Publ. 397, National Academy of Sciences- National Research Council, Washington DC, 1956; 135-145. (cit. from Boxenbaum H. et al. (5)).
29. Townsend JF, Luckey TD. Hormologosis in Pharmacology. J Am Med Assoc 1960; 173, 44-48.
30. Stebbing ARD. Hormesis- the stimulation of growth by low levels of inhibitor. Science of the Total Environment 1982; 22, 213-234.
31. Luckey TD. Hormology with inorganic compounds. In Heavy metal - toxicity, safety and hormology. Luckey TD, Venugopal B, Hutchenson (ed.). Stuttgart: Georg Thieme Publ 1975, pp 81-103.
32. Luckey TD.Activity spectrum of ingested toxicants. Comp Anim Nutr 1977; 2, 144-178
33. Calabrese EJ, McCarthy ME, Kenyon E. The occurence of chemically induced hormesis. Health Phys 1987; 52, 531-541.
34. Luckey TD. Hormesis with Ionizing Radiation. Boca Raton, FL: CRC Press 1980.
35. TD Luckey. Physiological benefit from low levels of ionizing radiation. Health Phys. 1987; 43, 771-789.
36. Furst A. Hormetic effects in pharmacology. Health Phys 1987; 52, 527-531.
37. Richet C. De l'action des doses minuscules de substances sur la fermentation lactique: Periodes d'acceleration et de relentissement. Archiv. Intern Physiol 1906; 4, 18-50.
38. Stebbing ARD. Hormesis- stimulation of growth in Campanularia flexuosa (Hydrozoa) by cadmium copper and other toxicants. Aquatic Toxicol 1981; 1, 227-238.
39. Stebbing ARD. Growth hormesis: a by - product of control. Health Phys 1987; 52, 543-547.
40. Sagan LA.What is hormesis and why haven't we heard about it before? Health Phy. 1987; 52, 521-525.
41. Sagan LA. On radiation, paradigm and hormesis. Science 1989; 245-574.
42. Hickey RJ, Bowers EJ, Clelland RC. Radiation hormesis, public help, and public policy: a commentary. Health Phys 1983; 44, 207-219.
Hickey RJ. More comments on radiation hormesis, epidemiology and public health (letter). Nuclear News May 1984; 46, 1159- 1160.
Hickey RJ, Clelland RC, Bowers EJ. More comments on radiation hormesis, epidemiology and public health. Health Phys 1984; 46, 1159-1160.
43. Liu ZS, Liu WH, Sun BJ. Radiation hormesis: its expression in the immune system. Health Phys 1987; 52, 579-583.
44. Congdon CC. A review on certain low- level ionizing radiation studies in mice and guinea pigs. Health Phys 1987; 52, 593- 597.
45. Feinendegen LE, Muehlensiepen, Bond VP, Sondhaus CA. Intracellular stimulation of biochemical control mechanism by low-dose, low-let irradiation. Health Phys 1987; 52, 663-669.
46. Jones GE, in Hood DW (Ed.). Organic Matter in Natural Waters. Alaska: University of Alaska Press 1970.
47. Davis HC, Hidu H. Effects of pesticides on embryonic developement of clams and oysters and on survival and growth of the larvae. Fish Bull. Fish Wildl Serv US 1969; 67, 393- 404.
48. Dunstan WM, Atkinson LP, Natoli J. Stimulation and inhibition of phytoplankton growth by low molecular weight hydrocarbons. Mar Biol 1975; 31 (4), 305-310.
49. Fabricant JI. Adaptation of cell renewal systems under continuous irradiation. Health Physics 1987; 52, 663-671.
50. Kondo S. Altruistic cell suicide in relation to radiation hormesis Int J Radiat Biol 1988; 53, 95-102.
51. James SJ, Makinodan T. T cell potentiation in normal autoimmune-prone mice after extended exposure to low doses of ionizing radiation and/or caloric restriction. Int J Radiat Biol 1988; 53, 137-152.
52. Heidrick ML, Hendricks LC, Cook DE. Effect of dietary 2- mercaptoethanol on the life span, immune system, tumor incidence and lipid peroxidation damage in spleen lymphocytes of aging BC3F1 mice. Mech Ageing Dev 1984; 27, 341-358.
53. Craig EA. The heat Shock Response, CRC Crit Rev Biochem 1985; 18, 239-280.
54. Stebbing ARD. The kinetics of growth control in a colonial hydroid. J Marine Biol Assos United Kingdom 1981; 61. 35- 63.
55. Stebbing ARD, Heath GW. Is growth controlled by a hierarchical system? Zool J Linnean Soc 1984; 80, 345-367.
56. Stebbing ARD Norton JP Brinsley MD. Dynamics of growth control in a marine yeast subjected to perturbation. J General Microbiol 1984; 130, 1799-1808.
57. Sacher GA, Trucco E. Biological Aspects of Aging. (Ed. W. Shock). 244-251 Columbia University Press, New York: 1962.
58. Sacher GA. Handbook of the Biology of Aging. (Ed. Finch CE, Hayflick L). . New York: Van Nostrand 1977, chap 24, pp 582-638.
59. Goodrick CL, Ingram DK, Reynolds MA, Freeman JR, Cider LA. Effects of intermittent feeding on growth and life span in rats. Gerontololy 1982; 28 (4), 233-241.
60. Young VR. Diet as a modulator of aging and longevity. Fed. Proc. 1979; 38, 1994-2000.
61. Endler PC. Die Ernährung des (...) Hochbetagten. Medizinverlag Maudrich, Vienna 1993.

62. Cotte J, Bernard A. Effets de dilutions hahnemanniennes de Mercurius corrusivus sur la multiplication en culture de fibroblastes intoxiqués par le chlorure mercurique. In: aspects de la recherche en homéopathie (Ed. Boiron J, Abecassis J, Belon P). Lyon: Editions Boiron 1983, pp 51-59. .

63. Cazin JC, Cazin M, Boiron J, Belon J, Gaborit L, Chaoui A, Cherrault Y, Papanayotou C. The Study of the effect of decimal and centesimal dilutions of arsenic on the retention and mobilization of arsenic in rats. Human Tox 1987; 6: 315-320.

64. Wurmser L. Action de doses infinitesimales de bismuth sur l'elimination de ce metal chez le cobaye. Hom Fr 1954; 45: 427-433.

65. Harisch G, Andersen M, Kretschmer M. Aktivitaetsaenderung von Succinatdehydrogenase und Glutathionperoxidase nach Verabreichung von homoeopathisch aufbereiteten Phosphorverduennungen im Rattenlebermodell. Dt Apotheker Zeitung 1986; 26: 29-31.

66. Aubin M. Elements de pharmacologie homéopathique. Etude de l'Aconitine. Hom Fr 1984; 72: 231-235.

67. Penec JP, Aubin M, Baronnet J, Manlhiot L, Payrau B, Scaglier D. Action de l'Aconitine sur le coeur perfuse d'anguille. Hom Fr 1984; 72: 237-243.

68. Auquiere JP, Moens P, Martin PL. Recherche de l'action de dilutions homeopathiques sur la vegetaux. II action du $CuSO_4$ 14 DH sur la moutarde blanche intoxiquée au $CuSO_4$ 1/1000 et 2/1000. J Pharmac Belg 1982; 37,2: 117-134.

69. Noiret R, Glaude M. Etude enzymatique du grain de froment intoxique par du sulfate de quivre et traite par differentes dilutions hahnemanniennes de la m'ème substance. Rev Belg hom 1976; IX (1): 461-495.

70. Haseman J. Patterns of tumor incidence in two year cancer bioassay feeding study in Fisher- 344 rats. Fundam appl toxicol 1983; 3(1):1-9.

71. Larue F, Cal JC, Dorian C, Cambar J. Influence du pretraitement par des dilutions infenitesimales de Mercurius corrosivus sur la mortalité induite par le chlorure mercurique. Nephrologie 1985; 6: 86

72. Cal JC, Larue F, Dorian C, Guillemain J, Dorfman P, Cambar J. Chronobiological approach of mercury- induced toxicity and of the protective effect of high dilutions of mercury against mercury- induced nephrotoxicity. Liver Cells Drugs 1988; 164: 481-485.

73. Weinberg A, Storer J. Ambiguous carcinogens and their regulation. Risk Anal 1985; 5(2), 151-156.

74. Lagache A. Echos du Sensible. Paris: Atelier Alpha Bleue Publisher 1988.

75. Lagache A. Evolution des modèles logigues dans la science contemporaine. In: M. Bastide ed. Signals and Images. Paris: Atelier Alpha Bleue Publisher 1991, pp 73-114.

76. Bastide M, Lagache A. The Paradigm of signifiers. Paris: Atelier Alpha Bleue Publisher 1992.

77. Bastide M, Lagache A. Essential Elements. Paris: Atelier Alpha Bleue Publisher 1993

REAPPRAISAL OF A CLASSICAL BOTANICAL EXPERIMENT IN ULTRA HIGH DILUTION RESEARCH. ENERGETIC COUPLING IN A WHEAT MODEL

W. Pongratz, P.C. Endler

SUMMARY

A. The effects of three different dilutions of silver nitrate log 24, log 25 and log 26, specially prepared according to the instructions for the preparation of "homoeopathic" remedies, following a historical protocol (1926) on germination and growth of coleoptiles of wheat seedlings were investigated. A typical effect pattern was found, with the dilutions log 24 and log 26 significantly enhancing development. The consistency with a previous historical study is discussed.

B. In experiments with dilution log 24, it was found that both unprepared solvent (water) as well as analogously "homoeopathically" prepared water is suitable for reference. The consistency with a previous multicenter study is discussed.

C. The effects of the dilutions, as described in study A, were investigated when the dilutions were kept in sealed glass vials that were brought in permanent contact with the water the grains were submerged with. The typical effect pattern already described in study A was found.

ZUSAMMENFASSUNG

A. Es wurden die Wirkungen dreier verschiedener Verdünnungen von Silbernitrat log 24, log 25 und log 26 auf Keimung und Koleoptilwachstum von Weizensamen untersucht. Die Verdünnungen wurden entsprechend der "homöopathischen" Vorschrift eines historischen Protokolls (1926) zubereitet. Ein typisches Wirkungsmuster, bei dem die Verdünnungen log 24 und log 26 die Entwicklung deutlich förderten, wurde beobachtet. Die Übereinstimmung mit einer früheren historischen Studie wird diskutiert.

B. In Experimenten mit Verdünnung log 25 wurde gefunden, daß sowohl nicht eigens präpariertes Lösungsmittel (Wasser) als auch analog "homöopathisch" zubereitetes Wasser als Kontrolle verwendet werden können. Die Übereinstimmung mit einer früheren multizentrischen Studie wird diskutiert.

C. Es wurden die Wirkungen der in Studie A beschriebenen Verdünnungen untersucht, wenn diese Verdünnungen in versiegelten Glasphiolen eingeschlossen waren, die in permanentem Kontakt mit dem die Samen bedeckenden zugegossenen Wasser waren. Das bereits in Studie A beschriebene Wirkungsmuster wurde erneut beobachtet.

INTRODUCTION

An increasing number of studies report different effects of diluted substances, depending on whether or not those were submitted to a "homoeopathic" agitation process during dilution (see the examples discussed in this book). There is clinical evidence that even in very high dilutions, when these were submitted to a special preparation process, specific information can be stored [1,2].

In the background of such more or less therapy-oriented studies, important work has been done in research on Ultra High Dilutions (UHD) on plant models [3-22]. In many of

19

P.C. Endler and J. Schulte (eds.), Ultra High Dilution, 19–26.

these studies [e.g. 14,15], but also in other fields of UHD research [23], it was a.o. claimed that, from a logarithmic range of dilutions (e.g. 10^{-24}, 10^{-25}, 10^{-26}), only certain dilutions are able to exert biological activity.

The aim of the study discussed here is the critical evaluation of the reliability of a test system which has been quoted as a basic model for the research on "homoeopathic" drugs, and has been discussed for decades [8,10,11,12,14,15,21].

The efficacy of a "homoeopathic" drug commonly used, namely silver nitrate, was investigated in the model system of the germination and the development of leaves of wheat grains. The study is based on the facts a. that silver nitrate in a common molecular dose inhibits life processes [24] and b. that inverse - i.e. stimulating - effects of specially prepared high dilutions of silver nitrate have been reported [8,10,11,12,14,15,21], especially from "homoeopathically" prepared dilutions log 24 and log 26 (but not log 25) [14].

In order to evaluate the respective protocol by Kolisko [14] critically, a study was performed to compare the effects of different steps of the dilution of silver nitrate, namely log 24, log 25 and log 26. Unprepared solvent (distilled water) was used as an additional control (A).

Furthermore, the dilution log 24 was tested against analogously "homoeopathically" prepared distilled water (B). Data from a respective study, organized by the biological editor of this book and performed in three independent research institutes (see discussion), have been published earlier [22].

Furthermore, an additional approach was added: it concerned the evaluation of the exposure to the test dilutions in a way that they were not mixed with the water the grains were submerged with. For this purpose, the test dilutions silver nitrate log 24, log 25 and log 26 were applicated in closed glass vials (C).

METHODS

Study A

Plants: Unbroken, unsorted grains of winter wheat (sort: Mephisto), grown without an application of herbicides or pesticides, were used.

Sites involved: All experiments were performed by one researcher, but at two different sites, at the Institut für strukturelle medizinische Forschung, Graz and the Ludwig Boltzmann Institut für Homöopathie, Graz, respectively.

Observed development: According to the protocol of 1924 [14], germination and the initial development of stalks (coleoptiles) were observed. It is reported that biological systems in such circumscript developmental transitions show maximal sensitivity [25,26, see the contribution of Endler et al.]. The parameters observed according to [14] seem to be adequate for a first quantification of the development of the plants.

Preparation of test solutions: The grains were observed under the influence of an aqueous solution $1:10^{24}$ part of weight of silver nitrate (Merck), specially prepared according to "homoeopathic" instructions. The stock solution had a concentration of 1:100 part silver nitrate of weight, as described in [14], and it was diluted in distilled water in steps of 1:10. The solutions, including the stock solution, were agitated according to standardized instructions [27]: at every step a sterile bottle is partly filled with the dilution, and is pushed down at short regular intervals (e.g. against a rubber impediment) to create mechanical shocks. The test dilution prepared in this way is called dilution silver nitrate D24. In addition to the dilution log 24, dilutions log 25 and log 26 part silver nitrate of weight were analogously prepared. Three analogously prepared sets of ranges of the dilutions were used. Untreated distilled water was used as an additional control.

Independent solution coding: All sets were applied blindly. The sets were coded at the Institut für strukturelle medizinische Forschung, Graz, and at the Institut for Plant Physiology of the University of Graz, respectively.

Exposition to probes according to [14]: The grains with the germination furrow downwards were put into glass dishes (diameter 11cm), each containing 20 ml of the respective probe. The number of germinated grains and the stalk length were measured after 5 days. Natural light was used. The experiments were performed inside in the summer season at a temperature of 20°C (comparatively long coleoptiles). The position of the dishes was rotated in the course of the experiment.

Avoidance of contamination by the probes: Before use, the dishes were treated with dry heat for 45 minutes and splashes of the dilutions into the surroundings, when they occurred, were removed with soapy water according to the instructions of Boyd [28].

Data base: Three sets of dishes for the treatment with dilution silver nitrate D24, D25 and D26, respectively, plus dishes for additional water control were always used for the experiments. 30 grains were put into one dish; the number of grains was equal in each experiment (for total numbers, see below).

Evaluation of the data: The number of germinated seedlings was compared to the number of non-germinated seedlings for the groups according to the treatment in a four-field table by the chi-square test. For a description of the data on stalk lengths, the statistical mean was used, and the lengths were compared by one way variance analysis. S.E. of the arithmetic mean was calculated. According to [14], the groups of seedlings treated with the different silver nitrate probes were mutually compared (see figures). Furthermore, they were compared to the water control (see tables).

Study B

Researchers involved: Two sets of experiments were performed by two independent researchers (Endler, Pongratz) from the Ludwig Boltzmann Institut für Homöopathie, Graz, at two different laboratory sites.

Preparation of test solutions was done as described for study A (dilution silver nitrate D24). Analogously prepared distilled water (dilution water D24) was used for control. Two analogously prepared sets of the dilutions were used.

Data base: 16 sets of dishes for the treatment with dilution silver nitrate D24 and water D24, respectively, were always used for the experiments. 20 or 30 grains, respectively, were put into one dish; the respective number of grains per dish was equal for both groups (for total numbers, see below).

The experiments were performed inside in the winter season at a temperature of 15°C (comparatively short coleoptiles). For further conditions, see study A.

Study C

Exposition to the probes: The dilutions silver nitrate D24, D25 and D26, as well as those for control, respectively, were sealed in glass vials. The coded vials were first all hung into a common water bath to make sure that they were not individually contaminated outside and then hung into the water submerging the corresponding sets of grains.

Ampoules used: The dilutions were sealed in hardglass vials with an optical transmission spectrum starting from about 350 nm, the optical transmission spectrum was limited by the properties of the submerging water to wavelengths less than about 2500 nm.

The experiments were performed inside in the summer season at a temperature of 20°C (comparatively long coleoptiles). For further conditions, see study A.

RESULTS

Study A

Experiments with a total of 890 grains treated with silver nitrate D24, 890 grains treated with silver nitrate D25 and 890 grains treated with silver nitrate D26 were performed. In addition, one group with 686 grains was treated with unprepared distilled water. The comparison of the germination rates as well as of the mean stalk length showed a certain pattern of the effects of the three different dilutions. All the parameters observed indicate that development is enhanced by the probes silver nitrate D24 and silver nitrate D26 as compared to the probe silver nitrate D25 (see Fig. 1 and Tab. 1).

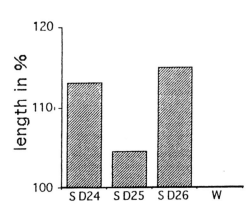

Fig. 1: The influence of the test dilutions silver nitrate D24, D25 and D26, compared to unprepared water, on wheat germination. Ordinate: mean stalk length. 100% refers to the stalk length under the influence of unprepared water. Abscisses: S D24 - D26, silver nitrate D24 - D26; W, unprepared water.

Tab. 1: Stalk lengths, study A

AgNO₃ D24	AgNO₃ D25	AgNO₃ D26	Water unprepared
60.1 ± 1.4	55.5 ± 1.4	61.1 ± 1.4	53.1 ± 1.4
**	-	*	

Tab. 1: Mean stalk length of wheat under the influence of silver nitrate D24, D25, D26 and unprepared water, respectively. Median (mm) ± SEM. **, $P < 0.01$; *, $P < 0.05$; -, not significant: the effect of the test substance compared to the effect of unprepared water.

The differences could be found both for the pooled data as well as for most of the 16 successive experiments. They were also found when the statistical median was used for the calculations instead of the statistical mean.

Study B

Experiments with a total of 400 grains treated with silver D24 and 400 grains treated with water D24 were performed by the two researchers. When the data were pooled, the germination rate of grains treated with the test dilution silver nitrate D24 was above that for

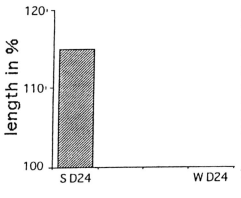

reference at about 5 % (P > 0.05), and the mean stalk length in the group treated with the test dilution was above that for reference at about 15% (P < 0.001, see Fig. 2).

Fig. 2: The influence of the test dilution silver nitrate D24, compared to water D24, on wheat germination. Ordinate: mean stalk length. 100% refers to water D24. Abscissa: S D24, silver nitrate D24; W D24, water D24.

Tab. 2: Stalk lengths, study B

AgNO$_3$ D24	Water D24
20.0 ± 2.5	17.4 ± 2.5
**	

Tab. 2: Mean stalk length of wheat under the influence of silver nitrate D24 and water D24, respectively. For further information, see Tab. 1.

Experiments with a total of 200 plus 200 grains were performed by Endler. Both the germination rate as well as the mean stalk length for the grains treated with the test dilution were above those for reference (however, only the difference in the stalk lengths was statistically significant - P < 0.01 -).

Experiments with a total of 200 plus 200 grains were performed by Pongratz. Both the germination rate as well as the mean stalk length for the grains treated with the test dilution were above those for reference (however, only the difference in the stalk lengths was statistically significant - P < 0.05 -). The differences were also found when the statistical median was used for the calculations instead of the statistical mean.

Study C

One experiment with 600 grains treated with silver nitrate D24, sealed in glass vials, 600 grains treated with silver nitrate D25, sealed in vials and 600 grains treated with silver nitrate D26, sealed in vials, was performed. In addition, one group with 600 grains was treated with unprepared distilled water.

The comparison of the germination rates as well as of the mean stalk length showed the same pattern of the effects of the three different dilutions as was found in study B. All the parameters observed indicate that the development is enhanced by the probes silver nitrate D24 and silver nitrate D26 as compared to the probe silver nitrate D25, when the dilutions were sealed in glass vials (see Fig. 3 and Tab. 3)

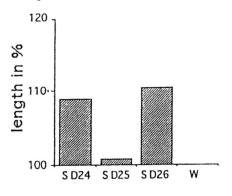

Fig. 3: The influence of the test dilutions silver nitrate D24, D25 and D26, sealed in glass vials, compared to unprepared water, sealed in a glass vial, on wheat germination. For further information, see Fig. 1.

Tab. 3: Stalk lengths, study C

AgNO₃ D24 in vial	AgNO₃D25 in vial	AgNO₃D26 in vial	Water unprepared in vial
65.1 ± 1.0	60.3 ± 1.3	66.1 ± 1.3	59.8 ± 1.4
*	-	*	

Tab. 3: Mean stalk length of wheat under the influence of silver nitrate D24, D25, D26 and unprepared water, respectively, when all test substances were applied in sealed glass vials. For further information, see Tab. 1.

The differences were also found when the statistical median was used for the calculations instead of the statistical mean.

DISCUSSION

The findings of study **A** show a variable growth of wheat seedlings under the influence of the three different "homoeopathically" prepared dilutions of silver nitrate log 24, log 25 and log 26, "D24", "D25" and "D26". This is consistent with the description by Kolisko (1926) [14]. All Kolisko - data, too, show more growth under the influence of silver D24 and D26 than under the influence of D25 (Fig. 4).

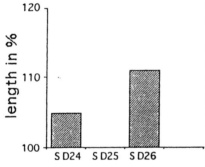

Fig. 4: The influence of the test dilutions silver nitrate D24, D25 and D26 in the experiments of Kolisko in 1924. Abscissa: 100% refers to silver nitrate D25. For further information, see Fig. 1. (Adopted from [14], modified.)

Furthermore, when in our study the silver nitrate dilutions were compared to the unprepared solvent, a statistically significant enhancement of growth was found under the influence of D24 and D26, but no difference at D25. This is consistent with our expectations derived from the Kolisko-study.The importance of the study discussed here is the critical proof of the reliability of a test system which has been quoted as a basic model for the research on "homoeopathic" drugs, and has been controversially discussed for decades.

In the study above, the effect of silver nitrate D24 was compared to that of D25 (according to Kolisko) and to that of untreated water (by laboratory convenience). Furthermore, we compared the effect of silver nitrate D24 to that of the analogously prepared solvent, i.e. water D24. In study **B**, it was found that the wheat seedlings show enhanced growth under the influence of silver nitrate D24 also when compared to water D24. This is consistent with the findings of a previous three-center study organized by the writers and performed at the Institut für strukturelle medizinische Forschung, Graz (W. Pongratz), the University Institute for Plant Physiology, Graz (E. Bermadinger), and the University Institute for Botanics, Vienna (F. Varga) (two further institutes contributed data based on different observation modes). In this study comprising 1840 grains, both the differences in germination rate as well as in stalk length were statistically significant ($P < 0.05$ and $P < 0.001$, respectively), the values for the group treated with the test dilution silver nitrate D24 being above those for reference water D24. An enhancing effect of silver nitrate D24 on the observed stage in wheat development could be proven in all three laboratories [22].

The findings of study **C**, where again silver nitrate D24, D25 and D26 were mutually compared and tested versus unprepared water as in study A, show the same typical pattern of variable growth of wheat seedlings under the influence of the three dilutions as described in study A, also when the dilutions were sealed in glass vials.

At present, we have only very limited insight into the physical properties of such test dilutions and into their physiological interconnections with the living system. However, with regard to study C, a non-molecular interconnection between the UHD and the living system such as an electromagnetic field can be postulated. (For further information, see the contribution by Endler et al.)

ACKNOWLEDGEMENTS

Thanks are especially due to Th. Kenner, M. Moser, E. and Y. Clar and to the Hom.Int. Organisation for their help in the initial phase of this study, which was started by the writers at the Institut für strukturelle medizinische Forschung, Graz; further to M. Haidvogl and to the Ludwig Boltzmann Society for their help during the elaboration phase, which was set at the Boltzmann Institute for Homoeopathy, Graz.

ANNOTATION

A more detailed description of the experiments discussed here is in preparation for print. The authors thank Z. Bentwich for his collaboration.

REFERENCES

1. Reilly D, Taylor M et al. Is homoeopathy a placebo response? The Lancet 1986: II(2): 881.
2. Kleijnen, J, Knipschild, P et al. Clinical Trials of Homoeopathy. Br Med J 1991; 302: 316-323.
3. Auquière JP, Moens P, Martin PL. Recherche de l'action de dilutions homéopathiques sur les végétaux I. J Pharmacie Belgique 1981; 36: 303-320.
4. Auquière JP, Moens P, Martin PL. Recherche de l'action de dilutions homéopathiques sur les végétaux II. J Pharmacie Belgique 1982; 37: 117-134.
5. Boiron J, Zervouacki: Action de dilutions infintésimales d'arséniate de sodium sur la respiration de cleoptiles de blé. Ann Hom Fr 1962,5: 738-742.
6. Boiron J, Marin J. Action de dilutions homéopathiques d'une substance sur la cinétique d'élimination de cette même substance au cours de la culture de grains préalablement intoxiquées. Ann hom franc 1967: 9: 121-130.
7. Boiron J, Marin M. Action d'une 15CH de sulfate de cuivre sur la culture de Chlorella vulgaris. Ann Hom Franc 1971; 13: 539-549.
8. Bornoroni C. Synergism of action between indolacetic acid (IAA) and diluted solutions of $CaCO_3$ on the growth of oat coleoptiles. Berlin J Res Hom 1992; 1 (4/5): 275.
9. Dutta AC. Plants' responses to high homoeopathic potencies in distilled water culture, ICCHOS News Letter II 1989; 3: 2-8
10. Endler PC, Pongratz W. Homoeopathic Effect of a Plant Hormone? A Preliminary Report. Berlin J Res Hom 1991; 1: 148-150
11. Jones RL, Jenkins MD. Plant responses to homoeopathic remedies. Br Hom J 1981; 70: 120-128
12. Jones RL, Jenkins MD. Comparison of wheat and yeast as in vitro models for investigating homoeopathic medicines. Br Hom J 1983; 72: 143-147
13. Jousset P (1902), cited from Netien G, Girardet E. Ann Hom franc 1963; 10:9.
14. Khanna KK, Chandra S. A homoeopathic drug controls mango fruit rot caused by Pestalotia mangiferae. Experientia 1978; 34: 1167-1168.
15. Kolisko, L.: Physiologischer Nachweis der Wirksamkeit kleinster Entitäten bei sieben Metallen. Goetheanum Verlag, Dornach, Schweiz 1926
16. Kolisko L. Physiologischer und physikalischer Nachweis der Wirksamkeit kleinster Entitäten. Stuttgart: Arbeitsgem. Anthroposoph. Ärzte 1961.
17. Netien G. Action de dilutions homéopathiques sur la respiration des coléoptiles de blé. Ann hom franc 1962; 4: 823-827.
18. Netien G. Action de doses infinitésimales de sulfate de cuivre sur des plantes préalablement intoxiquées par cette substance. Ann hom franc 1965.

19. Netien G, Graviou E. Modifications rythmologiques d'un materiel sain traité par une 15CH de sulfate de cuivre. Annales hom franc 1978; 4: 219-225.
20. Noiret R, Claude M. Activité de diverses dilutions homéopathiques de cuprum sulfuricum sur quelques souches microbiennes. Ann Hom Franc 1977; 91-109.
21. Noiret R, Glaude M. Attenuation du pouvoir germinatif des grains de froment traitées par $CuSO_4$ en dilutions homéopathiques. Rev Belg Hom 1979; 31: 98-130.
22. Pelikan W, Unger G. (Die Wirkung potenzierter Substanzen.) The activity of potentized substances. Experiments on plant growth and statistical evaluation. Br Hom J 1971; 60: 233-266.
23. Pongratz W, Bermadinger E, Varga F. Die Wirkung von potenziertem Silbernitrat auf das Wachstum von Weizen. In: Stacher A (ed.): Ganzheitsmedizin. Zweiter Wiener Dialog. Facultas-Universitätsverlag Wien 1991b: 385-389.
24. Projetti ML, Guillemain J, Tetau M. Effets curatifs et préventifs de dilutions homéopathiques de sulfate de cuivre appliquées à des racines de lentilles pré- ou post-intoxiquées. Cahiers de Biotherapie 1985; 88: 21-27.
25. Strasburger, E., Lehrbuch der Botanik. G. Fischer Verlag, New York - Heidelberg (1990)
26. Popp F-A (ed). Electromagnetic Bio-Information - 2nd Edition. Vienna: Urban & Schwarzenberg 1989. See Popp's contribution in this book.
27. Smith CW. Homoeopathy, Structure and Coherence. In: ZDN (ed.). Homeopathy in Focus. Essen. VGM Verlag für Ganzheitsmedizin 1990. See Smith's contribution in this book.
28. Homöopathisches Arzneibuch. Deutscher Apothekerverlag, Stuttgart and Govi Verlag, Frankfurt (1991)
29. Boyd WE. Biochemical and Biological Evidence of the Activity of High Potencies. Br Hom J 1954; 44:7-44.

IMMUNOLOGICAL EXAMPLES ON
ULTRA HIGH DILUTION RESEARCH

M. Bastide

SUMMARY

Immunological models seem to be very useful in demonstrating the activity of homoeopathic high dilution effects of hormones or mediators. Various experimental models have been used. We will describe successively: the action of highly diluted antigens on the secretion of specific antibodies; the action of very low doses or high dilutions of molecules which participate in the education of T or B lymphocytes such as thymic hormones (T lymphocytes) or bursin (B lymphocytes); the immunomodulating effect of very low doses of cytokines; the immunopharmacological activity of high dilution of silica.

ZUSAMMENFASSUNG

Immunologische Modelle scheinen sehr gut geeignet, die Aktivität homöopathischer (Hoch-)Verdünnungen von Hormonen und anderen Substanzen zu demonstrieren. Der Einfluß von hochverdünnten Antigenen auf die Sekretion von spezifischen Antikörpern, der Effekt von Hochverdünnungen von Molekülen, die die Bildung von Lymphozyten beeinflussen (Thymushormone und Bursin), sowie die immunopharmakologische Aktivität von Silicat werden beschrieben.

INTRODUCTION

Immunological models are very useful in demonstrating the activity of homoeopathic remedies or high dilution effects of hormones or mediators. The whole living organism is able to respond to homoeopathic remedies as well as to antigens; of course, the immunological response is not mechanistically comparable to the action of a homoeopathic remedy but the immune system is naturally stimulated by extremely low doses of antigens as low as 0.1 ng (10^{-10}g) (Shinoshara, 1989). In an in vitro model, human monocytes have been shown to present optimal phagocytosis using 10^{-14} M postin; the activity was still evaluable at 10^{-18} M (Leung-Tack , 1986). The evaluation of high dilution or very low dose effect on immunological models is an interesting methodological possibility. In addition, it permits us to open up a new field of application by using the homoeopathical preparation tested in therapeutics.

Various experimental models have been used. We will describe successively: the action of highly diluted antigens on the secretion of specific antibodies; the action of very low doses or high dilutions of molecules which participate in the education of T or B lymphocytes such as thymic hormones (T lymphocytes) or bursin (B lymphocytes); the immunomodulating effect of very low doses of cytokines; the immunopharmacological activity of high dilution of silica.

METHODS AND RESULTS

1. The effect of highly diluted antigen on the humoral immune response in mice.

The effect of high dilutions of antigen on specific humoral immune response has been studied by Weissman et al (1991, 1992). In this model, IgM and IgG antibodies were evaluated after treating the mice with high dilutions of KLH (keyhole-limpet-hemocyanin) a 60,000 kD protein considered as a strong immunogen. The mice were pretreated with high

P.C. Endler and J. Schulte (eds.), Ultra High Dilution, 27–33.

dilutions of KLH, 10^{-15}M (C-6), 10^{-17}M (C-7) et 10^{-33}M (C-15) for 8 weeks (3 injections i.p. per week).Then a classic immunization was performed using 1mg of KLH associated with incomplete Freund adjuvant injected in the foot pads.

Following the immunization, an increase in the primary IgM response was observed, probably polyclonal, in almost all the preconditioned groups. A significant increase in the KLH specific IgG response in the KLH - treated groups could also be found (Fig.1).

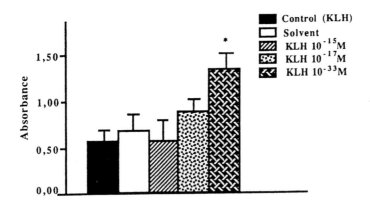

Fig. 1: The titer of IgG anti-KLH antibodies 20 days after immunization. *, $P < 0.05$ (ANOVA test, multiple measurement analysis of variance).

These results suggest that the immune system is able to recognize an antigen even though the solution does not contain any antigen molecules.

2. The action of high dilutions of molecules participating in the education of lymphocytes: thymulin and bursin

The immune system is able to protect the organism by the destruction of foreign compounds or living organisms like bacteria, parasites, viruses (foreign antigens). The immune response is based on the interaction between immunocompetent cells like lymphocytes. Antigen presenting cells cooperate with lymphocytes in order to produce cytotoxic cells and antibodies the function of which is to eliminate the foreign antigens. The efficiency of this system is related to features of the lymphocytes resulting from the "education" of the cells. The education organs, or central lymphoid organs, modify the immunological capacity of those cells born in the bone-marrow. These education organs are the thymus and the Fabricius bursa, the latter of which is only found in birds.

The thymus educates the T lymphocytes which can be divided into two groups: CD4 T lymphocytes or helper lymphocytes are able to cooperate with effector cells and CD8 T lymphocytes the function of which is to destroy target-cells bearing specific antigens by cytotoxicity. These target-cells carry viruses, bacteria or parasites or else tumor cells. These abilities of the TCD4 and TCD8 lymphocytes are acquired at the thymus by the action of various molecules (interleukin 1, interleukin 2, interleukin 4, g interferon etc..), thymic hormones (thymulin, thymosin, thymopoïetin), neuropeptides (substance P, vasointestinal peptide, etc..), prostaglandins, etc.

The Fabricius bursa, found only in birds, communicates with the cloaca. It is the B lymphocyte education organ which is replaced by bone marrow in mammals which have no B individualized education organ. This education gives the B lymphocytes their capability to recognize the antigen and to multiply through the action of the antigens and the cytokines liberated by the T lymphocytes. Then they secrete immunoglobulin antibodies specific to the antigen that has been recognized. These B lymphocytes then transform themselves into plasmocytes which secrete large amounts of antibodies. Amongst the molecules isolated from the Fabricius bursa is bursin which seems to contribute to the education of B lymphocytes. Bursin is a tripeptide able to regenerate the education function of the B in birds bursectomized in the embryonic stage (amputation of the Fabricius bursa).

The use of very low doses of thymulin in mice (a hormone present in the thymus) or of bursin in the bursectomized chicken (present in the Fabricius Bursa) allowed us to modify the immune system of the animals treated.

We observed a significant decrease of the cellular immune response against P815 mastocytoma murine cells when immunized C57BL/6 mice were treated by very low doses of thymulin (from 10^{-3} to 10^{-11}ng) (fig.2): the lowest dose (expressed as C9 in hahnemannian dilutions) was the most effective one (Bastide et al, 1985, 1987).

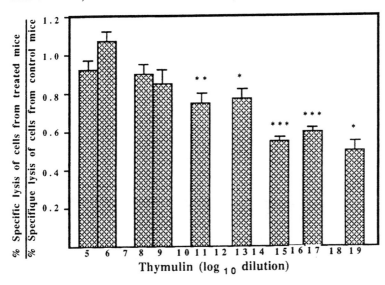

Fig. 2: Allospecific cellular immune response of mice treated with various dilutions of thymulin. Effector cells /target cells = 200.
*, P < 0.05;
**, P < 0.025;
***, P < 0.001
(Newman-Keuls test, multiple comparison).

Thymulin like other thymic hormones (thymosin or thymopoïetin) can act on the T lymphocytes modifying the helper function in humoral immune response: the helper lymphocyte TCD4 releases lymphokines which regulate the antibody synthesis of the B lymphocytes. Highly diluted thymulin (9C) was able to modify the humoral immune response in Swiss mice (Doucet-Jaboeuf et al., 1982, 1984) (fig.3).

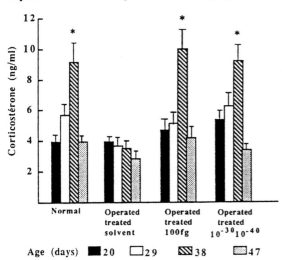

Fig. 3: Plasmatic corticosterone level before immunization (day 20) and following the 1st, 2nd and 3rd immunization (days 29, 38 and 47).
*, P < 0.05 (vs day 20, Student's t test).

We could also demonstrate that seasonal variations of the humoral immune response were amplified in mice treated with high dilutions of thymic hormones (Doucet-Jaboeuf et al., 1984).

Bursin is a molecule which contributes to the education of B lymphocytes like thymulin contributes to the education of T lymphocytes.

Three-day-old chicken embryos were surgically deprived of the tissues which give the Fabricius Bursa at the caudal bud. The egg shell was then closed and the eggs were incubated until hatching. It was shown that embryonically bursectomized chickens were unable to produce specific antibodies after immunization. Embryonic bursectomy resulted in a change in adrenocorticotropic and adrenocortical function. On the 6th and 9th day of incubation, in ovo administration of bursin at very low doses (1mg/ml, 100pg/ml, 100fg/ml) or at high dilutions (pool of 10^{-30}-10^{-40}M dilutions) was performed. No injection was done after hatching. The results obtained in two independent experiments showed that the adrenocorticotropic response after repeated antigenic challenge in the bursectomized chickens was very low. However, the 100fg and the 10^{-30}-10^{-40} M pool treated animals had a normal corticosterone secretion (fig 3) (Guellati 1990, Youbicier-Simo et coll., 1993). Specific antibody secretion against porcin thyroglobulin was restored in 100fg and 10^{-30}-10^{-40} M bursin treated chicks; the production of these antibodies by the bursectomized control chicks was very low. The specificity of the antibodies was checked (fig 4).

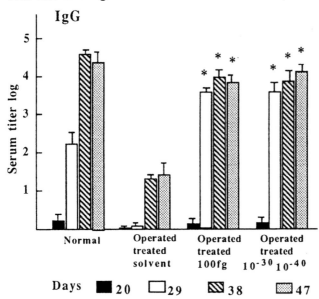

Fig. 4: IgM and IgG antibody levels before immunization (day 20) and after the 1st, 2nd and 3rd immunization (days 29, 38 and 47). * P < 0.05 (vs Operated treated by solvent; Student's t test).

3. The immunomodulating effect of very low doses of cytokines.

Immunocompetent cells and cooperating cells are able to communicate. Cytokines are released mainly by cells belonging to the immune system and able to act on other cells: they are also called interleukines as they allow cellular communication.

Some of them have been used at a dose equivalent to C4 or C9 as immunomodulating agents. It has been shown that homoeopathic dilutions of leucocytary interferon (C4 and C9) can modify the humoral and cellular immune response in immunocompromised mice (New Zealand Black or NZB mice) which show an early thymus involution followed by dysfunction of the immune system (fig. 5) (Daurat et coll. 1986a, 1986b, 1988; Daurat 1988).

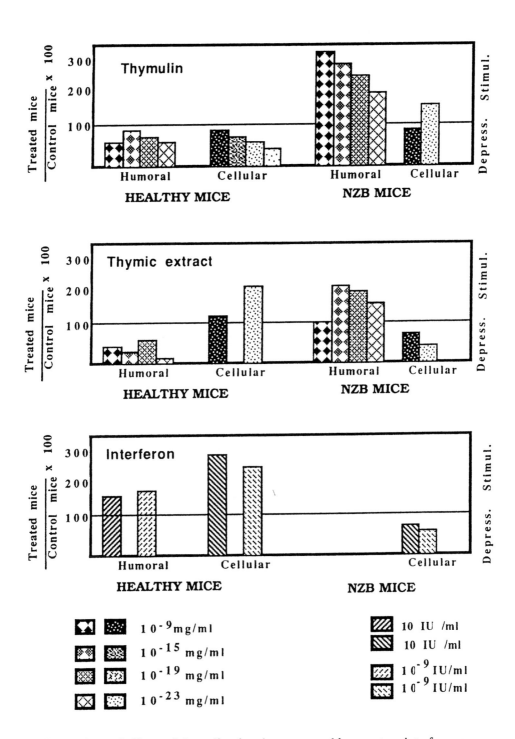

Fig. 5: Comparison of effects of thymulin, thymic extract and leucocytary interferon on the immune response of healthy and immunocompromised mice.

The effect of homoeopathic dilutions of leucocytary interferon on murine macrophages has also been tested (Carrière et coll. 1988a, 1988b, 1990). The C9 dilution (10^{-10} international units) caused a decrease in macrophage activity tested by chemiluminescence in 4-5 or 8-12 week old mice. An increase in macrophage activity was observed in 6-7 week old mice (puberty period). The same homoeopathic dilutions of interleukin 2 (C4 and C9) have also shown immunomodulating effects on specific humoral immune response when the treatment occurred for the primary response (stimulating effect). When the dilutions of interleukin 2 have been injected after repeated immunizations (secondary response), we observed a decrease in the antibody level (Daurat 1988). Immunostimulation models using low doses (picograms, fentograms) of cytostatic substances (plumbagin, azathioprin, colchicin, methotrexate) acting on macrophages or lymphocytes have been published (Wagner et coll., 1988).

4. The immunopharmacological effect of high dilutions of silica in mice.

High concentrations of silica are known to have a toxic effect on macrophages. It has been shown (Davenas, 1987) that homoeopathic dilutions of silica (C9) are able to modify the behaviour of macrophages. When the dilution C9 was administered to the mice in their drinking water, the peritoneal macrophages released a higher concentration of PAF-acether in vitro after stimulation. This acid arachidonic derivative released by the action of silica demonstrates that macrophages are sensitive to high dilutions of silica: furthermore, macrophages are able to secrete various cytokines like transforming growth factor b (TGF b), which acts effectively in wound healing.

An in vivo model has been used (Oberbaum et al., 1992) to demonstrate the action of silica in wound healing. The pathogenesis of silica expresses its capacity to release objects from the organism (splinters, for example). The proposed model used earrings which are fixed to mice's ears. The mice were treated by various dilutions of silica in drinking water compared to the control. Various lots of mice received C5 (10^{-10}), C30 (10^{-60}) or C200 (or 10^{-400}). The results were evaluated by measuring the hole in the ear by image analysis in the treated groups compared to the control group. The results demonstrate that the holes in treated groups are significantly smaller than the holes in the control group; the higher the dilution, the better the effect (fig 6).

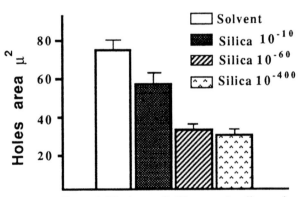

Legend:
- □ Solvent
- ■ Silica 10^{-10}
- ▨ Silica 10^{-60}
- ▨ Silica 10^{-400}

Fig. 6: Effect of dilutions of silica on the dimensions of holes in the ears of the mice (measured by image analysis).

DISCUSSION

In conclusion, the immunological models are very useful to demonstrate the activity of high dilution effects. From a mechanistic point of view, actually no explanations are valuable to understand the real activity of such dilutions.

A theoretical approach using another paradigm has been done concerning this activity: the diluted molecule could be the carrier of the information functioning in an analogical communication with the organism. In this case, the receiver directly understands the message as a signifier and is able to react. This paradigm is easily applied to homoeopathy (Lagache, 1988, 1991) and to immunology (Bastide et al, 1992). In fact, these two sciences are very close and this fact could explain the good results obtained by using high dilutions in the immunological models.

REFERENCES

Bastide M, Doucet-Jaboeuf M, Daurat V. Activity and chronopharmacology of very low doses of physiological immune inducer. Immunol today 1985; 6: 234-235.

Bastide M, Daurat V, Doucet-Jaboeuf M, Pélegrin A, Dorfman P. Immunomodulator activity of very low doses of thymulin in mice. Int J Immunotherapy 1987; 3: 191-200.

Bastide M, Lagache A. The Paradigm of Signifiers. Atelier Alpha Bleue, Paris 1992.

Carrière V, Dorfman P, Bastide M. Evaluation of various factors influencing the action of mouse ab interferon on the chemiluminescence of mouse peritoneal macrophages. Ann Rev Chrono-pharmacol 1988; 5: 9-12.

Carrière V, Dorfman P, Florentin I, Bastide M. Adherent LGL without NK activity modulates the chemiluminescence of mouse macrophages stimulated in vitro by a,b interferon. Intern. J. Immunopharmacol.1988; 10, Supp. 1: 23.

Carrière V, Bastide M. Influence of mouse age on PMA-induced chemiluminescence of peritoneal cells incubated with a,b interferon at very low and moderate doses. Intern J Immunotherap 1990; 6: 211-214.

Daurat V, Carrière V, Douylliez C, Bastide M. Immunomodulatory activity of thymulin and a,b interferon on the specific and the nonspecific cellular response of C57BL/6 and NZB mice. Immunobiology 1986; 173: 188.

Daurat V, Sizes M, Doucet-Jaboeuf M, Guillemain J, Bastide M. An immunopharmacological study on very low doses of mediators in mice. 6th International Congress of Immunology, Toronto, Canada, 1986.

Daurat V, Dorfman P, Bastide M. Immunomodulatory activity of low doses of interferon a,b in mice. Biomed & Pharmacother 1988; 42 : 197-206.

Daurat V. Modulation de la réponse immunitaire de la souris par des doses infrapharmacologiques d'immunomédiateurs. Thèse, Université de Montpellier I. 1988.

Davenas E, Poitevin B, Benveniste J. Effect on mouse peritoneal macrophages of orally administered very high dilutions of silica. Europ J Pharmacol 1987; 135: 313-319.

Doucet-Jaboeuf M, Guillemain J, Piechaczyk , Karouby Y., Bastide M. Evaluation de la dose limite d'activité du Facteur Thymique Sérique. C R Acad Sci 1982; 295, III : 283-287.

Doucet-Jaboeuf M, Pélegrin A, Cot MC, Guillemain J, Bastide M. Seasonal variations in the humoral immune response in mice following administration of thymic hormones. Ann Rev Chronopharmacol 1984; 1: 231-234.

Doucet-Jaboeuf M, Pélegrin A, Sizes M, Guillemain J, Bastide M. Action of very low doses of biological immunomodulators on the humoral response in mice. Intern J Immunopharmacol 1985; 7: 312.

Guellati M. Les relations immuno-corticotropes: effets de la bursectomie embryonnaire et du traitement substitutif par la bursine. Thèse Université du Languedoc 1990.

Lagache A. Echos du Sensible. Paris: Atelier Alpha Bleue 1988.

Lagache A. Evolution des modèles logiques dans la science contemporaine.In: M.Bastide ed. Signals and Images. Paris: Atelier Alpha Bleue 1991, pp 73-114.

Leung-Tack J, Martinez J, Sansot JL, Manuel Y, Colle A. Inhibition of phagocyte functions by a synthetic peptide lys-pro-pro-arg (postin). Protides Biol Fluids Proc Colloq 1986; 34: 205-208.

Oberbaum M, Markovits R, Weisman Z, Kalinkevits A, Bentwich Z. Wound healing by Homeopathic Silica Dilutions in Mice. Harefuah 1992; 123:79-82.

Shinoshara N., Watanabe M., Sachs D.H. and Hozumi N. Killing of antigen-reactive B cells by class II-restricted soluble antigen-specific CD8 cytolytic T Lymphocytes. Nature 1989; 336:481-484.

Wagner H, Kreher B, Jurcic K. In vitro stimulation of human granulocytes and lymphocytes by pico- and fentogram quantities of cytostatic agents. Arzneim.Forsch./Drug Res 1988; 38: 273-275.

Weisman Z, Topper R, Oberbaum M, Bentwitch Z. Immunomodulation of specific immune response to KLH by high dilution of antigen. Abstracts GIRI Meeting, Paris 1992.

Weisman Z, Topper R, Oberbaum M, Harpaz N, Bentwich Z. Extremely low doses of antigen can modulate the immune response. ProcVIII Intern Congress Immunol, Budapest, Hungary, 1992, 23-28 August, 532.

Youbicier-Simo BJ, Boudard F, Mekaouche M, Bastide M, Baylé JD. The effects of embryonic bursectomy and in ovo administration of highly diluted bursin on adrenocorticotropic and immune response of chicken. Int J Immunother 1993, in press.

FURTHER BIOLOGICAL EFFECTS INDUCED BY ULTRA HIGH DILUTIONS.
INHIBITION BY A MAGNETIC FIELD

J. Benveniste

SUMMARY

Several studies on biological effects of agitated highly diluted substances 1. on cell-lines, 2. on isolated hearts and 3. on mice in vivo are presented. In these studies, statistically significant differences have been observed between the effects of the respective test substances and those of their controls, that is, the dilution buffer in analogous preparation.

It was shown that treatment with a magnetic field suppressed the effect of an agitated high dilution. A mechanism is discussed for the transmission of information from the molecular mother substance, including intermolecular communication by oscillating electromagnetic fields and perimolecular coherent water separated from the substance molecule during the process of agitation.

ZUSAMMENFASSUNG

Es werden verschiedene Studien zur Wirkung verschüttelter hochverdünnter Substanzen auf 1. Zellkulturen, 2. isolierte Herzen und 3. Mäuse in vivo vorgestellt. In diesen Studien wurden signifikante Unterschiede zwischen den Wirkungen der einzelnen Testsubstanzen und der Wirkung ihrer Kontrollen, nämlich Verdünnungsbuffer in analoger Zubereitung, festgestellt. Es wurde gezeigt, daß die Behandlung mit einem Magnetfeld die Wirkung einer verschüttelten Hochverdünnung aufhob. Ein Mechanismus der Informationsübertragung von der molekularen Ausgangssubstanz wird diskutiert, der intermolekulare Kommunikation durch elektromagnetische Felder und perimolekulares kohärentes Wasser, das während des Verschüttelungsprozesses von dem Substanzmolekül getrennt wird, einschließt.

INTRODUCTION

Our laboratory is at present conducting several studies on the biological effects of highly diluted ligants. Beyond the "memory of water", these studies could have widespread consequences for biology, revealing fundamental and, until now, unknown aspects of molecular interaction and recognition (see also ref. 1-3).

METHODS AND RESULTS

1. In vitro experiments

1.1 Effects on cell-lines

In the wake of heavy metal poisoning, serious disorders, either inflammatory or strictly immunological, occur. The toxic effects of cadmium (Cd) on human cell-lines were therefore studied. When the cells are cultured in the presence of 5 to 10 μ M Cd, a mortality rate of generally 40 % to 50 % - but sometimes up to 100 % - is observed.

However, when they are pretreated with ponderal though non-toxic doses or with high-dilutions of Cd (dilution log 16-25 or 26-35) for a period of several days, a significant modulation of cellular activation and growth is observed, either directly, i.e. before the addition of toxic concentrations of Cd, or after it. Functional tests used to include coloration

35

P.C. Endler and J. Schulte (eds.), Ultra High Dilution, 35–38.

with trypan blue, flow cytometry, the incorporation of tritiated thymidine and MTT. Amongst other implications of this kind of results, they must be taken into account for the protection of the environment.

1.2 Cardiac effects. Inhibition by a magnetic field

One of the most productive models is the study of vasoactive amines on the isolated heart. Isolated guinea pig (n = 77) or rat (n = 49) hearts were perfused at constant pressure in a Langendorff system with highly diluted agonists. A significant time-dependent modification (p < 0,0001) of the guinea pig heart coronary flow was induced by histamine dilutions (log 31-41) but not by the diluted / agitated buffer (diluted histamine vs diluted buffer, p = 2 x 10^{-7}) (Fig. 1).

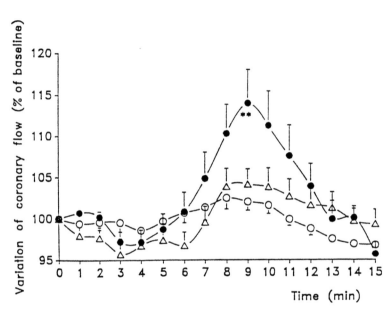

Fig. 1: Effect of highly diluted histamine (log 31-41) on the coronary flow of isolated perfused guinea-pig heart (full circles, n = 20), diluted vehicle (white circles, n = 24) and diluted histamine pretreated with a magnetic field (white triangle, n = 20). **p = 2 E^{-7} between histamine versus either magnetized dilutions or buffer.

Treating high dilutions of histamine with heat (70°C, 30 min) or with a magnetic field (50 Hz, 150 oersteds, 15 min) totally suppressed their effect whereas molecular histamine was unaffected. Coronary flow variations - also inhibited after exposure of the dilutions to a magnetic field - were observed in the rat heart treated with highly diluted serotonin. Modifications of maximal heart force and rate were also recorded on the rat and guinea pig heart during the injection of highly diluted histamine and serotonin respectively. These results were presented at the congress of the Federation of American Societies for Experimental Biology in Atlanta, USA, in April 91 (4) and in 92, the latter reporting the effects of highly antigen on sensitized guinea pig hearts (5). The effects of highly diluted serotonin (log 31-41) on the isolated rat heart have been recently observed by two independent teams of French pharmacologists.

2. In vivo experiments

The action of high dilutions of silica, a substance which, in poderal doses, is cytotoxic for macrophages, was studied on the synthesis, by mouse peritoneal macrophages, of an ether-lipid mediator of inflammation, paf-acether and its precursor, lysopaf-aceter (6). In the course of three series of experiments, of which two were blind, female C57 BL6 mice (n = 252) received orally, for a period of 25 days, $1,66 \times 19^{-11}$ M or $1,66 \times 10^{-19}$ M silica (final concentrations) or saline or lactose subjected to the same dilution (control groups). The perotoneal macrophages from silica-treated mice were stimulated in vitro by zymosan and their production of paf-aceter was amplified from 44,2 to 67,5 %, depending on the experiment, with respect to control mice. These differences were highly significant in all experiments ($p < 0,01$ to $p < 0,05$). There was no effect on lysopaf-acether synthesis. These results clearly demonstrate a cellular in vitro effect of high dilutions of silica.

These experiments, the results of which are indifferent to their being conducted blindly or openly, show that a specific biological activity can be induced or modulated by substances so highly diluted that the probability of a single molecule remaining in the volume used is less than 1×10^{-10}. No trace of the starting molecule was detected in high dilutions even using extremely sensitive methods (<1 ng/ml), such as electrodetection (serotonin) or spectro-fluorimetry (histamine). Another argument against a molecular basis for the activity of high dilutions is that such activity is eliminated by purely physical processes like heating (a dilution of a thermostable molecule) at 70°C or treating with a magnetic field. An artefact has been often suggested but not a single good hypothesis has so far been proposed. Controls used (buffers alone in the absence of ligand) are submitted to a dilution / agitation identical to that of the active solutions and, in order to observe the effect, the active substance must be present when beginning the dilution process.

In sum: 1) statistically significant differences have been observed between the effects of highly diluted substances and those of their controls, that is, the dilution buffer diluted in itself. 2) These effects have been obtained in several biological systems, underlining the ubiquitous nature of the phenomenon.

DISCUSSION

Hypothetical mechanism of biological effects at high dilution.

Oscillating fields and intermolecular communication.

If the active molecules are absent (or present in such low numbers that they cannot induce a response), we have to admit that the specific effect which we detect is of non-molecular origin. The physical basis of this phenomenon is at present unknown. Nonetheless, the suppressive effect of a magnetic field is compatible with a hypothesis proposed by Del Giudice and Preparata (7): an interaction between the electric dipoles of water and the radiation field of a charged molecule, an interaction which generates a permanent polarization of water, which thus becomes coherent, somehow like a laser. Results of the biological effects of magnetic fields are accumulating in literature (8-10).

Establishing the capacity of water (polarized ?) to trigger biological effects would, alone, represent a major advance. If water has such a capacity as an agent of transmission, it is because it belongs to the immediate environment of biological molecules (15,000 molecules of water for each protein molecule). It appears that during agitation we artificially separated the molecule from its message carried by the perimolecular coherent water. Moreover, this process is possibly linked to the mechanism of molecular communication, until now not elucidated (11). At the present time, we have no idea at all how molecules interact and transmit specific information from one to the other: the question as to the nature of the intimate mechanism(s) of molecular signalling, which are fundamental to all fields of biology, is not only without response, but has not even been posed. The essential molecular functions appear in fact to be determined by electro-magnetic mechanisms. One - if not the only - role of molecular structures is the carrying of electric charges which generate, in the aqueous environment, a field specific to each molecule. Those exhibiting such co-resonating or opposed fields ("electro-conformational coupling", 11) can thus communicate, even at a distance. Therefore a minute variation in the structure of molecules (plus or minus a P atom, a rearrangement of an amino acid ...), which even slightly modifies their radiating field, would allow their message to be received or not by a receptor, as in the FM waveband. Strictly, structural changes, or those of net charge, cannot take into account the sensitivity and specificity of these mechanisms of recognition and activation. And what about those agonist / antagonist couples, the respective structures of which have nothing in common ?

Theoretical and experimental advances are needed in order to establish the physical basis of these mechanisms; advances which could lead to the deciphering of the language of molecules. Considerable progress can be expected in the comprehension of fundamental biological activities and of traditional therapeutic approaches (be they homoeopathic or allopathic), as well as the opening of new pharmacological avenues.

REFERENCES

1. Davenas E, Beauvais F, Amara J, Oberbaum B, Robinzon A, Miadonna A, Tedesci A, Pomeranz B, Fortner P, Belon P, Sainte-Laudy J, Poitevin B, Benveniste J. Human basophil degranulation triggered by very dilute antiserum against IgE. Nature 1988; 333: 816-818.
2. Benveniste J, Davenas E, Ducot B, Cornillet B, Poitevin B, Spira A. L'agitation de solutions hautement diluées n'induit pas d'activité biologique spécifique. CR Acad Sci Paris 1991; série II: 461-466.
3. Benveniste J, Davenas E, Ducot B, Spira A. Basophil achromasia by dilute ligand: a reappraisal. FASEB J 1991; 5: A1008.
4. Hadji L, Arnoux B, Benveniste J. Effect of dilute histamine on coronary flow of guinea-pig isolated heart. Inhibition by a magnetic field. FASEB J 1991; 5: A1583.
5. Benveniste J, Arnoux B, Hadji L. Highly dilute antigen increases coronary flow of isolated heart from immunized guinea-pigs. FASEB J 1992; 6: A1610.
6. Davenas E, Poitevin B, Benveniste J. Effect on mouse peritoneal macrophages of orally administered very high dilutions of silica. Eur J Pharmacol 1987; 135: 313-319.
7. Del Giudice E, Preparata G, Vitiello G. Water as a free electric dipole laser. Phys Rev Lett 1988; 6:1085-1088.
8. Weaver JC, Astumian RD. The response of living cells to very weak electric fields: the thermal noise limit Science 1990; 247: 459-462.
9. Pool R. Electromagnetic fields: the biological evidence. Science 1990; 249:1378-1381; Is there an EMF-cancer connection ? 1990; 249: 1096-1098.
10. Smith CW, Best S. Electromagnetic Man. JM Dent and Sons Ltd, Avon, UK (1989)
11. Tsong TY. Deciphering the language of cells. Tr Biochem Sci 1989; 14: 89-92.

For further references, see the contribution of Citro et al.

A ZOOLOGICAL EXAMPLE ON ULTRA HIGH DILUTION RESEARCH. ENERGETIC COUPLING BETWEEN THE DILUTION AND THE ORGANISM IN A MODEL OF AMPHIBIA

P.C. Endler, W. Pongratz, R. van Wijk,
F.A.C.Wiegant, K. Waltl, M. Gehrer, H. Hilgers

SUMMARY

The present studies investigate the influence of extremely dilute thyroxine on the climbing activity of juvenile highland frogs and on two transitions in the metamorphosis of these animals.

The highly diluted thyroxine (10^{-30}, 1.125×10^{-30}M, "D30") to a small but statistically significant extent slows down the climbing activity of juvenile frogs, when once added dropwise to the aquaria water at the juvenile stage. It is also demonstrated that this dilution slows down climbing activity, when sealed in hardglass vials that were hung into the aquaria, without direct contact between the dilution and the organisms. In the experiments, water, analogously prepared and applied, was used as control.

It was further shown that preclimax metamorphosis (the transition from the two-legged to the four-legged tadpole) can be influenced by the test dilution in two typical ways: if the dilution was added at intervals of 48 hours from the two-legged stage on, it caused an inhibition of the observed transition. If the same dilution was added at intervals of only 8 hours from the two-legged stage on, it caused an acceleration of the very transition.

Further, when the dilution was added during and after preclimax metamorphosis at intervals of 48 hours, it also slowed down postclimax metamorphosis (the transition from the four-legged, tailed tadpole to the juvenile frog), expressed in terms of climbing activity; and when it was added at intervals of 8 hours, it also accelerated postclimax metamorphosis.

The method of constant indirect application, where the dilution was sealed in glass vials, showed an intermediate effect between the dropwise application at intervals of 48 hours and dropwise application at intervals of 8 hours: when constant indirect application was used, preclimax metamorphosis was slowed down and postclimax metamorphosis was accelerated.

Although the different protocols were applied in different experiments, the reproducability of several protocols by different researchers, even in different laboratories, suggests that the effect of extremely diluted thyroxine depends (1) on the stage of development where the first addition takes place, (2) on the frequency of application and (3) on the type of application.

ZUSAMMENFASSUNG

Die hier vorgestellten Studien untersuchen den Einfluß von extrem verdünntem Thyroxin auf die Kletteraktivität juveniler Hochlandfrösche und auf zwei Übergänge in der Metamorphose dieser Tiere.

Das hochverdünnte Hormon Thyroxin (10^{-30}, 1.125×10^{-30} M, "D30") verlangsamt die Kletteraktivität juveniler Frösche zu einem geringen, aber statistisch signifikanten Grad

P.C. Endler and J. Schulte (eds.), Ultra High Dilution, 39–68.

wenn es im Juvenilstadium einmalig tropfenweise dem Aquarienwasser zugesetzt wird. Es wird auch gezeigt, daß diese Verdünnung die Kletteraktivität verlangsamt, wenn sie in Hartglas-Phiolen versiegelt ist, die, ohne direkten Kontakt zwischen der Verdünnung und den Organismen, ins Aquarienwasser gehängt wurden. In den Experimenten wurde analog zubereitetes und zugegebenes Wasser als Kontrolle verwendet.

Weiters wurde gezeigt, daß die Präklimax-Metamorphose (der Übergang von der zwei-beinigen zur vierbeinigen Kaulquappe) durch die Testverdünnung auf zwei typische Arten beeinflußt werden kann: wenn die Verdünnung vom Zweibeinstadium an in Abstän-den von 48 Stunden zugegeben wurde, bewirkte sie eine Hemmung des beobachteten Überganges. Wenn diese Verdünnung vom Zweibeinstadium an in Abständen von nur 8 Stunden zugegeben wurde, bewirkte sie eine Beschleunigung ebendieses Überganges.

Wenn, weiters, die Verdünnung während und nach der Präklimax-Metamorphose zugege-ben wurde, verlangsamte sie auch die Postklimax-Metamorphose (den Übergang von der vierbeinigen, geschwänzten Kaulquappe zum juvenilen Frosch), wie sie sich durch die Kletteraktivität ausdrückt; und wenn sie in Abständen von 8 Stunden zugegeben wird, be-schleunigt sie auch die Postklimax-Metamorphose.

Die Methode der konstanten indirekten Applikation, bei der die Verdünnung in Glas-phiolen versiegelt war, ergab einen zwischen den Wirkungen von tropfenweiser Zugabe in Abständen von 48 Stunden und tropfenweiser Zugabe in Abständen von 8 Stunden liegen-den Effekt: wurde die konstante indirekte Applikation verwendet, wurde die Präklimax-Metamorphose verlangsamt und die Postklimax-Metamorphose beschleunigt.

Obwohl die verschiedenen Protokolle in unterschiedlichen Experimenten angewendet wurden, deutet die Reproduzierbarkeit einzelner Protokolle durch mehrere unabhängige Forscher, ja sogar in verschiedenen Labors, darauf hin, daß die Wirkung von extrem ver-dünntem Thyroxin (1) vom Entwicklungsstadium, in dem die erste Zugabe erfolgt, (2) von der Häufigkeit der Zugabe und (3) vom Typ der Zugabe abhängt.

INTRODUCTION

Much important work has been done to research into Ultra High Dilutions on animal mo-dels (Bastide et al. 1985, 1987; Bildet 1975; Cazin 1987; Daurat et al. 1988; Davenas et al 1987; Fisher et al 1987; Hadji et al. 1991; Harisch et al. 1987, 1990; Herkovits et al. 1988, 1989; Khuda-Bukhsh and Banik 1990; Lapp et al 1958; Larue et al. 1985; Schaefer 1984; Sukul et al. 1986, 1990 and others; see Righetti 1988; Roth 1991; Endler and Schulte - eds. - in press). The representative model described here is based on the development of amphibian larves and the influence of thyroxine.

Thyroxine (tetra-iodo-thyronine, T_4, a thyroid hormone) plays an important role in the regulation of the speed of the metamorphosis of amphibia. When thyroxine is added to the water of an aquarium with a final concentration of 1 or 2×10^{-8} part by weight (e.g. 20 nM /l), it induces or accelerates, respectively, the metamorphosis of amphibia (Giersberg 1972; Ganong 1974; Wehner and Gehring 1990). A thyroxine concentration of 10^{-7} part by weight or below causes an acceleration of development to such an extent that deformities appear in the animals (Pitt-Rivers and Trotter 1964). In previous experiments, L-thyroxine sodium pentahydrate (Sigma) at a concentration of 10^{-9} parts by weight in the basin water caused an acceleration of metamorphosis of 10-17%. Fig. 1, above, shows the cumulative frequency of animals having reached the four-legged stage in a control group (54 animals) and Fig. 1, below, shows the bias in the respective cumulative frequencies between this untreated con-trol group and a group (48 animals) exposed to an aqueous thyroxine dilution 10^{-9} from the two-legged stage (about Gosner's stage 31, see Gosner 1964) on.

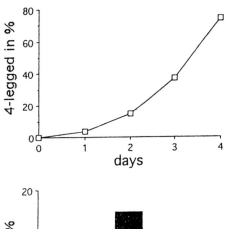

Fig. 1: The influence of an aqueous dilution 10^{-9} part L-thyroxine sodium pentahydrate by weight on preclimax metamorphosis of Rana temporaria. Abscissa: the number of days of duration of the experiment. For further details, see text.

In a further study, it is reported that hypophysectomized Rana pipiens larves, when immersed with different concentrations of thyroxine at an early non-feeding stage before gill appearance, reached only certain developmental stages and then remained at these stages. Immersion at a concentration of $0.002 \times 1{:}10^9$ part by weight DL-thyroxine/ liter sufficed to reach an early two-legged stage, whilst immersion at a concentration of about $0.2 \times 1{:}10^9$ part by weight was necessary to reach the four-legged stage, and at a concentration of $0.6 \times 1{:}10^9$ part by weight to reach the juvenile stage (Kollros 1961).

On the other hand, lack of thyroxine due to thyroidectomy causes a standstill of metamorphosis (Wehner and Gehring 1990, Pitt-Rivers and Trotter 1964).

Sensitivity of amphibian larves towards thyroxine is reported from very early stages on - before gill reduction - (Moser 1950, with regard to Rana temp.; Tata 1968; Prahlad and De-Lanney 1965; Chou and Kollros 1974). Precedent tail shrinkage can already be induced in early stages, this effect, however, can be achieved in two-legged tadpoles after a much shorter period (Chou and Kollros 1974). Larves from the non-feeding stages on up to the two-legged stages reacted with tail shrinkage to a thyroxine solution of $50 \times 1{:}10^9$ part by weight at $23°C$. Tail shrinkage occurred in all tadpoles with the latent period of response being 14.2 days in the animals at the larve's non-feeding stage and 4.9 days in the two-legged tadpoles. It is generally agreed that most larval tissues become reactive to thyroid hormones at a stage in larval development well before significant amounts of thyroid hormones are available (Dodd and Dodd 1976). However, in a strict temporal sequence, only different tissues are always concerned selectively, the hindlimb buds being the last tissue reacting in early tadpole development. Gill shrinkage induced by exogenous thyroxine has been reported for an urodele (Prahlad and De-Lanney 1965).

In general, thyroxine plays a more important role as an active hormone in premetamorphic tadpoles than it is thought to play in mammals (Galton and Cohen 1980). Early reactions to tri-iodo-thyronine (T_3) are more marked, being manifest already 2-4 days after fertilisation, than those to T_4 (Tata 1968). No literature has been found on the effects of thyroxine sodium pentahydrate.

The natural plasma level of iodine /thyroxine changes during and after spontaneous metamorphosis, with a slow increase during the two-legged stages and a rapid increase from the four-legged stage on, shortly before the reduction phase of the tail starts, followed by a rapid decrease during that reduction phase (Just 1976; Weil 1986). This high plasma level is interpreted to be due to an increasing synthesis of thyroid hormone (before tail resorption begins, Dodd and Dodd, 1976, 493) and increasing release into the circulation on the one hand, but also to be due to increased tissue saturation with thyroxine. Thus, tissue would become increasingly avid for thyroxine in the time before the plasma level increases (Dodd and Dodd 1976).

In general, as is known from experiments with radio-iodine, iodine begins to be trapped and stored already at the non-feeding stages, prior to the appearance of thyroid follicles and independent of TSH stimulation (Saxén 1958; Kaye 1961). There is an about 15-fold rise in the uptake during the transition from the two-legged to the four-legged stage (Kaye 1961). Uptake is then maximal at the time when the tadpole enters the four-legged stage (Dodd 1955; Hanaoka 1966). This is assumed to be the period of most active synthesis and storage of thyroid hormone, before the stored hormone is released to mediate climax (see above). After the tadpole has entered the four-legged stage, but already before tail shrinkage, due to thickening of the skin, shrinkage of the gills and cessation of feeding, the iodine absorption is reduced to a few percent (Dodd and Dodd 1976). Experiments have shown, that reactions can still be induced when iodine is injected, but not when it is added to the water containing the animals (Kaye 1961; Neuenschwander 1972) . However, at this time, there is already an accumulation of iodine in the gut (Lipner and Hazen 1962).

If the animals are treated with moderate doses of $NaClO_4$ in order to block their thyroid gland before the two-legged stages, metamorphosis does not proceed; if, as it was the case in a small experiment, they were treated at the stage when the hindlegs were already developed, metamorphosis continued (Bufo viridis, Stelzer 1990). Furthermore, it is known that during climax, when most of the physiological transformations occur, the animals are most sensitive to stress (Stelzer 1990).

Thyroxine is, however, not only able to enhance metamorphosis, but, when applied at a molecular concentration of both $1:10^6$ as well as $1:10^7$ part by weight (L-thyroxine sodium pentahydrate) in the basin water at the two-legged stage, it blocks the development of the animals due to deformation and death (data that were undeliberately found in previous experiments). Interestingly, body length increased to about 150% of that of the control group in the treated group, and tail length decreased to about 50%. Front limbs only occured in the control group, but not in the thyroxine-groups. Before the front limbs occured, on the sixth day after application, all tadpoles treated with thyroxine died. At that point of time, the tadpoles in the control group had already started to enter the four-legged stage. Thus, it can be stated that thyroxine in a molecular dose is able both a. to accelerate as well as b. to inhibit the development of forelegs.

No literature has been found on the dose-effect relationship of low concentrations of thyroxine on induced /accelerated metamorphosis. In any case, no effects are expected at a degree of dilution $1:10^{15}$, which corresponds to the accuracy of the measurement when in experiments on high dilutions aquaria water is checked for contamination with molecular thyroxine (K. Hagmüller, Institute for Zoology, University of Graz).

In several publications it has been suggested that highly diluted substances may influence metabolic /physiological processes, even in dilutions around or beyond Avogadro's number. In basic biological research, influences of such substances were observed on the immune system (Bastide et al. 1985, 1987; Poitevin et al. 1986, 1988; Davenas et al. 1987, 1988; Daurat et al. 1988; Benveniste et al. 1991), on certain enzymatic patterns in organs (Harisch et al. 1986, Petit et al. 1989) or on physiological parameters such as heart or ileum function (Hadji et al. 1991; Sukul et al. 1990). In clinical research, pathological processes are influenced by highly diluted substances (Reilly et al. 1986; Fisher 1989; Gibson 1980; see Kleijnen et al. 1991).

With respect to metamorphosis, already König (1923) has described that tadpoles are sensitive to highly diluted substances (silver nitrate).

This has led to the speculation that highly diluted thyroxine may influence the metamorphosis and the activity of amphibia. Our starting point to perform the studies discussed here was a preliminary investigation showing that a high dilution of thyroxine is able to enhance amphibian metamorphosis (Endler et al 1989).

In order to measure any influence of the test dilution, several conditions might be critical:

   a. the physiological background determined by the time in season the experiment is performed at,
   b. the stage of development the thyroxine dilution is added at,
   c. the number of additions,
   d. what type of later event will be scored for evaluating any influence, like the development of legs, loss of tail, climbing activity etc.

In the following studies our main approach was to add thyroxine in a high dilution ($1:10^{30}$) to frogs during metamorphosis at an early stage, i.e. two-legged. We chose one specific dilution, but varied the number of additions; the addition was done once every 48 h, or once every 8 h. At least two late events have been scored for the evaluation, i.e. the development of legs and climbing activity.

Two additional approaches were added to this main protocol. The first question was the evaluation of the addition at a time immediately before the scoring of the parameter. This has been done for the parameter 'climbing activity'.

The second question was the evaluation of the addition of the high dilution in a way that it was not mixed with the water of the aquarium. For this purpose it was applicated in a closed vial.

METHODS

The different studies described here were worked out by one researcher. The main experiments were performed blindly under the supervision by the zoological university institute Graz (G. Fachbach, department of developmental biology; G. Kastberger) and the Ludwig Boltzmann institute for homoeopathy (M. Haidvogl); furthermore, key experiments were independently and blindly repeated by other researchers in their own or in our laboratories (see Tab.1). For further details on the set-up of these control experiments, see Endler et al. 1993 b. Further information is in preparation for print.

Fig. 2 (next page) gives a schematic survey over the different transitions observed in the studies..

Fig. 3 (next page) gives a survey over the different methods of application of the test substance used in the studies.

Data from Fig. 2 and 3 are included in the comprehensive Tab. 1. The description of the protocols following below can be essentially derived from the list of variable conditions as presented in this Tab. 1. The test dilutions were prepared by Peithner KG, Vienna, by the Boltzmann Institute for Homoeopathy, Graz, and by VSM, Alkmaar, respectively. In the experiments, sets of dilutions from different origin were used in parallel, if that was possible.

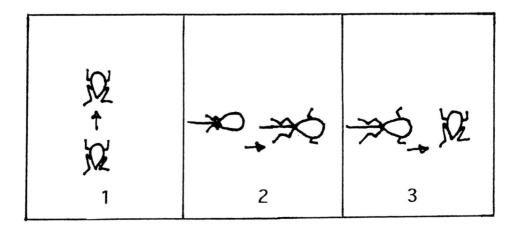

Fig. 2: Scheme of developmental stages of amphibia observed in the studies. 1, transition from the juvenile in the water to the juvenile at land; 2, preclimax transition from the two-legged to the four-legged tadpole; 3, postclimax transition from the tadpole to the air-borne frog. For further information, see text.

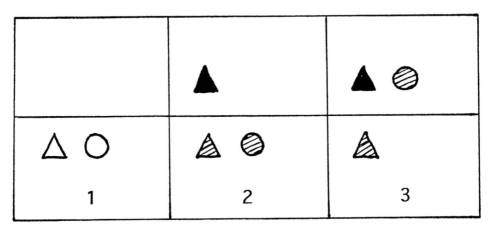

Fig. 3: Different methods of application of thyroxine D30. 1, short-term application; 2, middle-term application; 3, long-term application; white triangle, dropwise application once; white circle, application in sealed glass vial once; grey triangle, dropwise application at intervals of 48 hours; black triangle, dropwise application at intervals of 8 hours; grey circle, constant application in sealed glass vial. For comparison with the scheme of observed developmental transitions of the animals, see Fig. 2., for explanation, see text below.

The main data discussed in this paper were found with Rana temporaria from an Austrian highland biotope. For information on experiments with Rana temp. from the low site, see Endler et al. 1993b.

Tab.1

Type of protocol	1.1	1.2	2.1	2.2	2.3	3.1	3.2	3.3
Type of animals	R	R	R	R	R; B	R	R	R; B
Biotope	H	H	H;L	H	H,L;H	H,L	H	H,L;H
Starting stage	j	j	2	2	2	2	2	2
Type of treatment	d/o	v/o	d/48	d/8	v/c	d/48	d/8	v/c
Measurement stage	j	j	4	4	4	j	j	j
Parameter	c.	c.	4.	4.	4.	c.	c.	c.
Researcher	E,P	E,P,G,Wi	E,P,V; P	E,P	H,P,G; E,P; 68E,Wa	E,P	E	H,P,G; P; E,Wa
Laboratory	B,K	B gr	B,gr,UU; B	B	UV,gr,B; B,K; B	B,gr	B	UV,gr,B; B,K; B
N of animals	177	410	1139 288	864	469 1224 451	47	108	510 144 232

Tab. 1: Variable conditions in methodology. Type of protocol: for the respective chapters, see below. Type of animals: R, Rana temporaria; B, Bufo bufo. Biotope (place where the animals originally were taken from): H, highland site; L, low site. Starting stage (here, the first addition of the test substance was done): j, juvenile frog; 2, two-legged tadpole. Type of treatment: d, direct dropwise application of the test substance; v, indirect application in vial; o, application only once or during minutes, respectively; 48, application at intervals of 48 h; 8, application at intervals of 8 h; c, constant presence of the vial. Measurement stage (stage at which the final observation was made): j, juvenile frog; 4, four-legged stage. Parameter (observed parameter): c., climbing activity; 4., four-legged stage. Researcher (researchers involved): E, Endler; P, Pongratz; G, Gehrer; Wi, Wiegant; V, van Wijk; H, Hilgers; Wa, Waltl. Laboratory: B, site associated with the Ludwig Boltzmann Institute, Graz, Austria; K, site in the Koralpe region, Styria, Austria; gr, University greenhouse, Graz; UU, University of Utrecht; UV, University of Vienna. N of animals, total number of animals in the respective protocol or sub-protocol. For further information, see the text below and the list of contributors to this volume.

1. Short-term treatment

The immediate influence of the test dilution was studied on the climbing activity of non-pretreated juvenile highland frogs (see Fig. 2, studies 1).

1.1 Effect of the test substance, directly added at one point of time, on activity

Animals and staging: Rana temp. juveniles at the four-legged, tailless stage of the juvenile frog (about stage 46 according to Gosner's staging table) were taken from a highland pool 1600 m above sea level (Koralpe, Styria, Austria). For this population, metamorphosis takes place from August to November and is completed within days to weeks towards the end of the season. In comparison with Rana temp. from lower sites with metamorphosis earlier in summer, metamorphosis of the highland population takes place later in the season and proceeds slower.

Treatment: A few drops of the test dilution thyroxine D30 (for preparation, see below) as well as the control water D30 were added blindly to the water in the corresponding basins, containing the animals, followed by gentle shaking of the basins.

Observation of climbing activity: White plastic basins (l by w by h : 34 by 22 by 14.5 cm, Firma Miraplast, Austria) were each filled with 0.5 l of pure lake water. The juvenile frogs that spontaneously climbed out of the water and up the walls of the basins were counted. In some cases, photos and a video were prepared as documentation. The climbing activity of juveniles in the course of three minutes was investigated by always comparing two groups of animals in two different basins. All the animals that had climbed up the walls to a certain height were put back into the water before measurement, then both basins were shaken gently to provide the same starting conditions for the two groups. The number of climbers that had brought at least four of the five 'extremities' (including the stump of the tail) out of the water was counted (10 seconds and) 1, 2, and 3 minutes after last shaking the basins. This experiment was repeated 5 times within 20 minutes. In the control experiment (step A), the climbing activity was measured as described above without any treatment. In step B, the frogs were treated with dilution thyroxine D30 or dilution water D30. Immediately after this treatment, the climbing activity was observed as above (blind experiment). In step C (cross-over treatment), the water D30 - group from step B was treated with dilution thyroxine D30 and vice versa. The immediate observation was not performed. After two days, the climbing activity was observed within 20 minutes as above. With regard to the fact that the two groups according to the treatment with dilutions thyroxine D30 and water D30, respectively, could already be differentiated at the end of step B by the (blind) researcher, the later step C can only be discussed as an open experiment.

Further conditions: The positions of the basins were changed regularly to avoid any influence of spatial factors. Only very little indirect light was used. The room temperature was kept above 18°C. Any external shocks were avoided. If an animal had reached the edge of the basin, it was put back using a disposable wooden stick.

Preparation of testing solutions: The tadpoles were observed under the influence of tetra-iodo-thyronine sodium pentahydrate (Sigma) specially prepared in an aqueous solution $1:10^{30}$ part of weight. The stock solution had a concentration of $1:10^4$ part tetra-iodo-thyronine sodium pentahydrate of weight; it was diluted in steps of $1:10$. The diluted solution (10^{-5}, 10^{-6} etc.) was agitated according to standardized instructions (Homöopathisches Arzneibuch, 1991): at every step, using a disposable pipette, a sterile bottle is partly filled with the dilution, and is pushed down at short regular intervals (e.g. against a rubber impediment) to create mechanical shocks. The test solution prepared in this way was called dilution thyroxine D30. As reference solution, the solvent, pure distilled water, was prepared in an analogous way (dilution water D30).

Check for contamination: One set of the probes was checked for contaminations by T_4, T_3, TSH and iodine independently and blindly at a university institute (G. Passat, I. Med. Klinik, Graz). Hormone-active contaminations were excluded up to the accuracy of the measurement, i.e. the probes were in any case free from contamination within the range of classical pharmacological doses.

Data base: A number of small sub-experiments was performed. Two basins were always used for the treatment with dilution thyroxine D30 and water D30, respectively. Each basin usually contained 13 animals. The numbers of animals per basin were alike in each experiment.

Evaluation of data: As the results from the two researchers were in general comparable, the following evaluation was made both for the pooled data of the two researchers as well as for the data of each single laboratory. The cumulative frequency of climbers (F_c) was compared to that of animals remaining in the water. For every single of the five successive repetitions of the observation, the cumulative frequencies of climbing attempts in the two groups were evaluated as a 4-field table by a chi-square test at the specific measuring points in time. This differentiated evaluation was made for step 1 (reference), for step 2 (treatment) and for step 3 (cross-over treatment). Furthermore, the so-called 'survival analysis' (Lee /Desu, no year given; using the statistical package SPSS/PC) was used to determine the statistical significance of the bias in the number of animals remaining in the water between the two groups according to the treatment. This test sums up the data from all measuring points in time (minute 1, 2 and 3) and is more stringent than the chi-square test applied on the data from the single measuring points in time (survival analysis was independently calculated by F. Wiegand, Utrecht).

In order to investigate their reliability, the single repetitions of the climbing observation were compared with the Wilcoxon test at the measuring points in time.

Furthermore, in order to get a rough survey over all the respective data, the number of climbing frogs was added up for each step (A, reference; B, treatment; C, cross-over treatment) at the three different points of time (1, 2 and 3 minutes) after the start of the respective step in the experiment. These pooled data were evaluated with the chi-square test and survival analysis.

1.2 Effect of the test substance, sealed in glass vials and applied during a few minutes, on activity

Theoretically, in the thyroxine dilution used in the experiments described above in chapter 1.1, there will not be any molecules left from the original solution. However, in order to prevent the observed effects from being ascribed to a few remaining molecules, we decided to study the effect of thyroxine D30 when applied indirectly.

It has been shown in a previous study (van Wijk and Wiegant 1989), that such specially prepared solutions can influence a living system even if they are applied in sealed glass vials.

Animals and staging: Rana temp. juveniles were taken from the highland pool as described in chapter 1.1.

Exposition to the probes and observation of climbing activity: 10 ml of the probe dilution thyroxine D30 and of control, respectively, were sealed in the glass vials. The coded vials were first hung into a common water bath to make sure that they were not individually contaminated outside, and then hung into the corresponding basins.

In a first step (A) which can be taken as a zero-time-control, the climbing activity was measured without any treatment as described in chapter 1.1. In a second step (B), the frogs in basin 1 were treated with sealed dilution thyroxine D30, and the frogs in basin 2 with sealed water D30. In addition, in a third step (C), the frogs in basin 1 were treated with sealed water D30 and the frogs in basin 2 with sealed thyroxine D30. Climbing activity was observed after 10 seconds, 60, 120 and 180 seconds as above; further, additional measurements were made at 30, 90 and 150 seconds.

Further conditions, Preparation of testing solutions and *Check for contamination* were described in chapter 1.1.

Ampoules used: The dilutions were sealed in hardglass vials with an optical transmission spectrum starting from about 350 nm; the optical transmission spectrum was limited by the properties of the aquaria water to wavelengths less than about 2500 nm. The method of application of test substances in sealed ampoules had been used in therapeutical practice and by different researchers beforehand (van Wijk and Wiegant 1989, Smith 1990, Flyborg 1990, Hoffmann 1991).

Evaluation of the data: The number of climbers (F_c) was compared to the number of animals remaining in the water as described in chapter 1.1.

2. Middle-term treatment

The influence of the probe dilution was studied in the pre-climax metamorphosis (transition from the two-legged to the four-legged stage) of highland frog tadpoles (see Fig. 2, studies 2).

2.1 Effect of the test substance, directly applied at intervals of 48 hours, on pre-climax metamorphosis.

Animals: For the experiments with highland amphibia, Rana temporaria tadpoles were taken from the pool described in chapter 1.1. When the tadpoles were taken from the biotope in August or at the beginning of September, their metamorphosis from the two-legged to the four-legged stage proceeded quicker (within days) than when the experiment was performed at the end of September or in October (duration more than one week) (see Endler et al., 1991 a).

Staging: For the experiments we chose only those two-legged tadpoles which had just started to develop hind legs and where the hindlegs were not yet straddled to the extent that one could easily look through the angle between thigh and shank, comparable to stage 31 according to Gosner's staging table (Gosner, 1960). The tadpoles were observed until the forelegs preformed under the skin broke through and the animals thus had entered the four-legged stage. After a development time of one to some weeks, the forelegs broke through within a few minutes. Thus, this parameter seems adequate to define the final stage.

Exposition to probes: Two drops of the probe dilution thyroxine D30 and of control were added blindly to the corresponding basins, followed by gentle stirring, every other day about midday. The transition from the two-legged to the four-legged tadpoles was examined in basins as described in chapter 1.1 containing 5l of water each. Having reached the four-legged stage, the animals were transferred into aqua-terraria or into a natural biotope, respectively, in order to avoid drowning.

Further conditions: The positions of the basins were rotated in the course of the experiment. Indirect natural light was used. The temperature was kept above 18°C. The tadpoles were fed with cooked greens (lettuce). The experimental design was non-violent.

Preparation of testing solutions and *Check for contamination* were done as described in chapter 1.1.

Avoidance of contamination by the probes: Each basin was used only for one experiment. Metal and glass tools were treated with dry heat for 45 minutes according to the instructions of Boyd (1954).

Data base: A number of small sub-experiments was performed. Two sets of basins were always used for treatment with dilution thyroxine D30 and water D30, respectively. Each basin usually contained 18 animals; the number of animals per basin was alike in each experiment.

Evaluation of the data: As the results from the different researchers were in general comparable, the following evaluation was made both for the pooled data from the three researchers as well as for the data from each single laboratory. The cumulative frequencies of tadpoles that had reached the four-legged stage (F_a) were added up for each set and compared to that of tadpoles with only two or three legs. In a further step, the successive sub-experiments were treated as one experiment and the cumulative frequencies (F_a) were evaluated as a 4-field table by chi-square test at the specific measuring points in time. This was possible by normalisation of the duration of the experiment: four quartiles of the duration were defined for each experiment (see Endler et al. 1991a).

Furthermore, survival analysis was used to determine the statistical significance of the bias in the number of the remaining two- or three-legged animals between the two groups according to the treatment (see chapter 1.1).

2.2 Effect of the test substance, directly applied at intervals of 8 hours, on pre-climax metamorphosis

Animals: Rana temp. tadpoles were taken from the highland pool described in chapter 1.1.

Staging: The experiments were performed with two-legged tadpoles which already had developed their hind legs to the extent that these hind legs were straddled, i.e. the animals were more developed than those used in study 2.1 The tadpoles were observed until they entered the four-legged stage as described in chapter 2.1.

Exposition to probes: Two drops of the probe dilution thyroxine D30 and of control were added to the corresponding basins, followed by gentle stirring, thrice a day, i.e. every 8 hours.

Preparation of testing solutions, Further details of observation, laboratory handling, data base and *Evaluation* are described in chapter 2.1.

2.3 Effect of the test substance, sealed in glass vials continuously remaining in the basin water, on pre-climax metamorphosis

Animals: For different sub-studies, Rana temp. tadpoles were taken from the highland pool described in chapter 1.1, Rana temp. tadpoles from lower sites as described in chapter 2.1, as well as Bufo bufo from a comparatively high site with the metamorphosis in May /June, which is markedly later in season than the metamorphosis of lowland Bufo.

Staging: For the experiments we chose only two-legged tadpoles at stage 31 according to Gosner as described in chapter 2.1. For the Viennese experiment only, slightly older tadpoles were taken. The tadpoles were observed until they entered the four-legged stage as described in chapter 2.1.

Exposition to probes: 10 ml of the probe dilution thyroxine D30 and of control were sealed in the glass vials. The coded vials were first all hung into a common water bath as described in chapter 1.2 and then hung into the corresponding plastic basins (content 5 l). Only one net was used by each researcher to handle the animals in all the basins in order to provoke immediate contamination of all aquaria should the water in any of the basins become contaminated with a dilution.

Preparation of testing solutions and *Check for contamination* was done as described in chapter 1.1.

Ampoules used: The dilutions were sealed in hardglass vials as described in chapter 1.2.

Details of observation, laboratory handling, data base and evaluation are described in chapter 2.1. The temperature was kept about 18-21°C. Only some of the studies on animals from low site, due to circumstances, were performed at higher temperatures.

3. Long-term treatment

The influence of the test dilution, applied to tadpoles from the two-legged stage on, was studied in post-climax metamorphosis (the transition to the air-borne stage), expressed in terms of climbing activity of the frogs at the end of metamorphosis (see Fig. 2, studies 3).

3.1 Effect of the test substance, directly added at intervals of 48 hours, on post-climax metamorphosis

These studies investigated the influence of thyroxine D30, added dropwise at intervals of two days to the basins containing a. highland Rana temp. and b. Rana temp. from a lower site from the early two-legged stage on as described in chapter 2.1, on the post-climax metamorphosis of these animals, expressed in terms of the climbing activity. For details of the treatment during metamorphosis, see chapter 2.1.

The four-legged tadpoles were transferred into plastic basins containing 0.5 l of water as described in chapter 1.1. The day after the point in time when about 90% of the animals had reached the four-legged stage, the number of animals (by then mostly with reduced tail) spontaneously climbing out of the water was observed and evaluated as described in chapter 1.1.

3.2 Effect of the test substance, directly added at intervals of 8 hours, on post-climax metamorphosis

This study investigated the influence of thyroxine D30, added dropwise at intervals of eight hours to the basins containing highland Rana temp. The experiment started at the late two-legged stage as described in chapter 2.2.

For further details, see chapter 1.1. Observation C was repeated ten successive times for each basin.

3.3 Effect of the test substance, sealed in glass vials remaining constantly in the basin water, on post-climax metamorphosis

Animals and staging: For different sub-studies, Rana temp. and Bufo bufo, respectively, were taken both from the highland pools described in chapter 2.3 a. at the stage where the animals had just started to develop their hind legs as described in chapter 2.1 and b. at the stage where the hind legs were already straddled as described in chapter 2.2 . The animals were observed until they started to leave the water (airborne stage).

Exposition to probes during the pre-climax metamorphosis was done as described in chapter 2.3. The four-legged tadpoles were transferred into the basins containing 0.5 l of water. Ampoules containing the dilutions were hung both into the water filled basins as well as into the basins with a water depth of 1 cm. When about 60% of the animals had reached the four-legged stage, the number of animals (by then mostly with reduced tail) spontaneously climbing out of the water was observed and evaluated as described in chapter 1.1. Only one net was used by each researcher to handle the animals in all the basins in order to provoke immediate contamination of all aquaria should the water in any basin become contaminated with a dilution.

Preparation of testing solutions was done as described in chapter 1.1. *Glass ampoules* as described in chapter 1.2 were used.

Check for contamination: One set of the probes was checked for contaminations and hormone-active contaminations were excluded as described in chapter 1.1.

Data base: Two sets of basins for treatment with dilution thyroxine D30 and water D30, respectively, were used for the experiments. 18 tadpoles and 18 juveniles (36 in one laboratory), respectively, were put into one basin. The basins were placed in an alternating sequence.

RESULTS

As the results of the experiments in the different laboratories were in general comparable, the data were pooled for each of the studies described in this contribution. For details on the data from the independent laboratories, see Endler et al. 1993b.

1. Short term treatment

1.1 Effect of the test substance, directly added at one point of time, on activity

Several experiments were performed with a total of 177 animals (five repetitions led to 884 cases). In step A, before treatment, the increase of F_c is practically identical for both groups at all points of time. In the second step, B, dilutions thyroxine D30 and water D30 were added and the observation was repeated immediately. In Fig. 4, above, the increase of F_c is shown for the control animals in step B. Fig. 4, below, shows the bias in the cumulative frequencies of climbing animals between the test group and the control group. The F_c-values for thyroxine D30-animals are now below those for reference at about

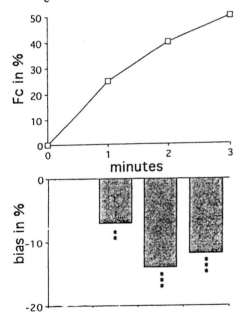

7-14%. The difference is statistically highly significant both for the chi-square test at all three points of time (P < 0.001) as well as for an overall survival analysis (P = 0.000).

Fig. 4: The influence of the test dilution, at one point of time added directly, on the climbing activity of highland Rana temporaria (pooled data of step B, see methods). Ordinate: above: Cumulative frequency of climbers. 100% refers to 442 observed cases in the control group. Below: Bias in the cumulative frequencies of animals climbing out of the water between the groups treated with thyroxine D30 and with the water D30 control, respectively. Abscissa: the number of minutes of duration of the respective observations 1-5 (see methods). ***, P < 0.001 (evaluation with chi-square-test). For further details, see text.

In Fig. 5, (next page) the data from the respective repetitions 1-5 (130 animals) were pooled for each step. In Fig. 5A, the two curves give the increase of climbing activity F_c for animals assigned for later treatment with dilution thyroxine D30 or dilution water D30. In this step A, before treatment, the increase of F_c is practically identical for both groups at all points of time. In Fig. 5B, the curves give the increase in step B, when dilutions thyroxine D30 and water D30 were added and the observation was repeated immediately for 5 times.

The F_c-values for the dilution thyroxine D30-group are now 5-10% below those for the group treated with dilution water D30. This difference is highly significant when evaluated with the chi-square test as well as with the survival analysis (P = 0.007). In Fig. 5C, the two curves give the increase after the cross-over treatment: in the five repetitions of the climbing observation, the F_c-values for the animals treated with dilution thyroxine D30 + dilution water D30 until step C remained below those of step A, whereas the F_c-values for the animals treated with dilution thyroxine D30 in step C only had decreased by 3-10%.

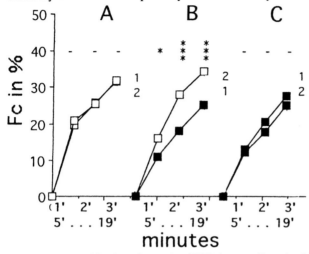

Fig. 5: The influence of the test dilution, at one point of time directly added, on the climbing activity of juvenile highland frogs (steps A, B, C, pooled data from the respective repetitions 1 - 5). Ordinate: cumulative frequency of climbers Fc. 100% refers to 325 cases per dilution in each step. Abscissa: the number of minutes of duration of the respective observations 1 - 5; step A, climbing activity before treatment: cumulative frequency of climbers assigned for later treatment with dilution thyroxine D30 (curve 1) and with dilution water D30 (curve 2); step B, immediately after treatment with the dilutions: black squares, animals treated with thyroxine D30; white squares, animals treated with water D30; step C, two days after cross-over treatment: all animals are finally treated with the dilution thyroxine D30. -, not significant; *. P < 0.05; ***, P < 0.001(evaluation with chi- square test). For further details, see text. (Adopted from J Vet Hum Tox, Endler et al., 1993c.)

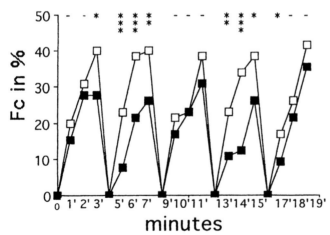

Fig. 6 is a more detailed description of step B (survey in Fig. 5B). It shows the F_c-values for dilution thyroxine D30- (black squares) and for dilution water D30- animals (white squares) immediately after the first treatment during the repetitive observations 1-5. at the end of the third minute. Only in the second repetition, the differences could be proven with survival analysis (P = 0.04).

Fig. 6: The influence of the test dilution, at one point of time directly added, on the climbing activity of juvenile highland frogs. Detailed description of step B (pooled cumulative frequencies in Fig. 5B). 100% refers to 65 animals per group. **, P < 0.01 (evaluation with chi-square test). For further details, see Fig. 5 and text. (From Endler et al 1993c.)

The F_c-values for dilution thyroxine D30- animals are below those for reference animals in all five repetitions. This difference is statistically significant (P-values between < 0.05 and < 0.001) at most, but not at all repetitions. When the five repetitions were compared with a Wilcoxon test, P is 0.043 at the end of the first, 0.068 at the end of the second and 0.042 at the end of the third second.

1.2. Effect of the test substance, sealed in glass vials and applied during a few minutes, on activity

Several experiments were performed with a total of 410 animals (five repetitions led to 2050 cases). In step A, no treatment was given in order to determine control values. The increase of F_c is practically identical for both groups at all points of time.

In step B, coded vials containing either thyroxine D30 or water D30 were hung into two respective basins. The observation of the climbing activity started immediately after the introduction of the vials into the water basin containing the juveniles. The increase of F_c is practically identical for both groups at all points of time.

In step C, the positions of the vials were changed and the observations were repeated. The F_c-values for thyroxine D30 are now about 7-8% below the reference values observed with water D30. The difference is statistically highly significant in the chi-square test; it could also be proven with survival analysis (P < 0.005).

In Fig. 7, the available data from the respective repetitions 1-5 were pooled for each step.

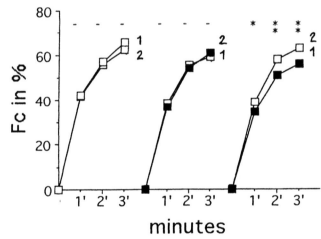

Fig. 7: The influence of the test dilution, sealed in glass vials, on the climbing activity of non pretreated juvenile highland frogs.
Ordinate: 100% refers to the whole number of observed climbers in the five successive repetitions of each step (A, B, C) of the experiment (1080 cases in each step). For details, see Fig. 5, 6 and text.

Fig. 8 (next page) is a more detailed description of step C. It shows the F_c-values for animals pretreated with sealed dilution thyroxine D30 for about 20 minutes, and then immediately treated with sealed dilution water D30 for further 20 minutes (white squares) and the F_c-values for animals pretreated with sealed water D30 for about 20 minutes and then treated with sealed dilution thyroxine D30 for further 20 minutes (black squares). The F_c-values for dilution thyroxine D30- animals are slightly below those for reference animals in most of the repetitions. However, this difference is statistically significant only at two points of time (in repetition 3 and 4). When the five repetitions were compared with a Wilcoxon test, the difference could not be proved, and also survival analysis did not prove the differences for the single successive repetitions.

To a small degree, (in the studies of all four researchers, see Tab. 1), climbing activity was slowed down by thyroxine D30 in step C, but not in step B.

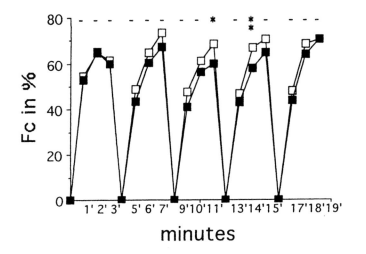

Fig. 8: The influence of the test dilution, sealed in glass vials, on the climbing activity of non - pretreated juvenile highland-frogs.
Detailed description of step C.
100% refers to 410 animals per group. For further details, see Fig. 6 and text.

2. Middle term treatment

2.1 Effect of the test substance, directly applied at intervals of 48 hours, on pre-climax metamorphosis

Several experiments were performed with a total of 1139 two-legged animals (starting stage: early two-legged stage, see methods). In Fig.9,above, the increase of the cumulative frequencies of four-legged animals Fa is shown for the control animals. In Fig. 9, below, the bias in Fa between the groups treated with the test dilution and the control dilution, which can be taken as a parameter for the level of metamorphosis, is documented for all quartiles. The Fa-values for thyroxine D30- animals are below those for reference at about 5-10%. S.D. of Fa-values both for thyroxine D30- and for water D30-animals is about 10%.

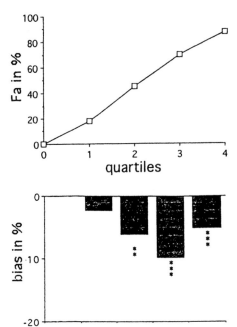

Fig. 9: The influence of the test dilution, when added at intervals of 48 hours, on pre-climax metamorphosis of highland tadpoles. Pooled data. Ordinate: above: Cumulative frequency of four-legged tadpoles in the control group. 100% refers to 570 animals in the control group. Below: Bias in the cumulative frequencies of four-legged tadpoles between the groups treated with thyroxine D30 and with the water D30 control, respectively. Abscissa: the number of quartiles of normalized duration of experiments (see methods).
***, P < 0.001; evaluation with chi-square test.. For further details, see text.

When all experiments are treated as one, the chi-square test as well as the survival analysis show that the F_a-values for thyroxine D30 - animals are statistically significantly below those for the reference animals (P < 0.01 in both tests).

In other words, the chance to pass preclimax metamorphosis is generally smaller for the group treated with dilution thyroxine D30 than for the water D30- group if the dilution is added at intervals of 48 hours.

The evaluation of these and of further recent data showed that especially experiments performed at the end of September or later (duration: more than one week) are suitable to prove this difference.

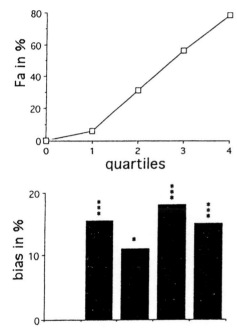

2.2 Effect of the test substance, directly applied at intervals of 8 hours, on pre-climax metamorphosis

One experiment was performed with a total of 144 animals (starting stage: late two-legged stage, see methods). In Fig. 10 (below), the bias in the cumulative frequencies of four-legged animals F_a between the groups treated with the test dilution and the control dilution, are documented. The F_a-values for thyroxine D30 - animals are above those for reference at about 10 - 20%. The difference could be proven with the chi-square test (P < 0.001 in the third quartile) as well as with the survival analysis (P < 0.001).

Fig. 10: The influence of the test dilution, when added at intervals of 8 hours, on the preclimax metamorphosis of highland Rana temp. 100 % refers to 72 animals in the control group. For further details, see Fig. 9 and text.

Two further experiments with a total of 720 animals (starting stage: middle two-legged stage) were performed while this paper was in print. The F_a-value for thyroxine D30 - animals was above those for reference at about 5 - 10%.

In other words, the chance to pass preclimax metamorphosis is generally bigger for the group treated with dilution thyroxine D30 than for the water D30- group if the dilution is added at intervals of 8 hours.

2.3 Effect of the test substance, sealed in glass vials continuously remaining in the basin water, on pre-climax metamorphosis

Rana temporaria from highland: Several experiments were performed with a total of 469 animals. In general, the differences showed that dilution thyroxine D30, even if not in direct contact with the water or the animals, but separated by the glass wall of the ampoule, slows down the development to the four-legged stage. However, this could not be statistically proved with the chi-square test or survival analysis.

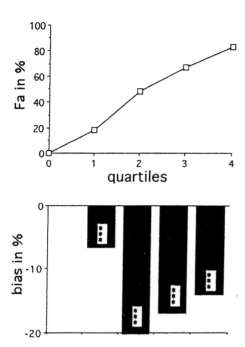

Bufo bufo: Several experiments were performed with a total of 451 two-legged animals. At one of the measuring points, the F_a-value for thyroxine D30-animals is below that for reference ($P < 0.001$ at the fourth quartile). The difference, however, could not be proved with the survival analysis.

Rana temporaria from low site: Several experiments were performed with a total of 1224 animals. At all measuring points, the F_a-values for thyroxine D30-animals are below those for reference ($P < 0.001$). The difference could also be proven with the Wilcoxon test and the survival analysis ($P < 0.001$) (Fig. 11).

Fig. 11: The influence of the test dilution, when applied in sealed glass vials, on preclimax metamorphosis of lowland Rana temporaria. 100% refers to 612 (quartiles 1 and 2, respectively) and 468 (quartiles 3 and 4, respectively) animals in the control group. For further details, see Fig. 9 and text.

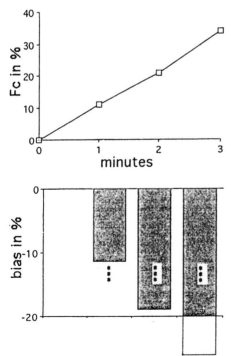

In other words, specially for Rana temp. from the low site and for Bufo bufo, but only to a very small extent for Rana temp. from the highland, the chance to pass metamorphosis is generally smaller for the group treated with dilution thyroxine D30, sealed in glass vials, than for the water D30-group.

3. Long-term treatment

3.1 Effect of the test substance, directly added at intervals of 48 hours, on post-climax metamorphosis

One experiment was performed with 47 animals (five repetitions led to 235 cases). In Fig. 12 (below), the bias in the cumulative frequencies of animals climbing out of the water F_c are documented (pooled data from repetition 1-5). The F_c-values for thyroxine D30- animals are markedly below those for reference at about 10-30%.(S.D. is about 5%.) The difference is statistically highly significant both when evaluated with the chi-square test as well as with the survival analysis ($P < 0.001$).

Fig. 12. For legend, see next page.

Fig. 12 (see above): The influence of the test dilution, directly added at intervals of 48 hours, on postclimax metamorphosis, expressed in terms of the climbing activity of animals treated since the two-legged stage. 100% refers to 117 cases in the control group. For further details, see Fig. 4 and text.

This is also true, if the data from the five successive repetitions of the observation of climbing activity are evaluated separately.

In other words, the chance to pass metamorphosis and to climb onto land is generally smaller for the group treated with dilution thyroxine D30, directly added dropwise at intervals of 48 hours, than for the water D30- group.

Rana temporaria from low site: One experiment was performed with 144 animals (five repetitions led to 720 cases). The F_c-values for thyroxine D30- animals are below those for reference at about 5-10 %. The difference is statistically highly significant when evaluated with the chi-square test, and shows $P < 0.05$ when evaluated with the survival analysis.

The difference can also be found, if the data from the five successive repetitions of the observation of climbing activity are evaluated separately.

In other words, the chance to pass metamorphosis and to climb onto land is generally smaller for the group treated with dilution thyroxine D30, added dropwise directly at intervals of 48 hours, than for the water D30- group.

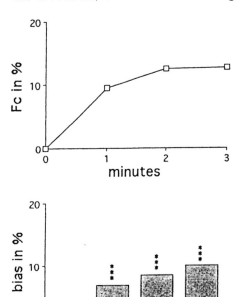

3.2 Effect of the test substance, directly added at intervals of 8 hours, on post-climax metamorphosis

One experiment was performed with 108 animals (ten repetitions led to 1080 cases). In Fig. 13 (below), the bias in the cumulative frequencies of animals climbing out of the water are documented on the ordinate (pooled data from repetition 1-5). The F_c-values for thyroxine D30- animals are above those for reference at about 6-10%. The difference is statistically highly significant both when evaluated with the chi-square test as well as with the survival analysis ($P < 0.001$).

Fig. 13: The influence of thyroxine D30, directly added at intervals of 8 hours, on postclimax metamorphosis, expressed in terms of the climbing activity of animals treated since the two-legged stage. 100% refers to 540 cases. For further details, see Fig. 4 and text.

This is also true for most, but not for all of the ten successive repetitions of the observation of climbing activity, when these were evaluated separately ($P < 0.05$ and < 0.01).

In other words, the chance to pass metamorphosis and to climb onto land is generally bigger for the group treated with dilution thyroxine D30, directly added dropwise at intervals of 8 hours, than for the water D30- group.

3.3 Effect of the test substance, sealed in glass vials remaining constantly in the basin water, on post-climax metamorphosis

Rana temporaria from highland: Several experiments were performed with a total of 510 animals (five repetitions led to 2550 cases). In Fig. 14 (below), the bias in the cumulative frequencies of animals climbing out of the water between the groups treated with the test dilution and the control dilution are documented. The F_c-values for thyroxine D30- animals are above those for reference at 10-15%. Standard deviation of the data from each laboratory is about 14%. The difference is statistically highly significant both when evaluated with the chi-square test (P < 0.001 in all measuring points) as well as with the survival analysis (P < 0.001).

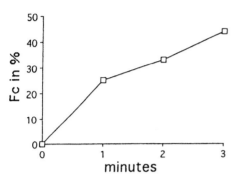

Fig. 14: The influence of the test dilution, applied in sealed glass ampoules, on post-climax metamorphosis, expressed in terms of the climbing activity of highland Rana temp. 100% refers to 1275 cases in the control group. For further details, see Fig. 4 and text.

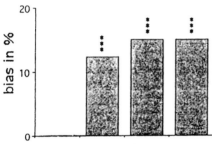

If the data from the five successive repetitions are treated separately, again the difference is statistically highly significant both when evaluated with the chi-square test (P-values < 0.001) as well as when evaluated with the survival analysis (P < 0.01 for the first repetition; P < 0.001 for each of the repetitions 2-5).

Bufo bufo: Several experiments were performed with a total of 232 animals (five repetitions led to 1160 cases). The F_c-values for thyroxine D30-animals are above those for reference at 8-15%. S.D. is about 24%. The difference is statistically highly significant both when evaluated with the chi-square test (P < 0.001 at all measuring points) as well as with the survival analysis(P < 0.001).

In other words, for highland Rana temp. and for Bufo bufo, respectively, the chance to pass metamorphosis and to climb onto land is generally bigger for the group treated with dilution thyroxine D30, sealed in hardglass vials and remaining in the basin water since the two-legged stage, than for the water D30- group.

DISCUSSION

These different studies investigate the influence of extremely dilute thyroxine on the climbing activity of juvenile highland frogs Rana temporaria and on two transitions in the metamorphosis of these animals. Studies on Rana temp. from lower sites are discussed in Endler et al. 1993b.

Studies on short-term treatment

The highly diluted agitated thyroxine (10^{-30}, "D30") slows down the climbing activity of juvenile frogs to a small but statistically significant extent, when once added dropwise to the aquaria water at the juvenile stage. Evidently, this inhibitory effect of thyroxine D30 is opposite to that expected from thyroxine in classical pharmacological doses (Zimbardo 1983).

It was also observed that, in some experiments this dilution slowed down the climbing activity, when it was sealed in hardglass vials that were hung into the aquaria, without direct contact between the dilution and the organisms. When the test dilution was sealed in glass vials, only in 50% of the experiments an inhibition of climbing activity was observed. The positive experiments had a protocol in which the animals had been pretreated with dilution water D30 for about 20 minutes. In other experiments where the test dilution did not exert its effect, the ampoules were brought near the non-pretreated juveniles for the first time. In order to evaluate the role of pretreatment with the dilution water D30 it will be necessary to vary this parameter in one and the same experiment.

In the experiments using the thyroxine dilution in sealed vials, also older juvenile animals were investigated. An effect of the test dilution could not be found when juveniles were used that had already reached the stage of the air-borne animal some days before the experiment was performed, but only when freshly air borne animals were taken. This may point towards a specific stage dependence on the effect of the test dilution, which might go hand in hand with the saturation of the tissue with thyroxine.

In the experiments, water, analogously prepared and applied, was used as control.

Studies on middle-term treatment

The present set of studies confirmed our preliminary results (Endler et al. 1989), namely that a high dilution of thyroxine is able to enhance amphibian metamorphosis (see chapter 2.2).

Furthermore, it was shown that, depending on the way of frequency of application, thyroxine D30 is also able to exert an inhibitory effect on amphibian metamorphosis.

Our research strategy was to follow both lines of experiments in order to demonstrate the bivalent effects that can be achieved with one and the same high dilution (for further discussion, see below).

It was shown that preclimax metamorphosis (the transition from the two-legged to the four-legged tadpole) can be influenced by the test dilution in two typical ways.

When the dilution was added at intervals of 48 hours from the two-legged stage on, it caused an inhibition of the observed transition. Evidently, this inhibitory effect of thyroxine D30 is opposite to that known from thyroxine in classical pharmacological doses (Wehner and Gehring 1990).

Comparing the results of the different experiments carried out using the same protocol, the differences were more marked in the experiments performed late in September or in October (duration: more than one week) than in the experiments performed in August or at the beginning of September (duration: some days). (see Endler et al. 1991a, 1993b). In the experiments, tadpoles at an early two legged stage (the angle between shank and thigh, but no open triangle between shank, thigh and tail is visible) were taken.

When the same dilution was added at intervals of only 8 hours from the two-legged stage on, it caused an acceleration of the very transition. The effect described was only observed in experiments where animals were treated from certain starting stages on (middle two-legged stage: an open triangle between shank, thigh and tail is visible; late two-legged stage: the hindlimbs are now straddled). The effect was not found in experiments where animals were treated from the early two-legged stage on, and it was less marked when animals at the middle two-legged stage were taken than when animals at the late two-legged stage were taken. Most of these experiments were performed in August and at the beginning of September.

When the dilution was applied in sealed glass vials that constantly remained in the aquaria, it caused little or no inhibiting effect in the observed transition in highland Rana temp. (metamorphosis late in the season), but a significant inhibition in Rana temp. from low site (metamorphosis early in the season). (For further discussion, Endler et al. 1993b.)

Studies on long-term treatment

Further, in experiments where the dilution was added from the two-legged stage on at intervals of 48 hours, it also slowed down postclimax metamorphosis (the transition from the four-legged, tailed tadpole to the juvenile frog), expressed in terms of climbing activity. In other experiments where it was added at intervals of 8 hours, it accelerated postclimax metamorphosis.

In a third group of experiments where the dilution was applied in sealed glass vials that constantly remained in the aquaria, it caused an acceleration of postclimax metamorphosis in highland Rana temp. (The effect on lowland Rana temp. is different; it is discussed in Endler et al. 1993b.)

These findings suggest that the method of application determines the physiological effect.

Thus, the method of constant indirect application, where the dilution was sealed in glass vials that remained in the basin water during the whole experiment, showed an intermediate effect between a dropwise application every 48 hours and a dropwise application every 8 hours: when constant indirect application was used, preclimax metamorphosis was slowed down and postclimax metamorphosis was accelerated. The initial inhibition, in all experiments with highland tadpoles being rather small, however, was observed only in experiments where animals at the early two-legged stage (stage 31) were used to start the experiment with, and it was not found when animals at the late two-legged stage were taken. The final acceleration, however, was observed both in experiments where animals at the early as well as at the late two-legged stage were taken. In highland Rana temp. the effect could be proven not only on the first test day (see methods) but also on the following days, whereas in Bufo bufo it could be proved only on the first test day. In order to prove the importance of the respective application method in the ultimate effect, experiments are needed in which the different methods of application are performed in one and the same experiment.

When, in our experiments, the climbing activity was studied in non-pretreated animals, an inhibition by the test dilution was observed (studies 1.1 and 1.2). This inhibitory effect is independent of the way the thyroxine D30 is applied (direct as well as indirect application). This observation suggests that it is the frequency of stimulation during a certain time in metamorphosis that leads to an acceleration during the postclimax phase.

In the study on long-term treatment, it became obvious that the sealed test dilution, although it had been present through the whole course of metamorphosis, did not immediately exerted its effect on the climbing activity, but only after the first repetition of the climbing experiment ($P > 0.05$ in the first of the repetitions and $P < 0.01$ in all follow-up repetitions of all researchers involved). This may point to the fact that a certain (experimental) stress is a condition for the effect of the test dilution.

Stage dependence on the effect of the test dilution

When it is supposed that stage dependence and application method are determinants for the ultimate effect, this might be linked to the (influence on) plasma levels of iodine /thyroxine.

Effects on climbing activity:

The natural plasma level of iodine /thyroxine changes during and after spontaneous metamorphosis, with a slow increase during the two-legged stages and a rapid increase from the four-legged stage on, shortly before the reduction phase of the tail starts, followed by a rapid decrease during that reduction phase (Just 1976; Weil 1986; see Fig. 15A,).

This high plasma level is interpreted to be due to an increasing synthesis of thyroid hormone (before tail resorption begins, Dodd and Dodd, 493, see Zelzer 1990) and increasing release into the circulation on the one hand, but also to be due to increased tissue saturation with thyroxine. Thus, tissue would become increasingly avid for thyroxine in the time before the plasma level increases (Dodd and Dodd, see Zelzer 1990).

Experiments were performed on the effect of thyroxine D30 on the climbing activity of highland Rana temp., combining the protocols of short-term, middle-term and long-term application (Endler et al. 1991b). When the test substance was added at intervals of 48 hours from the two-legged stage on, the effect on climbing activity at the juvenile stage (inhibition) was more marked than when the test substance was added at intervals of 48 hours from the four-legged stage on. In contrast, when the test substance was added at the juvenile stage, as well in observations after a few minutes as well as after 48 hours, the effect on climbing activity (inhibition) was again more marked (Fig. 15B, for details, see Endler et al. 1991b).

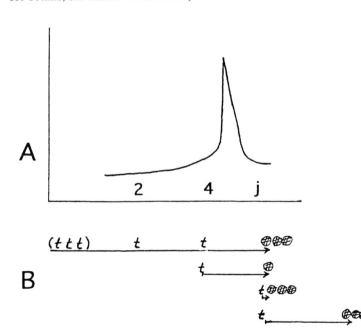

Fig. 15:

A: Changes in the plasma level of thyroxine during metamorphosis of frogs (schematized). Ordinate: Amount of iodine in the blood plasma (arbitrary units). Abscissa: course of time. 2, two-legged stage; 4, four-legged stage; j, juvenile stage.

B: Effect of thyroxine D30 on climbing activity of the juvenile with regard to the stage of the beginning of the treatment. Abscissa: course of time. ###, strong inhibiting effect; #, small inhibiting effect. t, application of a few drops of thyroxine D30 to the basin water.

Our speculation on these findings is that if the high dilution of thyroxine is presented during a stage with comparatively low physiological thyroxine plasma-level (before or after the transition from the four-legged tadpole to the juvenile frog), it is able to exert a stronger action than if it is presented during the time when the physiological thyroxine plasma-level is at a maximum. However, also the break in nutrition intake during the climax period must be considered here, as the pathway of the thyroxine dilution is yet unknown.

Effects on metamorphosis:

When the experiments on the influence of the test substance on the transition from the two-legged to the four-legged tadpole (study 2.1) were grouped according to the natural length in time of the observed transition, it was found that the effect of the thyroxine dilution (inhibition of development) is the more marked, the slower the natural development is. A tentative speculation on this finding is that, analogously to Fig. 15, the effect of thyroxine D30 increases, with a decreasing thyroxine level

Recommendations for further experiments

Fig. 16 gives a survey of the different effects of the test dilution on the amphibia dependent on the different methods of application. For a survey of the animals' stages in the respective studies see Fig. 2; of the methods of application of the test dilution Fig. 3; and of further details Tab. 2.

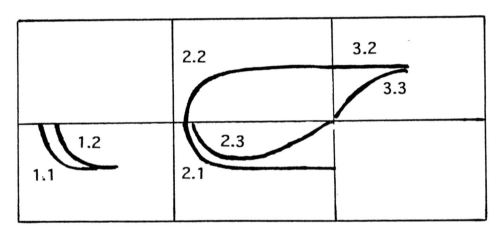

Fig. 16: Effects of thyroxine D30 dependent on different methods of application. 1, short-term application; 2, middle-term application; 3, long-term application (for explication see text). Ordinate: below the horizontal line, inhibition; above the horizontal line, stimulation (1, of activity; 2, of preclimax metamorphosis; 3, of postclimax metamorphosis). Abscissa: time (1, in minutes; 2, 3, in quartiles of duration of the experiment - see text -). The Fig. is arranged in correspondance with the symbols in Fig. 3.

Although data have been obtained from separate experiments with different protocols, we think that our data suggest that the activity of the test substance, when added dropwise to the basin water, diminishes in the course of time, so that a new addition every 8 hours causes an effect different to an addition every 48 hours.

The main experiments presented in this paper were performed blindly under the supervision of the university of Graz, Austria (mostly by the zoological university institute Graz

- codation: G. Fachbach, department of developmental biology; supervision of setup: G. Kastberger.-); furthermore, key experiments were independently repeated by other researchers in their own or in our laboratories (see Tab.1). These researchers, using our respective protocols, were able to confirm the senior author's results.

Thus, we think that with our model system, effects of the agitated high dilution tested could be proven. However, all studies that have only been performed in the central laboratory (study 2.2, 3.1, 3.2) should be independently repeated in another laboratory. Furthermore, we think repetitions of our studies should be generally suggested in high standard zoological /developmental laboratories.

By reasons of laboratory convenience, but also due to ethical aspects, specially the studies 2 on the transition from the two-legged to the four-legged Rana temp. as indicated in Tab. 2 seem adequate for this aim.

Tab.2

study 2.1 on highland animals in the early two-legged stage, performed at the end of September;
study 2.2 on highland animals in the late two-legged stage, performed at the beginning of September;
study 2.3 on animals from comparatively low sites, performed in May /June.

Tab.2: Study designs of special interest for further control experiments.

Unless further experiments clarify this point, a water temperature of more than 18°C should be looked at as a conditio sine qua non.

Evidently, the observations made in our laboratory model should be discussed with regard to general assumptions of homoeopathy. As compared to homoeopathic practice, the application of the thyroxine dilution at intervals of 48 hours can be interpreted as "thrice a week" and the application at intervals of 8 hours as "thrice a day" as according to clinical practice, whereas the application to the basin water including the animals could be interpreted as being in some way similar to the application to the body water of a human patient.

In the following, the observations made are discussed as a model system to achieve bivalent reactions of an organism to a defined stimulus with different intensity. However, this will need a repetition of the different protocols within one experiment, but it may shed light on physiological and medical aspects apart from the research on ultra high dilutions.

Bivalent effects of a thyroxine dilution. Laboratory background for some medical features

In this note, the data presented above are discussed with regard to typical features of homoeopathic medicine.

Thyroxine in molecular dose is known as a stimulator of metamorphosis and of activity (Wehner and Gehring 1990, Zimbardo 1983). In our studies, we showed that thyroxine in the 30th homoeopathic decimal dilution is able to cause both effects in the sense of those of molecular thyroxine (10^{-9}) as well as opposite effects. In order to arrive at a clear terminology, the termini "ortho-taxic" and "anti-taxic" are suggested (P. Fisher, personal communication) to describe effects in the sense of or opposite to the effect of a mother tincture.

With the help of an ("ultra-")high dilution (UHD), the possibility to influence given parameters of bio-assays both in the way comparable to the effect of a molecular concentration of the UHD's original mother substance as well as in the opposite way, is also described in other contributions in Part 1 of this volume. However, the amphibian model discussed here seems specially adequate to demonstrate such a bivalent effect of one test substance in one and the same bio-assay.

The experimental conditions of the respective experiments on highland tadpoles and frogs of Rana temporaria are summed up in Tab. 3. For further details, see chapters 2.1, 2.2, 3.1 and 3.2 of this contribution.

Tab. 3

A	B	
Metamorphosis from 2- to 4-legged swimming stage (preclimax metamorphosis). *Observed parameter: 4-leggedness*	Metamorphosis from 2-legged swimming to 4-legged climbing stage (pre- plus postclimax metamorphosis). *Climbing activity*	
↑ (acceleration 5-15%, 108 animals)		**1** Submersion with thyroxine 10^{-9}
↑ (acceleration 5-20%, 504 animals)	↑ (enhancement 5-10%, 108 animals)	**2** addition of thyroxine D30 at intervals of 8 hours
↓ (slowing down 5-10%, 1139 animals)	↓ (inhibition, 47 animals)	**3** addition of thyroxine D30 at intervals of 48 hours

Tab. 3: Effects of molecular thyroxine 1. at a concentration 10^{-10} and of UHD thyroxine D30, applied 2. at intervals of 1/3 day and 3. of 2 days, respectively, on the metamorphosis transitions A, B as described on top of the Tab. Flesh upwards: effect of tested molecular thyroxine dilution or analogous effect of UHD - ortho-taxic effect (acceleration, enhancement); flesh downwards: counter effect - anti-taxic effect (slowing down, inhibition).

The two possibilities discussed here are also consistent with observations on humans that have become fundamental for the knowledge on remedies in homoeopathic medicine.

Ortho-taxic effects

An effect of an UHD similar to that of the treatment or poisoning with the original molecular substance cannot only be found in our amphibia experiments, but is also the basis of Remedy Provings with UHDs (see Glossary, p. 221 and Walach's contribution). E.g. a Remedy Proving using thyroid substance was performed on healthy humans (Panos et al. 1964). The hypothesis was that the UHD of thyroid substance, among other symptoms, is able to provoke symptoms similar to those of molecular thyroid substance ("empirical homoeopathic intoxication study"). The aim of this study was to provoke symptoms in order to learn more about the remedy and to get to know the very symptoms that, according to the Law of Similarity (see Glossary, p. 221), might successfully be treated when they occur in patients with dilutions of thyroid substance. In that study, thyroid substance D30 was taken by voluntary participants at short intervals. In their double-blind experiment, Panos et al. found a list of symptoms that were typical for volunteers in the group receiving the UHD of thyroid substance, as for instance restlessness and hyper-agitation, which were obviously caused by the frequent application of the UHD.

Anti-taxic effects

There is a certain parallel between detoxification studies and our studies. In detoxification experiments, organisms are first intoxicated with a high dose of a toxin followed by the addition of a low dose or a highly diluted, agitated solution of the same toxin. In various studies, an enhanced detoxification has been observed in comparison with control in a variety of organisms (Lapp and Wurmser 1958; Cazin et al. 1987; see Law of Similarity, Glossary, and Poitevin's contribution). In our study the animals were, of course, not intoxicated with thyroxine. When thyroxine is applied in pharmacological doses, a stimulatory effect on metamorphosis has been found out; in contrast, when the UHD of thyroxine was added at intervals of 2 days, the speed of metamorphosis decreased. This is also in agreement with other observations described in the "homoeopathic" literature such as e.g. the reduction of hayfever symptoms when highly diluted agitated grass pollen was administered to patients suffering from hayfever (Reilly and Taylor 1986). Furthermore, in homoeopathic literature (Kent 1989, Barthel 1987) it is reported that agitated high dilutions of the iodine-containing substances such as thyroxine and thyroid substance can be useful remedies in the case of hyperactivity, when applied in an adequate way.

Obviously, however, the parallel of naturally occurring "intoxication" with thyroxine and "detoxification" with the UHD thyroxine D30 is valuable only within a certain range of enhanced thyroxine level (see the discussion section on stage dependence), probably with an optimal thyroxine level (blood /tissue-concentration), where the sensitivity of the animals towards the UHD is at a maximum.

In conclusion, we think that both the ortho- as well as the anti-taxic effects of thyroxine D30 are adequate to illustrate the biological laws underlying the ortho- and anti-taxic effects that have also been observed in "homoeopathy" for about 200 years.

Dependence on the frequency of application

The dependence of the effect of thyroxine D30 in middle-term treatment on the frequency of application, namely every 8 hours or every 48 hours, respectively, has been discussed above.In these experiments it was shown that more frequent dropwise application of the dilution leads to an ortho-taxic effect (stimulation of the transition from the two- to the four-legged stage), while less frequent application leads to an anti-taxic effect (inhibition). Recent experiments show that this is only true within a certain range of variation of the frequency of application. To our present knowledge, when the conditions suggested in Tab. 2 are used, an optimal ortho-taxic frequency interval is that of 4 to 8 hours and an optimal anti-taxic frequency interval is that of about 2 days.

Specificity of the originally diluted molecule

Apart from the studies described in detail in this paper, it was observed that, similar to thyroxine D30, also iodine (D30) and, surprisingly, the halogene chlorine (D30) were able to inhibit the transition from the two- to the four-legged tadpole when applied as described in the protocol 2.3. Highland Rana temp. were taken for these studies.

Bio-informational aspects

Using the method of continuous indirect application, one can rule out the hypothesis that an observed effect might be caused by some remaining thyroxine molecules from the thyroxine dilution interacting with the biological system. Therefore, the significant effect observed must be explained by non-molecular interactions between the homoeopathic drug and the organism. Here, too, general aspects of low energy bio-information might be elucidated. For respective biophysical studies, see Part 3 of this volume; for a comprehensive discussion, see the Prospects section, p. 245ff.

ACKNOWLEDGEMENTS

Thanks are due to P. Kokoschinegg (+) who called our attention to the general sensitivity of amphibia during metamorphosis, the "Institut für strukturale medizinische Forschung", Graz (Th. Kenner, M. Moser, E. Lehner, F. Muhry; E. and Y. Clar), where we had the possibitity to perform the preliminary experiments which then led to the studies presented here; to the Zoological Institute of the University Graz (H. Heran, K. Hagmüller, G. Fachbach, G. Kastberger, U. Neumeister-Stelzer), to M. Haidvogl from the Ludwig Boltzmann Institut für Homöopathie, who gave us the organisatorical frame for our studies, to G. Karmapa and D. Lama, and to many others (see Preface, p. VIIf). Important parts of our research work were co-financed by the HomInt-Group (1.1, 2.1), the Dolisos Laboratories (2.3) and by the Ludwig Boltzmann Society. The studies on iodine and chlorine mentioned in the discussion section were encouraged by the Blackie Foundation Trust.

A personal word: The authors submit the discussion of the amphibian model to the public hoping that it will be a challenge to scientific creativity to explore it further, but without any investigation harmful to the animals.

REFERENCES

Barthel H. Synthethic Repertory 3. ed. Heidelberg: Haug 1987.

Bastide M, Doucet-Jaboeuf M, Daurat V. Activity and chronopharmacology of very low doses of physiological immune inducers. Immunol Today 1985; 6: 234-235.

Bastide, M, Daurat V, Doucet-Jaboeuf M, Pèlegrin A, Dorfman P. Immunomodulator Activity of Very Low Doses of Thymulin in Mice. Int J Immunotherapy 1987; III: 191-200.

Bildet J. Etude de l' action de différentes dilutions homéopathiques de Phosphorus blanc (Phosphorus) sur l`hepatite toxique du rat - Thèse Pharmacie. Université de Bordeaux, Bordeaux II 1975.

Boyd WE. Biochemical and Biological Evidence of the Activity of High Potencies. Br Hom J 1954; 44:7-44.

Brewster WR, Isaacs JP, Osgood PF, King TL. Hemodynamic and metabolic inter-relationships in the activity of epinephrine, norepinephrine and the thyroid hormones. Circulation 1956; 13: 1.

Cazin JC, Cazin M, Gaborit JL, Chaoui A, Boiron P, Cherruault Y, Papapanayotou C. A Study on the Effect of Decimal and Centesimal Dilutions of Arsenic on the Retention and Mobilisation of Arsenic in the Rat. Hum. Toxicology 1987; 6: 315.

Chou HI, Kollros JJ. Stage modified responses to thyroid hormones in Anurans. Gen Comp Endocrinol 1974; 22: 255-260.

Daurat, V., Dorfman, P., Bastide, M.: Immunomodulatory Activity of Low Doses of Interferon in Mice. Biomedicine and Pharmacotherapy 42: 197-206, 1988

Davenas E, Poitevin B, Benveniste J. Effect on Mouse Peritoneal Macrophages of Orally Administered Very High Dilutions of Silica, Eur J Pharmacol. 1987; 135: 313.

Dodd JM. Studies on amphibian metamorphosis using [131]I. J Physiolog (London) 1955; 130: 11.

Dodd MHI, Dodd JM. The Biology of Metamorphosis. In: Lofts B. Physiology of Amphibia, vol III. London: Academic Press 1976, pp 467-599.

Endler PC, Pongratz W, Moser M, Kenner T. Effects of Highly Diluted Succussed Thyroid Hormone on Amphibia. Presentation Meeting ZDN, Essen 1989.

Endler PC, Pongratz W, Van Wijk R, Kastberger G, Haidvogl M. Effects of Highly Diluted Succussed Thyroxine on Metamorphosis of Highland Frogs. Berlin J Res Hom 1991a; 1: 151-160.

Endler PC , Pongratz W. Effects of agitated extremely diluted thyroxine on amphibia. Proc. Liga Medicorum Homoeopathica Internationalis, Köln, Mai 1991c.

Endler PC, Pongratz W, Kastberger G, Wiegant FAC, Haidvogl M. Climbing activity in frogs and the effect of highly diluted succussed thyroxine. Br Hom J 1991b; 80:194. Further data have recently been recommended for acceptance to the editor of the J Vet & Hum Tox.

Endler PC, Pongratz W, Van Wijk R. Transmission of hormone signal by water dipoles. Poster presented at the Meeting of the Americ Ass Advancment Sci, Boston 1993 (a).

Endler PC, Pongratz W, Van Wijk R, Wiegant FAC, Waltl K, Gehrer M, Hilgers H. On effects of agitated highly diluted thyroxine (10^{-30}) on the activity and the development of frogs (I) - Comprehensive report including detailed information on the substudies performed by different independent researchers 1990/91/92. Edited by the Ludwig Boltzmann Institut für Homöopathie, Dürergasse 4, A-8010 Graz, 1993 (b).

Fisher P, House I, Belon P, Turner P. The Influence of the Homoeopathic Remedy Plumbum metallicum on the Excretion Kinetics of Lead in Rats. Human Toxicology 1987; 6: 321-324.

Fisher P, Greenwood A, Huskisson EC, Turner P, Belon P. Effect of homoeopathic treatment on fibrositis (primary fibromyalgia). British Medical Journal 5 August 1989; 299: 365-366.

Flyborg K, Flyborg G. Die Flyborg-Therapie mit frequenzaktiviertem Wasser. Erfahrungsheilkunde 1990; 7: 424-427.

Gibson RG, Gibson SLM, Mac Neill DA, Watson-Buchanan W. Homoeopathic therapy in rheumatoid arthritis: evaluation by double-blind clinical trial. Br J Clin Pharmac 1980; 9: 453459.

Galton VA, Cohen JS. Action of thyroid hormones in premetamorphic tadpoles: an important role for thyroxine? Endocrinology 1980; 107 (6): 1820-1826.

Ganong WF. Lehrbuch der Medizinischen Physiologie. Berlin: Springer 1974.

Giersberg H. Hormone. Berlin: Springer 1972.

Gosner KL. A Simplified Table for Staging Anuran Embryos and Larvae with Notes on Identification, Herpetologica 1960; 16: 183.

Hadji L, Arnoux B, Benveniste J. Effect of dilute histamine on coronary flow of guinea-pig isolated heart. FASEB J 1991; 5: A1583.

Hanaoka Y. Uptake of ^{131}I by the thyroid gland during metamorphosis in Xenopus laevis. J Fac Sci Hokkaido Univ 1966; 16: 106-112.

Harisch G, Kretschmer M, von Kries U. Beitrag zum Histaminrelease aus peritonealen Mastzellen von männlichen Wistar-Ratten. Dt tierärztl Wschr 1987; 94: 515-516.

Harisch G, Kretschmer M. Jenseits vom Milligramm-Die Biochemie auf den Spuren der Homöopathie. Berlin: Springer 1990.

Herkovits J, Perez-Coll C, Zeni S. Reduced Toxic Effect of Cd on Bufo arenarum Embryos by Means of very Diluted and Stirred Solutions of Cd. Communicaciones Biológicas 1988; 7: 70 -73

Herkovits J, Perez-Coll C; Zeni S. Incremento en la toxicidad del Cd por empleo de soluciones preparadas con agitacion y microdosis de Cd. Medicina (Buenos Aires) 1989; 49:477-480.

Hoffmann M. Hautwiderstand und Lebensmittelqualität. Wetter-Boden-Mensch 1991; 3: 24-28.

Homöopathisches Arzneibuch. Stuttgart: Deutscher Apothekerverlag, and Frankfurt: Govi Verlag, Frankfurt 1991.

Just JJ. Protein bound iodine and protein concentration in plasma and pericardial fluid of metamorphosing anuran tadpoles. Physiol Zool 1972; 45: 143-152.

Kaye NW. Interrelationships of the thyroid and pituitary in embryonic and premetamorphic stages of the frog Rana pipiens. Gen Comp Endocrinol 1961; 1: 1-19.

Kent JT. Repertory of the Homoeopathic Materia Medica. New Delhi: Jain Publishers 1989.

Kollros JJ. Mechanisms of amphibian metamorphosis: Hormones. Amer Zool 1961; 8: 107-114.

König K. Über die Wirkung extrem verdünnter ("homöopathischer") Metallsalzlösungen auf Entwicklung und Wachstum von Kaulquappen. Zschft ges exp Med 1927; 56: 881-593.

Khuda-Bukhsh AR, Banik S. Assessment of Cytogenical Damages in X-irradiated Mice and Their Alterations by the Oral Administration of a potentized Homeopathic Drug, Ginseng -200. Perspectives in Cytology and genetics 1990: 7.

Kleijnen J, Knipschild P et al. Clinical trials of homoeopathy. Brit Med J 1991; 302: 316-323.

Lapp C, Wurmser L, Ney J. Mobilisation de l'Arsenic fixé chez le cobaye sous l'influence de doses infinitesimales d'arseniate, Therapié 1958; 13: 46.

Larue F, Dorian C, Cal JC, Guillemain J, Cambar J. Influence of the Pretreatment of Infinitesimal Dilutions of Mercurius corrosivus on the Mortality Induced by Mercuric chloride. Nephrologie 1985; 6: 86.

Lee E, Desu M. A computer program for comparing the samples with right-censored data, (Computer Programs in Biomedicine 2). No year given. p. 315.

Lipner H, Hazen S. Extra thyroidal iodine pump in tadpoles (Rana grylio). Science 1962; 138: 898-899.

Moser H. Ein Beitrag zur Analyse der Thyroxinwirkung im Kaulquappenversuch und zur Frage des Zustandekommens der Frühbereitschaft des Metamorphose-Reaktionssystems. Rev Suisse Zool 1950; 57 (Suppl 2): 1-144.

Neuenschwandner P. Ultrastruktur und Jodaufnahme der Schilddrüse bei Larven des Krallenfrosches (Xenopus laevis daud.) Z Zellforsch Mikrosk Anat 1972; 130: 553-574.

Panos M, Rogers R, Stephenson J. Thyroidinum, A Proving (Hyganthropharmacology), J Americ Inst Hom 1964; July-August: 197-207.

Petit C, Belon P, Got R. Effect of Homeopathic Dilutions on Subcellular Enzymatic Activity. Human Toxicology 1989; 8: 125-129.

Pitt-Rivers R, Trotter WR. The Thyroid Gland. London: Butterworth 1964.

Poitevin B, Aubin M, Benveniste J. Approche d'une analyse quantitative de l'effect d'Apis mellifera sur la dégranulation des basophiles humains in vitro. Innovation et Technologie en Biologie et Medicine1986; 7: 64-68.

Poitevin B, Davenas E, Benveniste J. In vitro immunological degranulation of human basophils is modulated by lung histamine and Apis mellifica. Br J Clin Pharmacology 1988; 25: 439-444

Prahlad KV, DeLanney LE. A study of induced metamorphosis in the axolotl. J Exp Zool 1965; 160: 137-146.

Reilly D, Taylor M. Is homoeopathy a placebo response? (Controlled trial on pollen in hayfever), The Lancet 1986; oct.19th: 881.

Righetti M. Forschung in der Homöopathie - Wissenschaftliche Grundlagen, Problematik und Ergebnisse. Göttingen: Burgdorf Verlag 1988.

Roth C. Literature Review and Critical Analysis on the Topic of "In- and Detoxication Experiments in Homoeopathy". Berlin J Res Hom 1991; 1: 111-117.

Saxen L. The onset of thyroid activity in relation to the cytodifferentiation of the anterior pituitary. Acta Anat 1958; 32; 87-100.

Schaefer W. Enzyme und metabolische Parameter des Glutathionsystems der Rattenleber nach Tetrachlorkohlenstoffvergiftung-Thesis. Tierärztliche Hochschule Hannover 1984.

Smith CW. Homoeopathy, Structure and Coherence. In: ZDN (ed.). Homeopathy in Focus. Essen. VGM Verlag für Ganzheitsmedizin 1990.

Sukul NC, Bala SK, Bhattacharyya B. Cataleptogenic Effect Prolonged by Homoeopathic Drugs. Psychopharmacology 1986; 89: 338-339.

Sukul NC, Zaghlool HA. Effect of two homoeopathic drugs, Agaricus muscarius and Nux vomica on the isolated ileum of rats. Sci.Cult. 1990; 56: 254-258.

Sukul NC. Mechanical agitation, the main factor in increasing the efficacy of Agaricus muscarius, a homoeopathic drug. Environment Ecol. 1992; 10(1): 7-10.

Tata JR. Early metamorphic competence of Xenopus larvae. Develop Biol 1968; 18: 415-441.

Wehner R, Gehring G. Zoologie. Stuttgart: Thieme 1990.

Weil MR. Changes in plasma thyroxine levels during and after spontaneous metamorphosis in a natural population of the green frog, Rana clamitans. Gen Comp Endocrinol 1986; 62 (1): 8-12.

Van Wijk R, Wiegant FA. Homoeopathic Remedies and Pressure-induced Changes in the Galvanic Resistance of the Skin. Alkmaar: VSM 1988.

Zimbardo PG. Psychologie. Berlin: Springer 1983.

Zelzer G. In vitro Untersuchungen zur intestinalen Phenylalaninresorption vor und während der Metamorphose bei Kaulquappen von Bufo viridis. Thesis, Universität Graz, Austria, 1990.

SCIENTIFIC PROVING OF AN
ULTRA HIGH DILUTION ON HUMANS

H. Walach

SUMMARY

A basic tenet of homoeopathy is that remedies which do not contain active drug molecules can have effects on the healthy human organism, by virtue of the specific preparation process of stepwise dilution and succussion. The claim that an ultra-high dilution of a homoeopathic remedy, Atropa belladonna C30 (100^{-30}), could produce effects different from placebo, was investigated in a pilot study. In a double-blind crossover trial, four weeks of Belladonna C30 were compared to four weeks of Placebo in 47 healthy volunteers. Data were collected daily. The number and types of changes were recorded into a predefined category system. Single-case evaluation showed differences between the two experimental phases for 21 subjects. Group evaluation showed only the tendency towards a difference between Placebo and Belladonna C30. The claim that homoeopathic high dilutions can produce symptoms other than placebo in healthy subjects should be put to further investigation.

ZUSAMMENFASSUNG

Es ist eine grundlegende Aussage der Homöopathie, daß Heilmittel, die keine aktiven Arzneimittelmoleküle enthalten, Wirkungen auf den menschlichen Organismus haben können, und zwar aufgrund des spezifischen Zubereitungsprozesses von stufenweiser Verdünnung und Verschüttelung. Die Annahme, daß eine Hochverdünnung eines homöopathischen Heilmittels, Atropa belladonna C30 (100^{-30}) von einem Placebo unterschiedliche Wirkungen hervorbringen könnte, wurde in einer Pilotstudie untersucht. In einem doppeltblinden cross-over-Versuch wurden 47 gesunde Freiwillige vier Wochen unter Belladonna C30 mit vier Wochen unter Placebo verglichen. Die Daten wurden täglich erhoben. Die Anzahl und Art der Veränderungen wurden in einem zuvor festgelegten Kategoriensystem festgehalten. Die Einzelfallanalyse zeigte zwischen den beiden Phasen des Experimentes bei 21 Personen Unterschiede auf. Die Gruppenanalyse zeigte nur die Tendenz eines Unterschiedes zwischen Placebo und Belladonna C30. Die Annahme, daß homöopathische Hochverdünnungen bei gesunden Probanden spezifische Symptome hervorbringen können, sollte weiter untersucht werden.

INTRODUCTION

Ever since its inauguration in 1796 by the German physician and experimental pharmacologist Samuel Hahnemann (1755 - 1843), homoeopathy has survived as a medical fringe discipline. Today, a renaissance in alternative medicine and homoeopathy has been recognized [1,2]. Although good scientific evidence is rare, as in many disputed fields of science, a recent review of clinical trials in homoeopathy uncovered 105 controlled trials, 81 of which indicated positive, 24 negative results[3]. However, no single study has been reported which experimentally puts to trial the very foundation of homoeopathic reasoning: the "Remedy-Proving" of a homoeopathic substance in succussed high dilution with healthy volunteers. It was with this method that Hahnemann began his work on homoeopathy: he administered all then known pharmaceutical substances to healthy volunteers, initially in crude doses, later in what he called "potentized" form - stepwise highly diluted. He then recorded the symptoms carefully and used the data in turn for therapy, applying the principle "like cures like": he treated patients who showed symptoms with agents which

P.C. Endler and J. Schulte (eds.), Ultra High Dilution, 69–79.
© 1994 *Kluwer Academic Publishers. Printed in the Netherlands.*

were able to produce similar symptoms in healthy subjects. He later tried to overcome to-xicological effects by stepwise diluting and succussing the drugs. This dilution process, called "potentisation", gradually reached a point at which it is most likely that no mole-cules are left in the solution. Thus, given Avogadro's number (6.023×10^{-23} molecules per mole of a substance), a one-molar solution stepwise diluted 12 times by a factor of 100 is highly unlikely to contain any molecule of solute in a liter of solution. This corresponds to a homoeopathic preparation of C12. Hahnemann did not know about these facts. Yet he thought that by the dilution process of stepwise succussing, "the dynamic power" of the remedy could be brought to the fore. These vitalistic conceptions, reminiscent of Paracelsus' notion of "arcana", can be understood in both systems-theoretic and semiotic ways [4,5].The question was addressed, whether so-called high dilutions or potencies, beyond Avogadro's number, can produce effects in healthy volunteers more and/or other than placebo, and if so, whether it is possible to investigate this phenomenon in a scientifically acceptable way[1].

METHODS
Substance and Preparation

The study was conducted as a strictly double-blind, randomized crossover trial with healthy volunteers. The proving substance was Belladonna 30C, a well-known homoeopa-thic drug, freshly prepared from the plant Atropa Belladonna, according to the methods of the German Homoeopathic Pharmacopoeia (HAB) by a well known homoeopathic manu-facturer (Deutsche Homöopathie Union, Karlsruhe)[2]. The highly diluted, potentized solu-tion was then spread out on sugar globules, such that 1 g solution was sprayed on 100 g of globules, which thereby get soaked, and are then allowed to dry. The dilution 30C is a standard dilution in many countries and was used by Hahnemann himself for therapeutical and drug-proving purposes [7]. It corresponds to a serial dilution of 100^{-30} or 10^{-60}, which is more than twice Avogadro's number. The placebo was absolutely identical in colour and taste, since it consisted of sugar globules of the same lot, simply soaked with 1g of ethanol per 100g of sugar globules. The manufacturer sent proving substance and placebo, which were both filled in identical 5g containers to a public notary who randomized the con-tainers, according to the study design.

Design

The study duration was 10 weeks altogether. For each week of the study a new container was used. The first week was a placebo run-in week, which was not counted. Then 4 weeks of Belladonna for group A, or 4 weeks of placebo for group B followed, and vice versa for the following 4 weeks. The last week was observation only. No washout was included, as the homoeopath in charge of homoeopathic remedy provings in Germany certified that it was unneccessary with this particular remedy. Each subject received 9 consecutively-numbered containers with 5 g of sugar globules each. Subjects were instructed to take 2-3 globules (approximately 1 mg) of sugar globules on Tuesday, Wednesday, and Thursday evening each week, and then to throw away the excess of globules so as to ensure that a new container was opened each week. The other four days of the week were observation only. The proving substances were given in a strictly double-blind method: the ran-domization plan was established by a computer program at a different site (Department of Biostatistics, University of Ulm) and deposited in single original with a public notary, who then randomized accordingly. For each subject, a single sealed copy of the individual sequence of proving substance and placebo was deposited with an emergency physcician who was instructed not to open the envelopes except in cases of emergency. The envelopes were collected at the end of the study and the seals inspected. The proving substances were mailed to the subjects by a third party (Dep. of Biostatistics, Ulm). The approval of the Ethical Committee of the Psychiatric University Hospital, Basel, was obtained. The subjects were told the basic aim of the study and that, according to homoeopathic theory, taking the substances might entail suffering from transitory light symptoms which would go away again. They were also informed that they might receive a homoeopathic remedy in high dilution, or placebo, or both. They were told neither the name of the remedy nor the precise potency, nor was information given as to the design. All participants gave their written consent.

Subjects

Volunteers were approached in lectures on homoeopathy and psychology in Freiburg, Germany, and in Basle and Zurich, Switzerland. Participation was strictly voluntary; there was no motivation other than interest in the subject. No payment or other form of reward was offered for participation. Volunteers were required to be and feel healthy, not to be under any medical, psychotherapeutic or other treatment. They were not allowed to take illegal or medical drugs at all, or drink stimulants like coffee or alcohol in excess, and they should lead a normal life without a lot of change, stress and travel. They were required to have a low level of physical complaints and to be psychologically healthy. This was screened employing three questionnaires: a questionnaire containing exclusion criteria items, 9 attribution and 2 knowledge-of-homoeopathy items, a questionnaire to assess physical complaints consisting of 10 subscales (Freiburg Complaint List - FBL), and a personality inventory consisting of 12 scales (FPI) [8,9].

Data

Data collection was performed on a daily basis with a diary specifically developed for that purpose. Subjects were asked to fill in the diary each day at the same time, preferably in the evening. The diary recorded information about intake of proving substance, intake of other drugs, life events, well being, mood, and changes observed. These data were obtained daily. Subjects were instructed to observe whether physical or psychological changes had occurred which were unusual or which represented a deviation from normal well being. In case changes had been observed, subjects were asked to classify the changes according to a predefined list of categories in the following dimensions: part of the body in which changes had occurred (body scheme with 102 numbered areas; e.g. 1 indicating head frontal, right, 2 head frontal, left, 5 right eye, 6 left eye, and so on), what kind of changes were noticed (66 categories), the time of day the changes occurred (23 categories)[3]. Additionally, the intensity and duration of each of the changes noted should be recorded, and free text for description was optional. The diary provided for the classification of 6 different changes per day. Subjects were instructed by a short manual how to use the classification system. A post-hoc-survey revealed that the system had been perceived as helpful and that the average time to fill in the diary was 7.6 minutes per day. All statistical analyses, except for the single-case statistics, were calculated on a VAX-computer using BMDP 1990 release [10].

The general hypotheses were that

a) per subject the two experimental periods are different
b) for the whole group, the quality of changes are different under the different conditions, and
c) the configuration or clustering of changes is different.

RESULTS

Study Population

66 healthy subjects volunteered to take part in the study: the median age was 25, 29 of them were female, 37 male, 48 of them were homoeopathy students, 18 came from other groups. 49 of them sent back diaries. Two subjects demanded breaking of the code, one of them in the placebo run-in-week, the other one on day 48, during the second week of Belladonna. One filled-in diary was lost in the mailing process, another was only partially completed and could be used only for a few of the analyses. Therefore, the database is n = 45.

An analysis comparing the dropped-out subjects with the remaining ones in the screening-, complaint-, and personality-variables shows only slight effects, which are not significant if corrected for multiple testing. The sample was physically and psychologically healthy, indicated by scores obtained with FBL and FPI questionnaires, i.e. stanine-values of FPI-scales were 5 or less, and mean values of FBL-scales were comparable to other healthy subjects. 56% of the subjects showed a good knowledge of homoeopathy, 44% some or little knowledge.

Single-Case Evaluation

For a first analysis on single-case basis, the question is, whether there were differences, between the two four-week experimental phases within each subject. Therefore, the categorial data of the changes perceived were re-translated into narrative form, and a qualitative judgement about group-membership was attempted. This failed.

Next, the changes reported in each phase were arranged according to a McNemar matrix: the frequency of changes noted in phase A was only compared with the number of occurences of those changes that were noted in both phases, and again with the number of those changes, which were reported in phase B only.

A McNemar statistic was calculated for each subject, according to the formula $Chi^2 = (A-D)^2/A+D$ with 1 DF, where A is the number of changes reported in phase A only, and D is the number of symptoms in phase B only (table 1, next page). This was done irrespective of group membership, i.e. no direction of the difference was assumed.

Group Evaluation

Next, a test for carry-over effects was calculated: the sum of differences between the two phases per group for all variables was compared. There was no significant carry-over effect (p = 0.7557, Mann-Whitney-U test).

Considering the group as a whole, the different categories were taken as dummy variables, and it was calculated how often a specific variable was counted during the placebo- and during the Belladonna-phase. Post hoc, subcategories which pointed in the same direction and were of the same contextual nature - e.g. all the subcategories of the left arm, or all subcategories of activation - were collapsed in order to reduce the number of variables and to increase the number of counts per category and phase. No variable was used twice, however.

The differences in count of a specific variable between placebo and Belladonna was then compared using Wilcoxon's test for dependent, non-parametric data. The results are given in Tables 2 - 4. P-values are given for larger differences.

Tab. 1

subject number of different changes reported in McNemar-Chi^2

no	phase A only	phases A and B	phase B only	p	
1	26	3	13	0.05	*
2	5	0	2	0.3	
3	34	8	5	0.0005	*
4	9	168	63	0.0017	* §
7	6	0	2	0.2	
8	3	2	6	0.4	
9	4	0	0	0.05	*
10	7	2	16	0.1	
11	14	2	7	0.2	
13	17	5	28	0.1	
14	25	6	32	0.4	
15	11	9	17	0.3	
16	12	4	9	0.6	
17	14	11	41	0.0005	* §
18	5	2	3	0.5	
19	34	9	5	0.0001	*
20	36	38	37	0.9	
21	4	0	0	0.05	* §
22	6	3	10	0.3	
23	6	4	1	0.1	
24	10	6	3	0.1	
25	34	9	25	0.3	
26	64	26	103	0.005	*
27	10	0	0	0.005	* §
28	5	0	0	0.05	* §
29	15	0	0	0.0005	*
30	25	10	42	0.05	*
31	11	5	9	0.7	
32	12	8	28	0.025	*
33	1	0	0	0.4	
34	4	0	3	0.7	
35	20	6	10	0.1	
36	36	15	24	0.2	
37	7	3	26	0.001	* §
38	6	0	6	0.8	
39	3	4	11	0.05	*
40	10	1	6	0.4	
41	8	0	0	0.05	* §
42	15	3	5	0.025	*
43	5	1	27	0.0005	*
44	40	0	0	0.0001	*
45	0	0	6	0.025	* §
46	27	1	6	0.0005	*
47	17	11	15	0.8	
48	9	3	17	0.2	

Tab. 1: Single-case evaluation of qualitative-categorical data: number of different changes reported by each subject in each experimental phase (4 weeks); p-value of McNemar statistics with 1 DF; *: p <= 0.05; §: paradoxical effects (explanation in the text).

Note that in this analysis only those changes are taken into account which occur in one of the two phases, and not in both. Therefore, this gives a fairly robust measure of divergence of the two phases per subject. The test was done to discern whether there was any difference at all, not whether it pointed in a specific direction. McNemar's test was employed because the data of each subject were dependent. Each single p-value can be considered independent of the other ones. Therefore, the p-values can be combined by what is known as the Fisher-Pearson method. K independent p-values are log-transformed, summed, multiplied by -2 and yield a Chi-square-value with 2k degress of freedom:

$$Chi^2 = -2 \times sum\ ln(p); (DF = 2k)$$
According to this formula for $Chi^2 = 280.78$ (DF = 90); $p < 0.0001$.

Visual inspection alone shows that, on an individual basis, the two experimental conditions are different for quite a few subjects reflected in individually significant McNemar-p-values. Some subjects exhibit a small, some a large difference between the phases. Most notably, in nine cases, paradoxical effects were observed (table 1, marked with "§"): the McNemar statistics indicated p values <= 0.05, but more changes were reported with placebo than with Belladonna.

Tab. 2

SITES	Belladonna	Placebo	TOTAL	Wilcoxon p-values
head frontal	87	189	276	0.13
nose	63	72	135	
ear right	12	4	16	0.18
ear left	12	20	32	
mouth	55	37	92	0.27
occiput	18	8	26	0.09
neck	9	5	14	
throat	36	54	90	0.19
back, upper part	10	13	23	
collar bone	0	2	2	
shoulder frontal	13	2	15	
shoulder backside	1	2	3	
shoulderblade	4	2	6	
thorax	6	1	7	
arm upper part	1	9	10	0.12
elbow	34	1	35	0.37
hands	14	16	30	
breast	34	22	56	
sides	2	4	6	
stomach	74	97	171	
genitals	15	20	35	
low back	15	8	23	
coccyx	7	11	18	
back	15	2	17	0.0039
nates	1	3	4	
hips	12	2	14	
leg left	18	11	29	
leg right	24	19	43	
TOTAL	592	636	1228	

Tab. 2: Frequency of changes at different body sites with Belladonna and Placebo.

Tab.3

TYPES OF CHANGES	Belladonna	Placebo	TOTAL	Wilcoxon p-values
accident	2	2	4	
nausea	10	13	23	
crawling	25	32	57	
itching	39	33	72	
warmth	9	30	39	
coldness	26	6	32	
other sensations	237	235	472	
blood	3	9	12	
secretion	23	19	42	
mucus	28	30	58	
pus	7	18	25	0.06
constipation	19	18	37	
diarrhoe	11	16	27	
urine	0	2	2	
menses	9	8	17	
sweat	5	1	6	
trembling	2	1	3	
twitching	1	9	10	0.12
lameness	8	0	8	0.06
inflammation	91	51	142	
skin	2	17	19	0.06
senses	5	9	14	
balance	8	16	24	
activation	119	78	197	0.05
negative feelings	86	161	247	0.06
positive feelings	23	19	42	
food	8	17	25	
sex	13	6	19	0.18
thoughts	22	32	54	
falling asleep	9	11	20	
sleeping	26	19	45	
dreams	13	19	32	
TOTAL	889	937	1826	

Tab. 3: Frequency of changes of different type with Belladonna and Placebo.

Tab. 4

TIMES	Belladonna	Placebo	TOTAL	Wilcoxon p-values
always	364	321	685	
night	119	67	186	0.02
evening	76	100	176	
early morning	16	25	41	0.08
morning	90	92	182	
breakfast	13	4	17	0.06
before noon	49	79	128	0.07
noon	6	22	28	0.06
afternoon	88	122	210	0.09
often in an hour	2	13	15	0.08
often	56	48	104	
TOTAL	879	893	1772	

Tab. 4 : Frequency of changes at different times with Belladonna and placebo.

As can be seen from the tables, effects, if present at all, were quite small. In general, there was no difference in the total number of changes reported with Belladonna or placebo. There were also "paradoxical" effects: subjects who reported more changes with placebo, where more changes with Belladonna would have been expected.

Responder/Non-Responder Analysis

From the single-case analysis, it was obvious that there were people who did report different changes in the different conditions and there were also people who did not. Those who did not respond, i.e. who could not be classified into the respective experimental group by the number or quality of changes reported into the respective experimental group, were defined as non-responders. Using screening-data, FBL and FPI, Mann-Whitney U-Tests and, if appropriate, t-tests were calculated, between Belladonna-responders and non-responders. The differences between the groups were not striking. But it was interesting to see that the differences emerged mainly in the variables of the FBL. The variables, which show a non-corrected difference with p </= 0.05, are reported in Table 5.

Tab. 5

Variable	Verum-Non-responder (n=15)	Verum-Re-sponder (n=32)	p=(Mann-Whitney/t)
homoeopathy stimulates	0.6	0.25	0.02
life-events present	0.13	0.0	0.04
pill	0.13	0.0	0.04
FBL1 well-being	20.23	16.58	0.04
FBL5 head/throat	19.30	15.72	0.05
FBL6 tension	20.0	15.71	0.03
FBL7 sensory symptoms	21.61	17.49	0.04

Tab. 5: Differences between responders and non-responders.

It is interesting to see that only one item out of 9 attribution items ("homoeopathy stimulates the body"), two items from the exclusion list (life events anticipated during the experimental period, taking of contraceptive pill), and 4 scales of the FBL, and none of the other variables show differences for the two groups. The data show a tendency for the responding group to exhibit less complaints as measured by the FBL and to fulfill all the criteria which were set for inclusion into the study.

DISCUSSION

Certainly, the issue whether a homoeopathic drug which does not contain any pharmaceutically active substance in the form of drug molecules, can produce any effects different from placebo, is highly controversial and must be handled with specific care. The data presented here are preliminary and do not permit any definite conclusion. Also, they are to my knowledge the first attempt to scientifically operationalize the rationale of a homoeopathic drug proving and to statistically evaluate the data obtained from it.

Single-case analysis shows that there seem to be differences between phases and within subjects. Some subjects report many changes, and also many different changes in the different phases. This is an interesting finding, for if Belladonna C30 had the same effect as placebo, one would expect the changes to be more evenly distributed over the phases. Obviously, the McNemar-statistics applied can only in a heuristic fashion capture effects, since the countings of changes within a phase are dependent. On the other hand, the statistics only considers those symptoms and changes which are particular to a certain phase, and information about those changes which appear in both phases, but in one more frequently than in the other, is lost. That makes the estimation conservative again.

The comparison of particular changes as dummy-variables reveals a specific problem of this kind of research: in order to validly depict the variety of symptoms which are purported to be possibly produced by a homoeopathic remedy, according to theory, one has to choose an open system of categories with a sufficiently large number of variables. On the other hand, this reduces statistical power considerably, since the number of variables compared to the counts of a single variable increases. This is appropriate for a pilot study. A few categories stand out, however, albeit only as tendencies. Sometimes they point in the same direction as homoeopathic theory would have it, sometimes they are paradoxical, i.e. clearly more symptoms are reported with Placebo than with Belladonna. Belladonna is said to be a remedy with special affinity to head, throat, ears, inflammatory processes with the triad of redness, swelling, pain. It is said to be effective in specific kinds of fevers and inflammation, and it should show symptoms in the area of general activation. Its specific time of action is said to be in the night. For further detail the reader is referred to one of the standard homoeopathic materia medicas [11,12]. However, in this study head localization was more prominent in the placebo-condition, as was the throat localization. Occiput, mouth, right-ear, and back are more outstanding with Belladonna, which would be expected. As can be seen from the single-case data, some subjects do have many symptoms when taking placebo, but no symptoms when taking Belladonna. Those paradoxical responders reveal quite a high level of placebo-symptomatology, with many head symptoms. These paradoxical effects could be curative, but this cannot be proven, since there is no baseline without intake. This is a major shortcoming of the design, which has to be considered for possible follow-up studies. It is interesting to find more changes in the categories warmth, pus, and negative feelings in the placebo condition, whereas coldness, inflammation, activation, and sexuality have more counts with Belladonna. The latter would be expected from clinical homoeopathic experience, the former is rather unexpected, and could again be a hint as to therapeutic actions. If this is so, namely that therapeutic effects, resulting in a reduction of symptoms with Belladonna, are mixed with experimental ones, yielding an increase of symptoms with Belladonna, effects might be confounded. This would entail that follow-up studies using only very healthy subjects should have more clear-cut results. The fact that Belladonna-non-responders show higher values in some of the complaint list scales is an indicator towards that direction, which should be looked into. It is interesting to find more symptoms following placebo during the day than during the night. It is known that the placebo-effect, and specially placebo-analgesia, is significantly reduced during the night [13]. We see nearly twice as many counts in the category "night" with Belladonna, which would be expected from both the homoeopathic materia medica and from the fact that the substances were taken in the evening and therefore, if active, should show their action predominantly during the night.

What is clear from the results is that placebo is able to produce a lot of symptoms and changes, numerically even more than Belladonna, when summed up. What we cannot infer from this design is whether these placebo-symptoms would be present in any case or are a specific effect of placebo as opposed to normal status. So if there are any effects of the homoeopathic drug at all, they are to be found neither in the quantity of changes, nor in single changes alone, unless with a very large number of subjects.

This is the first time to my knowledge that an experimental homoeopathic drug proving has been conducted in this highly-controlled manner with human volunteers, using a pre-structured way of data collection. There are interesting results from animal studies [14,15], showing that homoeopathic substances up to dilutions of C1000 could produce experimental catalepsy in animals, in accordance with homoeopathic practice and experience.

The results presented here seem to be promising. In the organization of the study, care has been taken to eliminate any kind of fraud or prediscovering of the random assignment. The study was strictly double blind and randomized. The data from the diaries seem to be valid, as a post-hoc-inquiry yielded favourable results. Therefore, one is prompted to conclude that the homoeopathic claim that potentized drugs even beyond Avogadro's number are able to produce changes in healthy human volunteers cannot simply be dismissed. To be sure, this is only a preliminary result. Follow-up studies have to determine reproducibility. Specifically, it should be considered whether placebo-changes differ from baseline-values. Another point to be taken into account is that one has to employ truly healthy subjects in order not to confound experimental and therapeutic effects, or else a way has to be found to examine these effects separately. One way of dealing with this problem may be the use of a single-case randomization design.

ACKNOWLEDGEMENTS AND ANNOTATIONS

This work has been carried out as Ph.D.-work at the Department of Clinical Psychology, University of Basel, Switzerland. I am grateful to Prof. V. Hobi for encouragement, advice and help in organizing the study. Prof. W. Gaus, Department of Biostatistics, University of Ulm, organized printing of the diaries, randomization and mailing of the test substances, and data entry. Prof. G. Haag, Prof. J. Fahrenberg and Dr. F. Potreck-Rose, Freiburg gave helpful advice as to the planning of the study. Dr. M. Stübler, Augsburg, gave homoeopathic advice as to dosage and remedy to be chosen. Statistical advice was given by Dr. Berres, Ciba-Geigy, Basel. The Deutsche Homöopathie Union funded the test substances, mailing and data entry. A scholarship was provided by Cusanuswerk, Bonn. The present work of the author is sponsored by the Robert-Bosch-Stiftung, Stuttgart.

Annotations by the author: 1. A different version of this paper is to appear in "Journal of Psychosomatic Research". - 2. A full report of this work in German can be found in ref. (6). Original raw data or print out of transcribed symptom lists are available on request. - 3. Details of the preparation procedure can be obtained from the author or the manufacturer: DHU, Postfach 410280, D-7500 Karlsruhe 41. - 4. A sample of the diary including the predefined categories and a German version of the instruction manual can be obtained from the author.

Annotation by the biological editor: The investigation of the effects of a substance in a Remedy Proving on healthy probands who were chosen at random is, ideally, followed by a more throughout investigation of the effects of that very substance on those persons who had shown the most marked reactions in the first step of the Proving. Such a (double-blind) follow-up study might be of interest also in the case of the Remedy Proving discussed in Walach`s contribution.

REFERENCES

1. Ullman D. The international Homeopathic renaissance. Berlin J Res Hom 1991; 1: 118-120.
2. Haag G, Walach H, Erbe C, Schrömbgens HH. Verbreitung und Verwendung unkonventioneller medizinischer Verfahren bei niedergelassenen Ärzten - Ergebnisse einer Fragebogenumfrage. Zeitschr Allgemeinmed (in press)
3. Kleijnen J, Knipschild P, ter Riet G. Clinical trials of homoeopathy. Br Med J 1991; 302: 316-323.
4. Walach H. Homöopathie als Basistherapie. Plädoyer für die wissenschaftliche Ernsthaftigkeit der Homöopathie. Heidelberg: Haug 1986.
5. Walach H. Homoeopathy as Semiotic. Semiotica 1991; 83: 81-95.
6. Walach H. Wissenschaftliche Homöopathische Arzneimittelprüfung. Doppelblinde Crossoverstudie einer homöopathischen Hochpotenz gegen Placebo. Heidelberg: Haug 1992.
7. Hahnemann S. Organon der Heilkunst. 6th edition, ed. by R. Haehl. Reprint. Stuttgart: Hippokrates 1979.
8. Fahrenberg J. Die Freiburger Beschwerdeliste FBL. Z Klin Psych 1975; 4: 79-100.
9. Fahrenberg J, Hampel R, Selg H. Das Freiburger Persönlichkeitsinventar FPI. Revidierte Fassung FPI-R und teilweise geänderte Fassung FPI-A1. 5th Ed. Göttigen: Hogrefe 1989.
10. Dixon WJ, Brown MB, Engelman L, Jennrich RI, editors. BMDP Statistical Software Manual. Berkeley, CA: University of California Press, 1990
11. Allen TE. Encyclopedia of Pure Materia Medica. New York & Philadelphia: Boericke & Tafel, 1874-80
12. Hahnemann S. Reine Arzneimittellehre Erster Theil. Dresden: Arnold, 1811.
13. Hildebrandt G, Pöllmann L. Chronobiologie des Schmerzes. Die Heilkunst 1987; 100: 340-358.
14. Sukul NC, Bala SK, Bhattacharyya B. Prolonged cataleptogenic effects of potentized homoeopathic drugs. Psychopharmacology 1986; 89: 338-339.
15. Sukul NC, Klemm WR. Influence of dopamine agonists and an opiate antagonist on Agaricus-induced catalepsy, as tested by a new method. Arch Int Pharmacodyn Ther 1988; 295: 40-51.

PHYSIOLOGICAL EFFECTS OF HOMOEOPATHIC MEDICINES IN CLOSED PHIALS; A CRITICAL EVALUATION

R. van Wijk, F.A.C. Wiegant

SUMMARY

In different series of experiments using the medicine test according to Voll, the observation was made that homoeopathic remedies in closed glass phials may influence a physiological characteristic of the human organism. The phial containing the remedy may be placed in close proximity to the patient, or it may be included in a testing circuit designed to measure the electric conductivity of the skin. In both cases, the remedy is not in direct contact with the skin and the phial seems to be permeable to the "information" which characterizes the remedy. Different series of experiments showed a significant deviation between an agitated dilution of sulphur and placebo. It is assumed that the test subject (the tester) is somehow affected by the patient in the fine-tuning of his muscular activity and in this respect influenced by the introduction of the remedy in the Voll testing process.

ZUSAMMENFASSUNG

In verschiedenen Experimentserien wurde mit Hilfe des Vollschen Medikamententestes die Beobachtung gemacht, daß homöopathische Heilmittel in geschlossenen Glasphiolen ein physiologisches Parameter des menschlichen Organismus beeinflussen können. Die das Heilmittel enthaltende Phiole kann entweder in nächster Nähe des Patienten plaziert sein, oder sie kann in ein zur Messung der elektrischen Leitfähigkeit der Haut entworfenes Testsystem integriert werden. In beiden Fällen ist das Heilmittel nicht in direktem Kontakt mit der Haut und die Phiole scheint für die "Information", die das Heilmittel charakterisiert, durchlässig zu sein. Unterschiedliche Experimentserien zeigten einen statistischen Unterschied zwischen einer verschüttelten Verdünnung von Schwefel und Kontrolle. Es wird angenommen, daß der Tester selbst in irgend einer Weise durch den Patienten in der Fein-Einstellung seiner Muskelaktivität beeinflusst wird und er selbst im Vollschen Testprozess auf diese Weise durch das Heilmittel beeinflußt wird.

INTRODUCTION

Research on the medicine test; why and how?

Homoeopaths have traditionally ignored the medicine test according to Voll. Instead, they have relied exclusively on the classic approach of Hahnemann, Kent and others. Because of this one–sided understanding and judgement of the application of homoepathic remedies, only a few questions raised in fundamental studies can be (partially) answered. The reason for studying the medicine test is the intriguing observation that homoeopathic remedies in closed phials (may) influence a physiological characteristic of the human organism (Voll, 1980). The phial containing the remedy may be placed in close proximity to the patient, or it may be included in a testing circuit designed to measure the electric conductivity of the skin. In both cases, the remedy is not in direct contact with the skin and the phial is thought to be permeable to the "information" which characterizes the remedy (Voll, 1971; Kramer, 1971). For homoeopaths, the principle should be of interest since it may elucidate the interaction of homoeopathic remedies and the human organism.

The application of this medicine test is developed as part of electroacupuncture diagnostics (Ferhenbach, 1988). It is developed worldwide and is part of the training in

81

P.C. Endler and J. Schulte (eds.), Ultra High Dilution, 81–95.
© 1994 *Kluwer Academic Publishers. Printed in the Netherlands.*

electro-acupuncture. The medicine test has met much criticism which is based on two major arguments or questionmarks. The first questionmark is usually based on doubts that such an influence on the human organism could be possible.

The second one is usually based on doubts with respect to the diagnostic value. An answer to the second question is of interest to electroacupuncture as a medical discipline. The purpose of the present research may be defined as answering the question whether any statistically significant changes in the electrophysiological properties of the skin occur after exposure to a phial containing a remedy. Several researchers have made a start by investigating the electrodermal changes which occur during the medicine test. Beisch and Bloesch (1979) demonstrated changes in conductivity with the administration of different remedies. And Van Wijk and Wiegant (1989) observed statistically significant differences between "active" remedies and control preparations. The latter study also considered the significance of the amount of pressure applied in the medicine test. Starting from the normal diagnostic test conditions we developed in the course of time a condition allowing blind testing of the medicine test.

Recently the results of this study have been described by Van Wijk (1992). This article summarizes the results of these investigations. The new data provided here offer a clear insight into the mechanism of the medicine test according to Voll. They can thus contribute to the discussion of the value of the EAV method for medicine.

This discussion here stresses the significance of the results in terms of the implications for basic homoeopathic research.

Conditions for the experiments

Experiments described in the literature and our own pilot experiments suggest a range of requirements which should be made in future experiments in order to achieve a greater degree of reliability and credibility compared to earlier studies. The key requirements are the following.

(a) The planning, execution and analysis of experiments should be carried out by different experts and should never be entrusted to one person or to a group of people who have insufficient experience in the area under study.

This requirement was met in our case. The team consisted of a study leader, testers, test subjects, one person supervising the automatic recording and one responsible for statistical aspects. The testers were professionals and taught training and refresher courses for EAV practitioners.

(b) If possible, experiments should be conducted double–blind.

This important condition was also met for all experiments. Both the tester and subject were unaware of each other's actions, while the results were recorded automatically.

(c) Experiments should proceed according to a set procedure.

The prescribed procedure was followed at all times during the experiments. It was impossible for the tester and the subject to influence the course of the investigation in a way not provided for in the procedure.

(d) Detailed statistical and data analysis should be carried out.

The processing of the data and statistical information was carried out by the usual calculation methods, with the help of Statistical Package for the Social Sciences (SPSS).

Information from the preparatory phase

Anyone familiar with the epistemological foundations of natural science and its practical application is aware of the problems which occur when observations are introduced into the discussion which appear to have a new character and which cannot be satisfactorily explained in theoretical terms. In an advanced field of science a model usually precedes a goal–orientated experiment. This means that the results of the experiment can make a statement about a specifically tested aspect. In such a case the theory should not be unnecessarily complicated and in the early phase should consist of some simple working hypotheses. Thus most scientific laws were initially established by accurate observation and measurement, which were only later integrated into a comprehensive theoretical description.

In the problem under consideration we need to establish the accurate observations first. As mentioned earlier, the first question which the experiments seek to answer is whether there are human reactions which are dependent purely on a homoeopathic remedy and are not explicable by the normal senses.

This question forms the basis for the formulation of a scientific hypothesis, and appropriate experiments should in principle provide an answer. There is as yet no need to build such experiments on far–reaching and comprehensive theoretical explanations of posited phenomena.

The numerous observations already recorded on the theme under consideration allow the formulation of working hypotheses which have a high correspondence with reality.

Thesis 1: There is no homoeopathic information in the sense of a clearly defined type of information. This means we must clearly reject undebatable and untestable notions such as that the EAV testing method reveals a new kind of information hitherto unknown to physics. We are assuming that this phenomenon can be explained within the framework of established biological principles and theories.

Thesis 2: The EAV testing method is a complicated biophysical phenomenon. We are assuming that different biological organisms, under specific circumstances, react with a physiologically determined signal, which is identified by the EAV tester as conductivity loss. We are assuming here that a physical cause produces a biological reaction.

Thesis 3: The EAV test is a highly efficient secondary aid for amplifying and illustrating a primary physiologically determined signal. This thesis relies on the indications that several and very experienced testers have no need of such an instrument and experience the signals direct with their bodies.

Complexity of the EAV test

The uncertainty about the impact of experimental and external conditions and above all the limited accuracy of the earlier studies made it difficult to develop problem–orientated tests. It was therefore necessary to move to an optimal set-up in stages. In this way we also wanted to avoid the mistakes and imperfections of earlier studies. It was our intention to improve the chances of success by, first of all, trying out several test set-ups. We distinguished between two types of experiments, namely those carried out under the usual diagnostic conditions, and those under artificially created circumstances.

(a) Problems with a natural testing process.

The tester tries to reduce the conductivity loss at specific points on the skin of the test subject. Experience has shown that in many cases a partial and temporary change does occur. This restricts the possibility of reproducible changes. In principle there are no problems with regard to double–blind testing (that is, blind with regard to tester and subject). Significant effects have been observed previously during the natural testing process under non–blind conditions (Van Wijk and Wiegant 1989) as well as under blind conditions (Van Wijk 1992).

(b) Problems with artificially created circumstances.

In these cases the aim is to create a situation in which the EAV tester records changes in a reproducible manner. However, the variation and optimization in the process of neutralization present a virtually insurmountable problem for the EAV testers or researchers. It should by no means be assumed that effecting an optimal combination would be easy. But if it succeeds, the benefits are self–evident, for the stimulus conditions can be easily altered and the experiments can be carried out double-blind.

In this paper we present data on the second approach, i.e. from experiments carried out under artificially created conditions.

METHODS

Measuring procedure and experimental model for testing the medicine test. DC measuring equipment

A device called the ACU–1, designed by P.J. Romeyn of Electronische Instrumenten voor de Geneeskunde Company, based in Zeist, The Netherlands, was used for the measurements with direct current. This device has been used in previous studies (La Rivière, 1988; Van Os–Bossagh, 1989; VanWijk and Wiegant, 1989).

The ACU–1 is connected to a two–channel Linseis recorder. The pressure on the active electrode is recorded by means of a measuring electrode with a pressure reader. The pressure reader is connected to the recorder's second channel. In this way current and pressure are recorded at the same time. The measurement set–up has previously been described in more detail (Van Wijk and Wiegant, 1989; Van Wijk, 1992).

Carrying out the measurements

The measurements were taken at the so–called Ting points on the hands. The skin under the active electrode is moistened with tap water. The usual method involves the measuring of the pressure–induced loss of electrical conduction at those skin points.

The analysis of conduction and conductivity loss is illustrated in Figure 1.

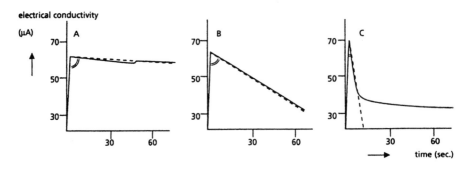

Fig. 1: Analysis of conductivity loss: three examples. Two types of calculation were applied (see next page).

1. In terms of degrees. The indicated angle is 90° or less. The angle is no absolute measure since the speed of recording determines the slope.

2. In terms of μA per time unit. The slope of the loss in conductivity is 0 μA per time unit or more.

In example A the conductivity loss is less, the decrease in terms of μA/sec is low and the angle in degrees is large. Example C shows the reverse situation, while example B represents an intermediate condition.

After a sharp increase in conductivity with the application of pressure on the moistened point, a maximum value was reached. This value was called the initial conductivity. Subsequently a more or less sharp loss in conductivity can occur. If no loss occurs, conductivity remains at the maximum level and shows an angle of 90 degrees with the sudden increase. The loss was expressed in μA per time unit.

Blind testing of the medicine test by means of an experimental model

An experimental model was devised in which an indicator drop was artificially induced in a subject who, in the preliminary treatment, showed no more symptoms of an indicator drop. An experimental model of this kind offers many advantages for the study of the medicine test, for we will know in advance which potentized remedies are capable of suppressing toxin–induced conductivity loss. It thus becomes possible to prepare valid control samples with which to conduct blind testings. The development of such an artificial testing system for inducing and then suppressing conductivity loss has been described previously (Van Wijk and Van der Molen, 1990; Van Wijk 1992). In short: the pre–treated indicator drop–free subject was given diphenyl orally. Indicator drop induced in this way was then modulated with the help of sulphur potencies.

Placing sealed phials sulphur D12 in the medicine holder produced a suppression of conductivity loss. The specificity of potentized sulphur D12 was the subject of a previous investigation (Van Wijk and Van der Molen, 1990). The conductivity loss in question was that which occurred after subjects who showed no further loss through the choice of remedies by the tester were given 0.1 μg of diphenyl orally. The suppressive effect of sulphur D12 was observed when the phial was placed in the holder and disappeared with the removal of the phial from the holder. It was shown that the phial could be placed in and removed from the holder many times without affecting the results. This opened the way to setting up blind testings. The central question was to what extent the tester was able to identify phials of sulphur D12 on the basis of observed conductivity loss.

Choice of EAV tester

There are several courses specifically aimed at testing homoeopathic preparations with the help of the EAV method. Obviously it cannot be assumed that all EAV testers possess the same skills or that patients react in exactly the same way. Experience, practice, the circumstances of the task and individual constitution all play a significant role. An example: it has been found that cutting out visual contact between tester and test substances easily leads to irritations of the tester.

Less experienced EAV testers often found it difficult to keep to the researcher's instructions. As an experiment series usually takes between three to four hours and requires heightened concentration from the tester, tendencies towards tiredness were inevitable.

Our subjective impression was that beginners are less sure of their measurements and take longer to look for active preparations than EAV testers with many years of experience. Only members of the latter group, which may be described as experts, were considered for the study. They were also expected to be prepared to work in an experimental situation. Two EAV testers, who may be considered to be representative of the expert group, took part in the study.

Experimental situation

The location of the people involved in the experiment are shown in Figure 2.

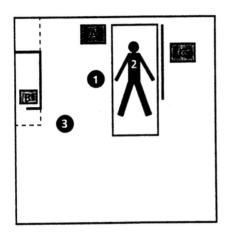

Fig. 2: Experimental situation. Location 1: Tester; Location 2:Test subject; Location 3: Experimenter. A: EAV device, B: Phial holder, C: Data recorder, apparatus.

Three people were involved in the experiment: a tester, a test subject and an experimenter, each playing a unique role.

(a) The *tester*, the expert on the measuring process. He measures the subject's skin points, and in doing so uses the EAV device which indicates conduction levels. He stands next to the treatment table on which the subject is lying. He can neither observe the movements of the experimenter, nor the recording of the data (with the Linseis recorder). Both are screened off.

(b) The *test subject*, the person (patient) whose skin points are tested, is lying on the treatment table. He cannot observe the movements of the experimenter or of the recording of data either.

(c) The *experimenter*, the test leader, divides up the experimental session in terms of space and time and supervises the experiment. In the blind testing (blind with regard to the tester, the test subjects and recording equipment), he chooses remedies (verum or placebo) and places the phials in the holders.

The recording of the results was done automatically. During the blind testing the distance between phial holder (with the experimenter) and the EAV device was around 1.25 m.

Procedure

The test was conducted according to the procedure outlined below.

(1) The tester makes measurements of the subject in order to find a point showing conductivity loss which is suitable for correction.

(2) The tester tries out phials of his choice, with the aim of finding a phial which appears capable of influencing conductivity loss. This process is continued until a phial or a combination of phials is found which achieves an optimal suppression of conductivity loss.

(3) The subject swallows 1 ml of liquid which contains 0.1 µg of diphenyl.

(4) The measurements of conductivity loss continued after a few minutes and the increased loss is recorded at three skin points (indicated in Figure 3).

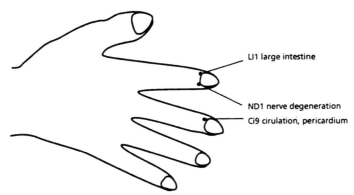

LI1 large intestine

ND1 nerve degeneration
Ci9 cirulation, pericardium

Fig. 3: Location of skin points measured in blind testing.

(5) A measurement is taken of the extent to which diphenyl–induced conductivity loss is again suppressed when sealed phials of sulphur D12 are placed in the holder. The tester also has at his disposal a phial containing a placebo (potentized alcohol), which he uses to control the difference in reaction between sulphur D12 and placebo. If the tester indicates that he can make a comparison, blind testing is started.

(6) In blind testing the experimenter places one of the two phials (verum/placebo) in the holder. The tester takes measurements at the three skin points and then repeats them. The phials are then removed and after a few seconds one of the two phials is again placed in the holder. The experimenter chooses the phials at random. After a few phials verum/placebo have been tested blind, an open measurement is made for control purposes.

RESULTS AND CONCLUSIONS
Blind testing of the medicine test by means of diphenyl–induced conductivity loss

Five experimental sessions were held on the basis of the above procedure. For analysis blind verum values were compared with blind placebo values. Open verum values and open placebo values were also compared. The data were compared in two ways, on the basis of the average of the four groups and using frequency distributions.

Table 1 (page 89) shows the average values of changes in conductivity of all sessions. A significant deviation between placebo and verum values in open measurements was observed. The blind values ranged between the two extremes. Thus, the difference between the averages of the blind verum and placebo values was smaller than that of the open measurements. The results in terms of frequency distribution patterns of conductivity loss are shown in Figure 4. The frequency distributions of verum and placebo values of initial conductivity in open experiments were normal and the overlap between the two groups was large (not shown). In contrast, the values of conductivity loss for the two groups were clearly separated (Figure 4, A and B, see next page).

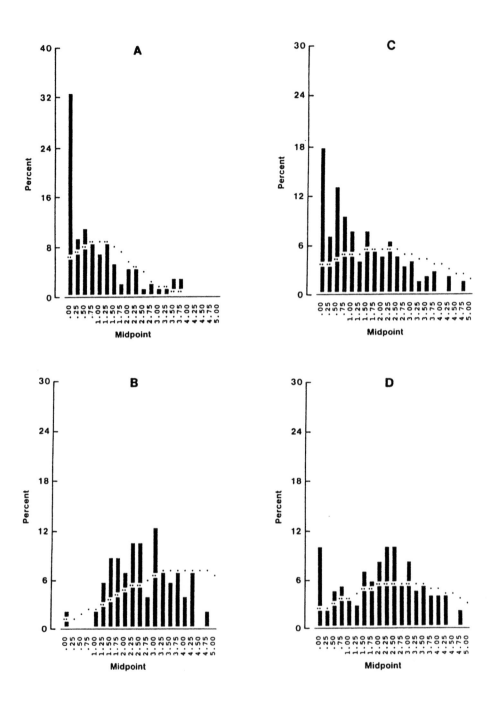

Fig. 4: Frequency distribution of the conductivity loss (Z.A.); results of the open and blind trials. All experiments. A: Open verum, B: Open placebo, C: Blind verum, D: Blind placebo.

Tab.1

	Open test		
	Verum (n=137)	**Placebo (n=92)**	**p**
I.V.	54.01 ± 9.9	59.07 ± 10.2	0.000
Z.A. (Rate)	0.97 ± 1.1	4.06 ± 2.10.000	

	Blind test		
	Verum (n=379)	**Placebo (n=371)**	**p**
I.V.	55.88 ± 10.1	57.40 ± 10.0	0.038
Z.A. (Rate)	1.95 ± 2.1	2.77 ± 2.10.000	

Tab.1: Total set of data. For details, see text (above).

The results of the blind experiments (Figure 4, C and D) differed considerably from those of the open experiments. The frequency distribution patterns showed two peaks for blind verum as well as for the blind placebo values, which correspond to the two peaks obtained for open verum and placebo values. As values were clustered in two areas in blind measurements, part of the measurements must therefore be regarded as either false–positive results (in which "verum–like" values were obtained with a placebo placed in the holder) or as false negative (in which "placebo–like" values were obtained with a verum). The cross–tabulation of the data of all open and blind experiments with respect to verum/placebo are shown in *Table 2, A and B* respectively. We compared the false and correct values and calculated the deviation with the chi–square test. Since overlapping between the two groups was found, a limit of 1.5 μA/min was used in order to applicate the chi–square test. As shown in Table 2, A and B, the results are statistically highly significant.

Tab. 2.

(a) Open trial			
	Placebo	**Verum**	**Row**
Total (%) No effect 50.7	87	29	116
Effect 49.3	5	108	113
Column	92	137	229
Total (%) 100.0	40.2	59.8	
Chi–Square	D.F.	Significance	
118.62275	1	0.0000	

(b) Blind trial			
	Placebo	**Verum**	**Row**
Total (%) No effect 60.0	278	172	450
Effect 40.0	93	207	300
Column	371	379	750
Total (%)49.550.5100.0			
Chi–Square	D.F.	Significance	
68.21130	1	0.0000	

Tab. 2: Cross tabulation of total data set of open and blind trials. Limit value 1.5 μA/min.

These findings led to two important conclusions. The first concerns the possibility of distinguishing in blind tests with the medicine test between the sealed phial containing sulphur D12 and the placebo. We recorded a statistically significant deviation between an average placebo and verum in several experimental sessions.

The deviation between verum and placebo observed in open tests is also important. When the deviation is so large that the distribution between open verum and open placebo values hardly overlaps, the number of correct and false positives and correct and false negatives can be determined in blind tests, and on this basis the effect of blind measuring can be determined. In this way a significant deviation was also revealed between verum and placebo values obtained in blind measuring.

The second conclusion concerns the factors which determine the percentage of false results. We observed that within a series of three measurements (three points) the variation was extremely small, the results were thus either correct or false as a group. Apparently the variation in values did not occur between individual measurements, but when a new phial was tested.

It seems therefore justified to conclude that we are dealing here with a very sensitive and tester–selective means of observing and recording. This sensitive form of observing and recording clearly also involves major uncertainties since the tester may play an important role in measuring the value of conductivity loss. To gain a better understanding of this process, research is focused on the analysis of the recording of conductivity.

Factors determining loss of conductivity

Conductivity is determined by several factors, including moistening of the test area, the amount of the pressure applied, the duration of the pressure applied, and the previous history (that is the time span between two applications of pressure or measurements) (Van Wijk and Wiegant, 1989). But our results have shown that even when these factors are constant, conductivity does not become constant but continues to show some variation. This is the result of the physiological regulation of sweat gland secretions, which thus influence conductivity. Physiologically and pressure–related conductivity changes differ at different locations on the skin. Even so, measurements taken at the same location on the same person and applying the same pressure exhibit a high degree of reproducibility. The effect of pressure can thus be most accurately assessed by examining particular points.

Several experiments have been performed to study the influence of pressure on changes in conductivity using fine–mechanical devices in which the pressure–registering electrode could be moved vertically with the help of a micrometer screw and a rotating switch.

The finger of the subjects could be positioned in such a way that in a downward movement, the pressure electrode experienced resistance in exactly the opposite direction (the tangential pressure is thus at a minimum). A more detailed description of the experimental set–up has been published elsewhere (Van Wijk, 1992). In these experiments, three aspects were studied:

(a) the stability of the conductivity without pressure and the effect of previous pressure;

(b) changes in conductivity at various amounts of constant pressure, that is, without an increase in pressure as it is being applied; and

(c) changes in conductivity as increasing pressure is applied.

From the obtained data (described extensively in Van Wijk, 1992; c. 7), the following conclusions may be drawn:

(a) the sudden application of pressure results in an increase in conductivity, with the amount of pressure applied being directly correlated to conductivity;

(b) after pressure has been applied and maintained at a constant level, conductivity does not remain constant but falls;

(c) the application of greater amounts of pressure does not produce further rises in conductivity; irrespective of the pressure applied (up to 280 g), conductivity rises to its level at 120 g;

(d) conductivity loss is reduced as existing pressure is increased.

(e) maximum unresponsiveness is produced with the application of 120 g for around 5 seconds, the rate of conductivity loss falls considerably under these conditions; and

(f) unresponsiveness disappears very quickly, within 1 to 3 seconds, the rate of conductivity loss then also begins to rise again.

The findings indicated that a particular initial application of pressure determined the conductivity of a skin point, even when the pressure was subsequently increased. Conductivity, which had been falling, rose again to about the initial value, regardless of the amount of pressure increase. Loss recurred, albeit at a reduced rate.

We examined this phenomenon further by applying additional increases after the initial increase in pressure. A few representative examples from one subject in one experiment are provided here to illustrate this phenomenon. As shown in Figure 5, in this experiment pressure was increased in stages and the stable periods between the stages were varied.

At first the pressure was quickly increased after the initial application. After the second pressure had been applied for 5 seconds, it was increased again and then increased for a fourth time. The results show that conductivity changed in jumps (Figure 5A). At the first increase, conductivity fell quickly, but at every subsequent increase conductivity rose. Each time the rise was temporary, however, and conductivity started falling again when the pressure applied was kept constant.

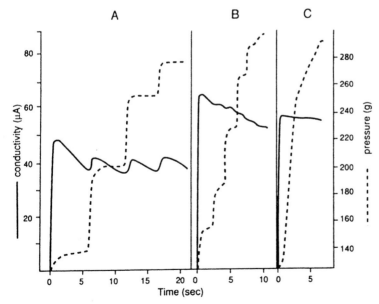

In the second approach (Figure 5B), pressure was increased in stages as before. But in this instance the period between increases was shortened to 2 seconds. Under these conditions conductivity loss occurred more gradually.

In the third approach (Figure 5C), pressure was gradually increased in such a way that conductivity remained constant.

Fig. 5: For legend, see next page.

Legend to Fig. 5: Example of the application of pressure from one subject in one experiment:

A: Increase of pressure at an interval of 5 seconds.

B: Increase of pressure at an interval of 2 seconds.

C: Gradual increase of pressure.

It can be concluded from these results that the phenomenon of pressure tolerance is of great importance for the study of changes in conductivity during the medicine test. It is possible that the suppression of conductivity loss must be seen as the result of a measurement in which a gradual increase of pressure occurs. If this assumption is correct, then the variations in the loss values should have something to do with the method of measurement, i.e. of pressure application, and with the behaviour (condition) of the tester rather than with an altered condition in the trial test person. Testing this hypothesis is described in the following paragraph.

Application of pressure as a determinant of conductivity loss

We tested the hypothesis that changes in conductivity follow from changes in the way pressure is applied, in experiments in which diphenyl–treated subjects were measured with and without sulphur D12 in open (not blind) experiments. We also recorded the application of pressure in several blind tests in which sulphur D12 was compared with a placebo. The recording of conductivity and pressure took place simultaneously.

Application of pressure in open tests

The measurements of the pressure and conductivity began after a subject had been normalized by the tester, that is, after one or several phials had been measured and no loss of conductivity was observed in the subject. The measurements were repeated and the application of pressure recorded at every measurement. Figure 6 A shows the average pattern of ten pressure curves (next page).

After a subject had taken 0.1 µg of diphenyl, a number of combined measurements of pressure and conductivity were also taken, both when sulphur D12 was present in the measuring circuit and when it was absent.

The average pressure curves when no sulphur D12 was present are shown in Figure 6B. The average development in the presence of sulphur D12 is shown in Figure 6C.

A comparison of the pressure and conductivity under these three conditions (Figure 6A to 6C) shows that pressure was applied differently in the three situations. In a normalized subject the pressure was gradually increased after the initial application. After the subject had taken diphenyl the pressure was quickly applied to the final level. In the situation that sulphur D12 was present the pressure was gradually increased. In these situations it was shown that, when the pressure was quickly applied and then not increased, a sudden conductivity loss occurred (Figure 6B). When the pressure was gradually increased after the initial application, conductivity fell slowly (Figure 6A and 6C).

From the conclusions described in the previous paragraph and the findings presented in Figure 6 we conclude that the deviations between placebo and sulphur D12 results can thus be ascribed to differences in the application of pressure.

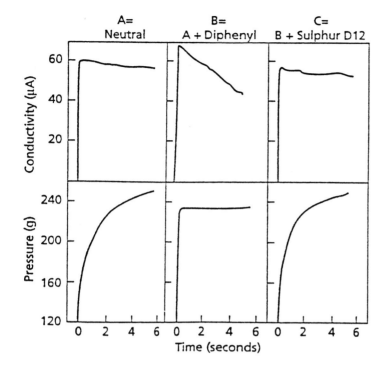

Fig. 6:

Simultaneous recording of pressure and conductivity in a subject treated by a tester (10 experiments).

A: Normalized subject.

B: Normalized subject with diphenyl.

C: Normalized subject with diphenyl and Sulphur D12.

Application of pressure in blind tests of sulphur D12 and placebo

In some blind tests the pressure and conductivity were recorded simultaneously. The curves produced are shown in Figure 7.

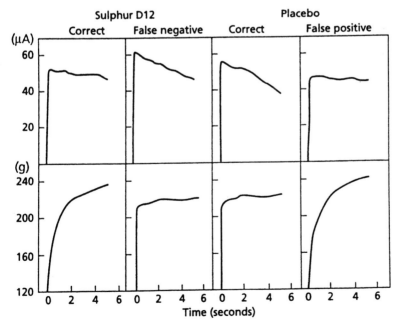

Fig. 7:

Simultaneous recording of pressure and conductivity in blind testings (8 experiments).

The conductivity loss was slow with a secondary increase in pressure. When pressure was quickly applied a sudden loss occurred. The correlation was independent of the correct/false scores in the blind tests. The false scores with sulphur D12 showed no secondary increase in pressure, while the false scores with placebo did.

These experiments show that the results of the medicine test are directly linked to the way in which pressure increases. The variation in the increase of pressure can only be explained by another pressure of the electrode against the finger of the subject or vice versa, of the subject's finger against the electrode. Since the tester holds the electrode in one hand and the subject's finger in the other (or with the subject's finger positioned), the change in pressure must be attributed to the tester. Changes in the tester's muscle tone always occur when he applies pressure. Many muscles in the body are involved in the application of pressure. The close coordination during the conscious application of this pressure is the outcome of a fine regulatory process, of the activation/deactivation of muscles which are involved in the relevant muscle groups.

This conclusion is of great importance for the interpretation of the results of the blind tests. Since these showed a significant deviation between sulphur D12 and placebo, we must assume that the tester is somehow disturbed by this remedy in the fine–tuning of his muscular activity. The question may be asked what role the test subject plays here. We cannot deduce from our results that the subject plays no role at all. The tester's reaction pattern is built up in direct contact with the subject. We may speculate that during the time when the tester is looking for remedies which will suppress conductivity loss in the subject, he is actually looking for a condition in which he applies less pressure. When he finds such a remedy, we may therefore also assume that it is one which affects his muscle tone condition during his contact with the subject.

In summary we may conclude that the tester uses himself as an intermediary in selecting remedies which influence him in his specific contact with the subject.

DISCUSSION

Recommendations for further scientific research

It is in the nature of things that the medicine test has many critics. But the comments by most of them prove that they have not bothered to acquaint themselves with the phenomenon in any great detail. Their counterargument is usually based on doubts that such a thing could be possible. The supporters of the EAV medicine test in turn do not offer an explanation, but merely the expectation that an explanation will be forthcoming in the future. "Where would science be today if we had neglected all phenomena which could not be explained at the time of their discovery. Their evidence must lie in the realms of physics and physiology" (Fehrenbach, 1988).

At times the proponents of EAV even claim that the method will have a pivotal role in future developments in homoeopathy and medicine. Since EAV primarily uses potentized remedies, it has made homoeopathy quantifiable and has thus removed its subjectivity. "No EAV without homoeopathy, but also no homoeopathy without EAV" (Bergher, 1988).

From our study we cannot conclude to what extent the experimental conditions during our research influence the reliability of the measurement. With respect to the serious errors which are made when conducting a blind test, the question can be raised whether these errors also occur in diagnosis during medical practice. Due to our study we suggested that the interaction between tester and test subject is an important factor in distinguishing phials with a different content in medical practice as compared to the experimental condition. In medical practice the anamnesis, the search for patient-specific characteristics and the motivation to help the patient are involved in establishing the interaction between the tester and the test subject. This means that the percentage of errors in blind tests cannot be automatically extrapolated to the medical practice. Electroacupuncturists think that in medical practice there are fewer errors because of an increased tester–patient interaction.

The need of scientific research has thus become increasingly urgent. Moreover, some practitioners have got themselves into deep water because of their claims for the diagnostic value of the measurements. As yet these claims – and no matter how fervently the practitioner believes in their accuracy and validity – are not justified. Despite years of experience there are as yet no formal studies of medicine testing. It is therefore important that the sytem's practicioners treat its diagnostic value as hypothetical or experimental postulates which still await verification by experiments.

This study, as well as other studies, clearly demonstrate the difficulties of preparing optimal protocols and of carrying out this research. In the meantime it is proposed that the diagnosis of a patient is performed independently by more than one tester. This procedure, although time–consuming, is for the good sake of the patient, but not only for his. When the data of these diagnoses are carefully and systematically registered, they are also useful to underscore the diagnosis of the medicine test.

ACKNOWLEDGEMENTS

The authors wish to thank C. van der Molen for his general help; P.J. Romeyn for his technical help and A.W.J.M. Bol for his help in the organization of the experimental sessions and analysis of data. This study is part of the Homint research project.

REFERENCES

Beisch K und Bloesch D. Ein Wirksamkeitsnachweis homöopathischer Medikamente am Beispiel der Nosoden – eine regelphysiologische Studie im Testgang der EAV. Uelzen, Germany: Med Lit Verlagsgesch 1979.

Bergher FJ. Axiome, Definitionen, Hypothesen der Elektroakupunktur nach Voll. Aerztezeitschrift für Naturheilverfahren 1988; 29(4): 295 – 307.

Fehrenbach H. Die Electroakupunkur nach Voll. Erfahrungsheilkunde 1988; 37(4): 233 – 337.

Kramer F. Beitrag zur Beweisführung der Medikamententestung. In: Electroakupunktur – Anderthalb Jahrzehnte Forschung und Erfahrung in Diagnostik und Therapie. Uelzen, Germany: Med Lit Verlagsgesch 1971, pp. 241 – 251.

La Rivière BM. Weerstandsmeting aan de huid in relatie tot electroacupunctuur. Verslag van onderzoek bij de Vakgroep Biomedische Natuurkunde en Technologie. Erasmus Universiteit, Rotterdam 1988.

Van Oss–Bossagh P. Electroacupunctuur.Verslag van onderzoek bij de Vakgroep Biomedische Natuurkunde en Technologie. Erasmus Universiteit, Rotterdam 1989 .

Van Wijk R and Wiegant FAC. Homeopathic remedies and pressure–induced changes in the galvanic resistance of the skin. Alkmaar, The Netherlands: VSM Geneesmiddelen bv. 1989.

Van Wijk R and Van der Molen C. Biological effects and physiological characteristics of potentized high dilutions of Sulphur; development of a model for fundamental studies in homeopathy. In: Homeopathy in Focus, Schlebusch KP ed., Essen: VGM Verlag 1990, pp. 122 – 147.

Van Wijk R. Homeopathic medicines in closed phials tested by changes in the conductivity of the skin: a critical evaluation; blind testing and partial elucidation of the mecnanisms. Alkmaar, The Netherlands: VSM Geneesmiddelen bv. 1992.

Voll R. Medikamententestung, Beweise für die Richtigkeit und Probleme der Grundlagenforschung der Medikamententestung. In: Electroakupunktur – Anderthalb Jahrzehnte Forschung und Erfahrung in Diagnostik und Therapie, Uelzen, Germany: Med Lit Verlagsgesch 1971, pp. 200 – 215.

Voll R. The phenomenon of medicine testing in electroacupuncture according to Voll. Am J Acupuncture 1980; 8(2): 97–104.

"We have finally de-mystified this field of research"

PART 2: PHYSICS

"High dilution effects have been demonstrated to exist in different areas of research, but have not been accepted as a fact by the scientific community so far. One of the reasons for the usual disregards seems to be that there is no theoretical model to understand how a dilution containing no molecules can produce any effect ..."

H. Walach, annual meeting of the GIRI, autumn 1992

OUTLINE OF EXPERIMENTAL PHYSICAL METHODS TO INVESTIGATE SPECIFIC STRUCTURES OF ULTRA HIGHLY DILUTED SOLVENTS

J. Schulte, P.C. Endler

SUMMARY

The mechanism of information storage in homoeopathic high dilutions that are supposed to cause biological and medical effects may be studied using methods known from experimental physics. Various prominent experimental techniques have been applied, here the most common ones will be briefly discussed.

ZUSAMMENFASSUNG

Die Mechanismen der Informationsspeicherung in homöopathischen Hochverdünnungen, die möglicherweise biologische und medizinische Wirkungen hervorbringen, können mittels aus der Physik bekannter Methoden untersucht werden. Verschiedene experimentelle Techniken wurden angewandt; die bekanntesten werden hier kurz diskutiert.

INTRODUCTION

General

For a long time, homoeopathy has been a complementary method to mainstream medicine without being accepted as a scientific branch in medicine. Along with industrialization, some branches of medicine have been highly developed, and are expected to become serious sciences with their own diversity, while homoeopathy is still struggling with justifying itself as a serious science.

In the research on homoeopathy, quite a number of experimental investigations try to show that there are some effects which may be traced back to homoeopathic influences. Clinical and human physiological studies as well as experiments on biological systems were discussed in Part 1 of this volume, as, for instance, modern in vitro experiments and in vivo studies which are expected to provide unbiassed information of the efficacy of homoeopathic drugs, since no placebo response is expected. In vitro experiments, however, have the advantage that standard laboratory equipment and methodology can be used to a higher degree than in in vivo studies [2]. Unfortunately, results obtained from in vitro experiments show systematic effects only, and are only valid within a limited range [3]. An investigation on the possible nature of information storage in homoeopathic potencies can be done by applying methods known from experimental physics. Most physical methods applied in homoeopathy, ultra low dose and ultra-high dilution (UHD) research, respectively, focus on structural changes of the solvent caused by the process of preparation of the dilutions (a stepwise dilution with mechanical agitation between the dilution steps, the so called "potentisation").

In the early days of homoeopathy, one assumed that the potency of the homoeopathic drug originated in the "power" or subtle "dynamical" force of the mother tincture which was diluted. This dynamical force of the mother tincture was supposed to be transferred and amplified through the potentisation process (see Glossary, p. 221)

The picture of homoeopathy has developed to a scientifically based view of potentised drugs, i.e. to succussed solvents with a memory function in their structural composition. Most of the contributions within this book refer to the specific structuring of water due to

99

P.C. Endler and J. Schulte (eds.), Ultra High Dilution, 99–104.
© 1994 *Kluwer Academic Publishers. Printed in the Netherlands.*

potentising, but the imprinting of information on water has been studied as well, by using acoustic, magnetic and electromagnetic methods [4].

Historical physical review

In 1948, Wüst suggested that the efficacy of potentisation may be caused by the drug's spatial and electrical influence on the potentisation medium [5]. Within the same year, H. Schoeler published his habilitation thesis on "Die Wissenschaftlichen Grundlagen der Homöopathie" (The Scientific Principles of Homoeopathy) [6]. 1953 Gay and Boiron speculated on specific polymer structures of solvents established during the potentisation process [7]. Later Bernard and Stephenson [8] formulated the so-called "imprint theory" in order to explain nuclear magnetic resonance (NMR) experiments by Boericke and Smith [9].

A common hypothesis on the imprint theory which has become prominent suggests that a specific structure is imprinted on the solvent by the drug. Once the structure is formed, it seems remarkably stable due to self-reproducibility, even at ultra-high dilution where no drug in the solvent can be expected anymore. During the past decades, several hypotheses have been developed for explaining the imprint mechanism. Most of these hypotheses are based on the structuring through electromagnetic interactions [10]. Other imprint theories are based on cluster theories [11], isotopic effects and coherent states [12] (see Anagnostatos's, Berezin's, Del Giudice's, Popp's and Schulte's contributions).

METHODS AND DISCUSSION

Miscellaneous Physical Methods in Homoeopathy

From the miscellaneous physical methods in the literature on homoeopathy we would like to mention the two most prominent ones, since they are most frequently cited. There is the study on ice crystals produced with homoeopathic potencies. For certain potencies in the frozen state certain crystal structures have been found, which are claimed to imply the reproduction ability of potencies. The tracing of radioactive phosphor and iodine in a potentisation sequence has been studied by further researchers [13-16]. The experiments are very interesting, but the results presented and their interpretation are rather questionable for there are too many parameters involved which may cause misleading results. In order to screen the sensitive parameters, these experiments require much more technical effort than an ordinary laboratory can provide. It might be that, for those experiments, new experimental measurement techniques have to be developed (where the methods themselves have to be proved and verified first).

Electric Conductivity and Dielectricity of Aqueous Solutions

Due to impurities (minerals and other ions), natural water has a fairly high electric conductivity. In extremely purified water there are not enough ions serving as a carrier for the electric current, thus the electric conductivity of pure water is low. Some experimental groups have investigated the change in electric conductivity and dielectricity (the response to an externally applied electromagnetic field) when potentising with purified (distilled) water. When potentising a probe with purified water, the well-known and unspectacular decreasing of electric conductivity can be observed first. Further potentisation with distilled water leads to the region of high and highest homoeopathic dilution where the electric conductivity becomes very small, and the experimental technique becomes unreliable because of large error bars of the measured current value [17]. The same can be said about measurements on the dielectric constant. Both experimental techniques focus on bulk properties of the probe only. However, at high dilution microscopic properties become more important. Unfortunately, the effect or contribution from microscopic properties just vanishes under the large error bars of the measured bulk value. Most prominent papers in literature do not even show error bars [18]. The method has recently been taken up by another working group [30,31].

Surface Tension

Starting in 1923, experiments investigating the surface tension of homoeopathic dilutions have been carried out measuring the climbing height of the dilutions at filter stripes. It has been claimed that specific steps of dilution reach a specific height [19]. Similar experiments have been carried out using capillary tubes [20]. With a decreasing concentration of ions, the surface tension decreases in a similar way to the electric conductivity. However, in the range of homoeopathic potencies, the effects are becoming so small that, with this technique, the effects are "potentised" into the error bars. The modern technique of measuring the surface tension could not verify a correlation of surface tension and homoeopathic potentisation [21].

Raman Spectroscopy

Raman spectroscopy requires a monochromatic light source, the light is analyzed after scattering at the probe. Due to Rayleigh scattering (elastic scattering), the scattered light loses intensity, but does not change its frequency. Within the probe, the molecules get excited by the monochromatic light, and the scattered light superposes with the excitation frequencies of the probe (Smekal-Raman effect, inelastic scattering). The difference in the frequency between the light source and the contribution due to Raman scattering gives the characteristic vibrational and rotational frequencies of the probe. Thus, the probe may be studied in its vibrational and rotational structure. Measurements in the low frequency region resolve large (global) structures, which cannot be covered by other optical methods. Unfortunately, the intensities at low frequency vibrations are very small, and only very few vibrational modes can be resolved. Experiments by Luu [22] have shown characteristic changes in the structure when comparing homoeopathically potentised water and just succussed water. The reproducibility of Luu's results has been recently shown by Weingärtner [21]. However, one has not yet been able to show the remarkable characteristics of homoeopathic dilutions with Raman spectroscopy, i.e. the periodicity in its efficacy related to the structure.

Nuclear Magnetic Resonance Spectroscopy (NMR)

The NMR spectroscopy uses the effect due to the interaction of the nuclear magnetic dipoles (spin) of the probe molecules and an external electromagnetic field. As in most cases the spin states of the molecules are highly degenerated (different spin states lead to the same energy eigenvalue), a very strong magnetic field is applied to the probe in order to separate individual states. The nuclear magnetic precision states depend on the structural characteristics of the probe, it is a finger print for structural changes within the probe. If an electromagnetic field is applied to the probe, the precision frequency (Lamor frequency) of the nuclear magnetic moments (spins) of the molecules show resonance effects (absorption) depending on the external electromagnetic field. Also, the spins are not only coupling to the external electromagnetic field but also to spins of other atoms and molecules (spin-spin coupling), and to the vibration modes of neighbouring molecules or of the solvent. All coupling results in a characteristic absorption of the external field, i.e., very detailed spectra concerning coupling and structure in the probe. The frequency of the external field is in the MHz range. This method is technically well-developed, easy accessible, sensitive and accurate, especially at the microscopic level. The interpretation of the spectra requires some experience and knowledge about atomic and molecular physics. Boericke and Smith [23] applied this method studying 87% alcohol, potentised sulfur D12, and non-potentised sulfur D12. They have found characteristic, reproducible differences in the NMR spectra, and they have come to the conclusion, that by potentisation the macro- molecular structure changes. Unfortunately, they did not give any explanation in which the structure has changed, although the NMR data were available. Periodic changes in the NMR spectrum of remedies of different potentisation have been found by Young [24]. These experiments as well as the experiments by Sacks on sulfur potencies, and D30 remedies that have shown significant differences in the NMR spectra, have to be re-evaluated and verified by other groups. Lasne et al. [25] studied the relaxation of the nuclear moments of sulfur, potassium, iodine and histamine in CH1 to CH30 dilution. Significant and reproducible differences to the just potentised CH1 and CH30 water has been found throughout the range of dilution in a total of 580 T2 measurements. Excellent methodological work has been done on NMR investigations by Weingärtner [26] and the working group of Poitevin [27]. The latter, e.g., reported an increase of the T1 time of relaxation, and an increase of the T1/T2 ratio in a study on lactose/ silicea.

UV-Spectroscopy

In addition to spectroscopical measurements in the low frequency region [28], Ludwig performed UV-spectroscopy investigations on ultra-high dilutions. E.g. in the UV region, 190-220 nm spectra of ultra-high dilutions of Atropa Belladonna (D30 and D200) were measured and compared with analogously prepared pure solvent. Differences between these samples in the UV-absorption both at D30 as well as at D200 were shown [29]. This important work should be further elaborated.

X-ray Spectroscopy

Promising preliminary results on X-ray spectroscopy of a homoeopathic remedy have been reported by Gutmann [30].

Differential (Micro)calorimetry

The differential calorimetric and microcalorimetric methods are, at present, applied to UHDs and promising first results have been achieved [31]. The authors, however, stress that more rigorous standardization is necessary in future experiments.

ANNOTATION AND ACKNOWLEDGEMENTS

For further details on some respective experimental methods see Smith's contribution in this volume and [32]. The authors of the contribution presented here wishes to thank F. Gross (Institute of Theoretical Physics, University of Graz) and M. Righetti [33] for their valuable help. A very sound survey on the topic has recently been given by Weingärtner [34].

REFERENCES

1. Dwarkanath SK, Stanley MM, Hahn Glean 1979; 46: 37-41.
   Chowdury H, Br Hom J. 1980; 69: 168-170.
   Chowdury H. Hom Rays 1986; 11: 251-152.
   Sharma RR. Molecular Homoeopathy, New Dehli: Cosmo Publication 1984.
   Day Ch. Br. Hom J 1986; 75: 11-14.
   Chandrasekhar K, Sarma GHR. J Reprod Fert 1974; 38: 236-237.
   Chandrasekhar K, Sarma GHR. Ind J Zootomy 1975;16: 121-127.
   Chandrasekhar K, Sarma GHR. Ind J Zootomy 1975; 16: 199-204.
   Prasad S, Chandrasekhar K. J Res Ind Med Y Hom 1978; 13 (4): 81-91.
   Prasad S, Chandrasekhar K. Ind J Exp Biol 1978; 16: 289-293.
   Prasad S, Chandrasekhar K, Br Hom J 1978; 67: 265-275.
2. Shrivastava JN, Kushwaha K, Hahn Glean 1984; 51: 371-374.
   Ray PG, Mukherjee SK. Ind J Exp Biol 1977; 15: 338-339.
   Sarma JB, Moses GJ. Proc 3rd Int symp Pl Path, IPS, ICAR&IARI, New Dehli 1980, pp 177-178.
3. King G., Experimental Investigation for the Purpose of Scientific Proving of Efficacy of Homoepathic Preparations. Thesis, Tierärztliche Hochschule Hannover, FRG, 1988.
   Majerus M. Kritische Begutachtung der wissenschaftlichen Beweisführung in der homöopathischen Grundlagenforschung (francophone literature). Thesis, Tierärztliche Hochschule Hannover, FRG, 1990.
4. Flyborg K, Flyborg G. Die Flyborg-Therapie mit frequenzaktiviertem Wasser. Erfahrungsheilkunde 1990; 7: 424-427.
   Kokoschinegg P. Informationsstrukturen von Wasser und ihre biologische Bedeutung. In: ISMF et al. eds, Heidelberg: Haug 1993.
   Smith CW, this volume.
   Ludwig W. Physikalische Grundlagenforschung in bezug auf Informationsspeicherung in lebenden Systemen und homöopathischen Medikamenten. Strukturierung von Wasser und Alkohol. Erfahrungsheilkunde 1991; 4: 293-295.
   Citro M, Endler PC, this volume.
5. Wüst J., Über den physikalischen Nachweis der Wirksamkeit potenzierter Lösungen nach R. Beck, Regensburg 1948.
6. Schoeler H., "Die Wissenschaftlichen Grundlagen der Homoeopathie", habilitation thesis, FRG 1948
7. Gay A, Boiron J. Demonstration physique de l'éxistence réelle du remède homéopathique. Edit Lab PH., Lyon 1953.
8. Bernard SP, Stephenson J. J Am Inst Hom 1969; 62: 73-85.
9. Boericke GW, Smith RB. J Am Inst Hom 1965; 58: 158-167.
   Boericke GW, Smith RB. J Am Inst Hom 1966; 59: 263-280.
   Boericke GW, Smith RB. J Am Inst Hom 1967; 60: 259-272.
   Boericke GW, Smith RB. J Am Inst Hom 1968; 61: 197-211.
10 Jussal RL, Dua RD, Mishra RK, Meera S, Agarwal A. (Electric conductivity.) Hahn Glean 1984; 51: 245-250 (1984).
   Jussal RL, Dua RD, Mishra RK, Meera S, Agarwal A. (Surface tension.) Hahn Glean 1982; 49: 114-120.
   Kumar A, Jussal RL. (Optical properties - optical density, refractive index - .) Br Hom J 1979; 68: 197-204.
   Khan MT, Saify Z. (Dielectricity.) Proc Congress LMHI, Rotterdam 1975, pp. 157-169.
   Jussal RL, Dua RD, Mishra RK, Meera S, Agarwal A. (Spectroscopy - IR, UV, Raman -.) Hahn Glean 1983; 50: 358-366.
   Gautam RS, Tewari KP, Roper NK, Mishra RK. Hahn Glean 1977.
   Boiron J, Vinh C, Proc Congress LMHI, Athens 1976, pp. 459-474.
   Beyr G. Kybernetische Denkmodelle der Homoeopathie. Heidelberg: Haug-Verlag 1982.
   Righetti M. Forschung in der Homoeopathie. Göttingen: Burgdorf 1988.
   Jones R, Jenkins MD. Br Hom J 1981; 70: 120.
   Luu C, Luu DV et al. J Molecular Structure 1980; 40 (1): 41-54.
   Luu C, Luu DV et al. J Molecular Structure 1982; 81: 1-10.
   Luu C, Luu DV et al.. Proc Congress LHMI, Lyon 1985.
11. Schulte J, Acta Medica Empirica 1990; 7: 418-423. See also Schulte, this volume.
12. Popp FA, Br Hom J 1990; 79: 161-166. See also Popp, this volume.
13 Khan MT, Saify Z. J Am Ins. Hom 1975; 68: 97-104.
   Verma HN. Adv Hom 1981; 1: 70-71, 73.
   Brigam N (1949), referenced after Coulter HL. Homoeopathic Science and Modern Medicine. Ritchmond:North Atlantic Books 1981.
14. Bonet-Maury P, Feysine A, Voegeli L. Ann pharm franc 1954; 12: 654.
15. Boiron J, Braise J. Ann hom franc 1965; 7: 586.
16. Ducasson P et al. Ann hom franc 1973; 15: 129.

17. Jussal RL, Dua RD, Mishra RK, Meera S, Agarwal A. Hahn Glean 1984; 51: 245-250.
    Jussal RL, Dua RD, Mishra RK, Meera S, Agarwal A. Hahn Glean 1982; 49: 114-120.
18. Jussal RL, Dua RD, Mishra RK, Meera S, Agarwal A. Hahn Glean 1983; 50: 358-366.
19. Koliko L. Physiologischer und physikalischer Nachweis der Wirksamkeit kleinster Entitäten (1923-1959).
    Stuttgart: Arbeitsgemeinschaft anthroposophischer Ärzte 1961.
20. Kumar A, Jussal RL. Br Hom J 1979; 68: 197-204.
21. Weingärtner O. Therapeutikon 1988; 5: 310-320.
22. Luu-d-Vingh C. Les dilutions homoeopathiques. Control et etude par spectrographie Raman-Laser. Thesis
    University Montpellier 1974.
    Luu-d-Vingh C. Etude des dilutions homoepathiques par spectroscopie Raman Laser. Lyon: Lab Boiron
    1976.
    Luu C, Luu DV et al. J Molecular Structure 1982; 81: 1-10.
    Luu C, Luu DV et al. Proc Congress LHMI, Lyon 1985.
23. Boericke GW, Smith RB. J Am Inst Hom 1965; 58: 158-167.
    Boericke GW, Smith RB. J Am Inst Hom 1966; 59: 263-280.
    Boericke GW, Smith RB. J Am Inst Hom 1967; 60: 259-272.
    Boericke GW, Smith RB. J Am Inst Hom 1968; 61: 197-211.
24. Young TM. J Am Inst Hom 1975; 68: 8-16.
25. Lasne Y, Duplan JC. et al. Comm Lab Boiron. Lyon 1976.
26. Weingärtner O. NMR-Spektren von Sulfur-Potenzen. Therapeutikon 3: 438.
    Weingärtner O. Experimentelle Studien zur physikalischen Struktur homöopathischer Potenzen. In: Albrecht
    H, Franz G (eds.): Naturheilverfahren, Zum Stand der Forschung. Berlin: Springer 1990.
    Weingärtner O. Homöopathische Potenzen. Wunsch und Wirklichkeit bei der Suche nach der therapeutisch
    wirksamen Komponente. Berlin: Springer 1992.
27. Demangeat JL, Demangeat C, Gries P, Poitevin B, Constantinesco N. Modifications des temps de relaxation
    RMN à 4 MHz des protons du solvant dans les très hautes dilutions salines des silice/lactose. J Med Nucl
    Biophy 1992; 16,2: 135-145.
28. Ludwig W. Eigenresonanzen homöopathischer Substanzen. Erfahrungsheilkunde 1987; 36: 952-953.
    Ludwig W. Spektroskopische Messungen an Patienten und Homöopathika im Niederfrequenzbereich. In:
    Ecolog '88. Umweltzentrum für ökologische Strukturforschung Schloß Türnich, FRG-W-5014 Kerpen-
    Türnich.
29. Ludwig W. Physikalische Grundlagenforschung in bezug auf Informationsspeicherung in lebenden
    Systemen und homöopathischen Medikamenten. Strukturierung von Wasser und Alkohol.
    Erfahrungsheilkunde 1991; 4: 293-295.
30. Resch G, Gutmann V. Wissenschaftliche Grundlagen der Homöopathie. Berg am Starnberger See: Barthel
    &Barthel (O Verlag) 1986. Scientific Bases of Homoeopathy. Berg am Starnberger See: Barthel &Barthel
    1987. (Reference is made to the work of E. Moser.)
31. Anagnostatos GS, Pissis P, Vyras K. Physico-chemical Study of Homepathic Solutions. Abstr Omeomed
    Meeting, Urbino 1992.
    Delinick A. Abstr ECRH Meeting, Vienna 1993.
    Cabaner C, Bastide M. Isothermal microcalorimetric analysis of homeopathic dilutions. Abstr GIRI
    Meeting, Paris 1992.
32. Bischof M., Rohner F. Wasser. In: Zentrum Dokumentation Naturheilverfahren and FFGB (eds):
    Dokumentation der besonderen Therapierichtungen u. nat. Heilweisen in Europa, Vol 2. Essen: Verlag für
    Ganzheitsmedizin 1992.
    Endler PC. Aspects of information storage and structures in water. Br Hom J 1989; 78: 253-254.
33. Righetti M., Forschung in der Homoeopathie. Göttingen: Burgdorf 1988.

CONSERVATION OF STRUCTURE IN AQUEOUS ULTRA HIGH DILUTIONS

Jürgen Schulte

SUMMARY

Theory on homoeopathy has developed a variety of models trying to explain the retaining of traces of drug molecules in aqueous solutions, as well as to explain the interaction with living systems. Some current promising theories are based on a concept of clustering of solution molecules, either directly or implicitly as a seed of the retaining trace. The fundamental concepts of current knowledge in cluster science is presented with respect to aqueous solutions. The basic ideas of some prominent, related theories, such as clustering, flickering clusters, isotope effect, clathrates, non-local electrodynamic fields and coherent states are outlined and discussed.

ZUSAMMENFASSUNG

Modelle und Theorien in der Homoeopathie versuchen die im Potenzierungsmedium verbleibenden Spuren von Drogenmolekülen zu erklären. Ebenso gibt es einige Ansätze, die sich dem Problem der Wechselwirkung von Homoeopathika mit lebenden Zellen widmen. Neue vielversprechende Theorien bauen direkt oder implizit auf das Konzept von Molekülen in Clusterformation auf. Das Konzept der Molekül-Cluster und deren Problematik wird an Hand von wässerigen Lösungen diskutiert. Von den heutzutage gängigen Theorien über Informationsspeicherung und -übertragung werden einige bekannte Theorien über Clustering, Flickering Clusters, Isotopizität, Clathrate, nichtlokale elektrodynamische Felder und kohärente Zustände vorgestellt und in ihren Grundlagen diskutiert.

INTRODUCTION

In physics, physical chemistry and quantum chemistry, water is known as an extremely complex substance in all of its states of aggregation. Water shows various qualities in its liquid state including the forming of distinct long-range structures, and localized and delocalized structures, commonly called clusters. Although the forming of clusters in liquid water and isolated water clusters in vacuum is presumed, they have become the subject of scientific research since the past few years only. The research was stimulated by the discovering and production of isolated clusters with modern experimental methods [1] using mass spectroscopy of a cluster beam in a carrier gas.

Within the scope of this paper I'll give a brief review on clusters, water, high diluted aqueous solutions, and structures in water, which is followed by theoretical comments on the transfer and storage of information into the structure of high and highest diluted aqueous solution, and the prospects of indirect paths of investigation used in physics to explain the interactions.

MAIN PART

1. The concept of atomic and molecular clusters

A cluster in this paper is defined by its phase rather than by its size. Although there are various names for atomic and molecular compounds with a size of 10^{-10}m - 10^{-7}m radius, throughout this paper those compounds are called clusters. The structures of clusters are

P.C. Endler and J. Schulte (eds.), Ultra High Dilution, 105–115.

different from the structures that may be found in the crystal phase and liquid phase of the same substance. Eg., a ring of 8 sulfur atoms may be found as a stable structure in liquid sulfur, and it seems to be a cluster, but it is not, for it is characteristic of liquid sulfur.

One characteristic that all clusters have in common is that a specific cluster has a characteristic unique structure, and that during growth the structure is not predictable due to complete structural reorganization, and in the same cluster size there are clusters with completely different internal structure (cluster isomers). Thus, the biophysical or biochemical information carried by a specific cluster of molecules is unique, eg. like that of the DNA. The DNA, a cluster of amino acids, carries well-defined genetic information. In Fig. 1 the geometric structure of $Na^+(NaCl)_n$ clusters of various sizes is shown, n is the

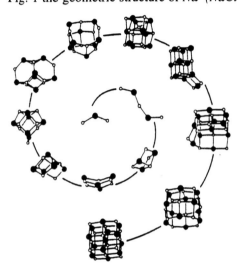

number of $Na^+(NaCl)$ units being connected by interaction bonds, n defines the cluster size [2]. The spiral begins with a single molecule, then successively an additional molecule is bonded, but not in the usual crystal growth pattern. The cluster structure has been rearranged completely with every single growth step. At the end of the growth spiral the cluster structure converges more and more to the $Na^+(NaCl)$ crystal structure. After the (phase) transition to the crystal structure, no completely reorganizing of structure is found. The structure is fixed for further growth. In a spectroscopy experiment on these clusters a different spectroscopic pattern can be found for every cluster. After the transition from the cluster phase to the solid phase the spectroscopic pattern will no longer change.

Fig. 1: Quantum chemical calculated configurations of $Na^+(NaCl)_n$ clusters. During the first phase of growth there are deviations from the crystal structure of $Na^+(NaCl)$ [26].

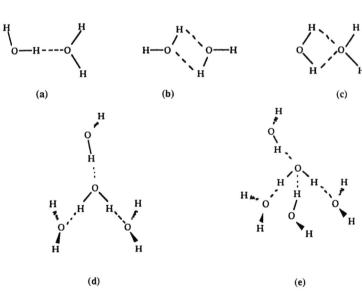

Fig. 2: Linear (a), cyclic (b) and bifurcate (c) water dimer, tetramer (d) and tetrahedral pentamer (e).

The example of unpredictability and structural reorganization shown in Figure 1, intuitively suggests that electrical, optical, and chemical properties of, for instance, a compound of

$Na^+(NaCl)_5$ clusters are very different from $Na^+(NaCl)_8$ or small compounds or crystals of the same substance.

It is exactly this complete structural rearrangement that makes clusters interesting in physics, chemistry, and material science, and suggests the cluster phase as vulnerable or receptive to information.

2. Water clusters and clusters in aqueous solutions

What about clusters in water, and what makes them form? Before outlining this question, some general comments on water are given, which are followed by a brief review on the physical properties of water molecules.

The seas on earth are covering an area of $361 \times 10^6 \, km^2$ (with an average depth of 3.8 km), that makes 71% of the total surface of the earth. A considerable amount of water is in the ground, and carried by rivers, lakes, the polar caps and the air. In plants 90% of their weight is water, and the portion in animals and human body counts 65%. Water is everywhere. It is the basis of life. It is part of us, and it is obvious that we should be respectful when dealing with water, especially once the mechanism of transferring and storing of information in water is understood.

Figure 2 shows some plausible simple structures of water molecules. Quantum mechanical calculations of their energies have shown [2] that the linear structure (Fig. 2a) is more stable than the cyclic structure (Fig. 2b) and the bifurcate structure (Fig. 2c). In the linear structure due to their repelling positive charges the distance between the oxygen atoms is larger. This result might imply favouring of linear chain forming in water, linear polymerization (which should not be confound with polywater [3]). In 3 dimensions these chains may branch out and produce quite different structures (Fig. 2d, 2e). Note that in Fig. 2 the quantum mechanical calculated linear structures have been calculated at 0 Kelvin temperature only. In an ensemble of 10^{23} molecules and at finite temperature (300 K) those structures might not represent the most stable ones. Using combinations of the bonding types shown, and using other bonding types not considered here, various stable structures may form in liquid water.

Recently, a small network structure has been found in an isolated water cluster with 20 water molecules. The structure with 20 molecules has an empty cavity, but it is also found with 21 water molecules, where the cavity contains one single molecule. In those clusters (clathrates) the water molecules form an open pentagonal dodecahedron, which is considerably stable against growth and decay [4]. Besides earlier implications clusters of those sizes have been verified in nozzle beam experiments (Fig.3) [5]. The mass spectroscopy plot shows a considerable frequency of water 21 clusters compared to neighbouring cluster sizes. Note that with the quantum mechanical calculated structures in Fig. 2 it is not possible to build up the high symmetric structure of the 20 or 21 water dodecahedron cluster.

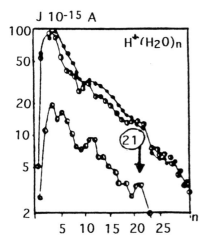

Fig. 3: Mass spectrum of water ions by Alexandrov *et al.* [8]. The logarithmic ion current is plotted over cluster size.

3. Models on the structures in aequeous solutions

Within the large number of models on the structure of water two main streams are essential: continuum models and mixing models. Continuum models assume that under normal (standard) conditions all water molecules are completely bonded by hydrogen bonds, i. e., there is no distinct different formation in the total of the liquid water (model) probe. Specific properties of water are considered due to lengths and angular among hydrogen bonds. The mixing type models consider water as a mixture of a small number of distinguishable formations (associative structures and non bonded single water molecules) like clusters, cells and clathrates.

Within the conception of clusters a mixture of 3 dimensional netted associate formation and single molecules is assumed, shown in Fig. 4. Here, clustering of water molecules is considered as cooperative phenomenon, accordingly if a hydrogen bond is formed, immediately several other bonds are formed with statistical distribution, but still correlated through their electronic states. Vice versa, if a bond in a cluster breaks, the cluster decays. The lifetime of those clusters is 10^{-11}s - 10^{-10}s, which is short enough to correspond to the relaxation time and to the permitivity of water, and, in addition, long enough for verifying

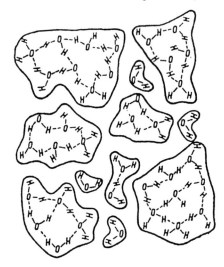

the existents of clusters in water with Raman spectroscopy [6]. Forming and decaying of clusters is considered as structural flickering (flickering clusters). Spectroscopic (Raman) investigations have shown that there are at least two or three coexistent flickering structures [7], i.e., structures that reproduce themselves while decaying.

Fig. 4: Cluster model of liquid water (flickering clusters).

Within the cell or clathrate model water is considered as a system with network structure, the cavities of which contain single molecules or are empty. The cavities are supposed to change some macroscopic physical properties of water. From a physical point of view the clathrate models are much closer to an imprint theory. The clathrate model is quite general and not restricted to aqueous solutions only. The solvent (water or a water-alcohol mixture) encases the probe (drug) molecule, or cluster, with a shell like shape [8]. The shell itself may be considered as filled cluster shell, commonly called clathrate. Also isolated empty clathrate exists. The clathrates themselves may be encased by other clathrates. With the succussion process inner and outer clathrates may separate and serve as a new seed for other clathrates. Once the basic structure of the original clathrate is formed the replication process of clathrate may go on without the probe (drug) molecules being present. The structure of the clathrate is considered as the drug's finger print. Unfortunately, those types of models cannot explain different medical effects caused by molecular related drug molecules. For instance, soluted sodium chlorine causes the same basic clathrate as soluted potassium chlorine does. The reason for forming an equivalent clathrate can be found in the bond length of the clathrate molecules that may vary within a certain range while maintaining the same geometric structure. Of course the enclosed volume differs, but with the replication process those details may disappear.

This brief review on water molecules and known cluster structures has formed an idea of basic structures recently discovered. However, very little is known about their stability and the forming mechanism in liquid water, or about transformation of information into water [9]. Experimentally little is known on this subject. From the theoretical point of view most molecular models and calculations are insufficient for they include polar pair interactions only, often taking into account the very next neighbour only. Few works deal with 3 body interaction. Quantum mechanical calculations require good initial conditions and in most cases they strongly depend on them, i.e. an assumption about a structure is put in, and an improved version of it comes out. Also full quantum mechanical calculations are restricted to a number of few molecules only, i.e.far away from a liquid.

What is it that arouses our suspicion to find specific structures and changes of structure in liquid water? Recalling the growth of $Na^+(NaCl)_n$ each individual cluster structure has its characteristic physical and chemical properties, and it may change dramatically with the cluster size. Extreme change in the physical property of liquid water can be found in its electrodynamics permitivity e [10]. The molecular dipole moment remains constant with varying temperature, while its permitivity underlies considerable changes, which makes water different and special from other natural liquid substances (Table 1).

Molecular dipole moment	T ($^\circ C$)	e
1.85	20	80.36
1.85	100	55.5
1.85	400 (10^3 bar)	16.5

Tab.1: Molecular dipole moment and permitivity of H_2O at various temperature.

The high polarity of the water molecules (Fig. 5) causes the water molecules to form hydrogen bonds (a polarized hydrogen atom of one water molecule forms a weak bond with an oxygen atom of a different water molecule). (Annotation: hydrogen bonds play an important role in molecular genetics, where they control the unification of the two branches of the DNA molecule [11].) Hydrogen bonds are the origin of specific local structures formed in liquid and solid water [10]. Figure 5 shows that in a water molecule the hydrogen atom cores are separated from their shell electrons to a considerable extent. Therefore, this compound is no longer electrical neutral, but is highly polar with the positive charge residing at the hydrogen core atoms. The hydrogen atom, i.e. in this case the bare hydrogen nucleus, a proton, may act on another polarizable molecule, or a polarized atom of another molecule, respectively, which might be a water molecule again. In this way water molecules may form arbitrary structures by means of hydrogen bonds.

There are, however, some implications on a constraint flexibility of the hydrogen bonds responsible for various structures to be formed and to decay in liquid water. On time average certain structures show up more stable than others. Experimental work on this is still in its initial stage, in theory there are no results yet that might give a sufficient insight into the nature of those structures. Much less can be said about a mechanism of information transfer. Some promising recent theories on the mechanism of information storage and transfer in liquid water will be briefly outlined after the following section which gives a simplified picture of the mechanism of information transfer in liquid water.

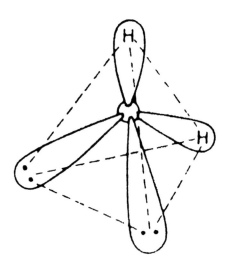

Fig. 5: Geometrical proportion of nucleus and electron charge distribution in a water molecule.

4. Information transfer and storage

In the preceeding paragraphs a brief summary on clusters, water and forming of structures has been given, which will be the starting point for the following theoretical issues and hypothesis on storing information in water.

Structurally, water is a highly sensitive system. The sensitivity is due to local interaction as well as to the collective field of actions of the properties of all its constituents in the field of interaction of all others (non-local interaction). What does it mean? A local structure is not given by a local field of the local particles only. Even looking at an isolated pair of molecules only, it is insufficient just to add up energy contributions of a common equilibrium structure. The calculated final energy would not give the correct equilibrium energy. There are non-linear, i.e. non-additive contributions to the equilibrium energy. Their origin lies in fields coming from a residuum interaction, the so-called zero-point interaction [12].

Assuming a single water molecule in far distance to a reference molecule in liquid water, its absolute contribution to the local field of the reference molecule may be small, but in water there are many other molecules that form a pair with the reference molecule. Therefore, the reference molecule feels the field of other molecules, and will react to them, i.e. it adjusts its energy, which again all other molecules will feel, and again will react and adjust. This may occur simultaneously with several other molecules, which is called a 2-, 3-, 4-, ... , n-body collision, a many-body or collective interaction *(Annotation: An exact calculation of the dipol n-body interaction problem in clusters is given in detail in [13].)* Collective interaction can be seen in the well-known collective motion of soliton water waves as well as in collective charge motion in modern plasma physics. Modern theoretical physics teaches us that collective effects cannot be estimated with linear thinking.

We have pointed out that in a collective system each individual depends on another one - groups on singles, and groups on other groups. What makes the system complicated is the process of their interaction (communication), e.g. the polar interaction caused by the fields of their charges, the zero-point field associated to every single atom, and the gravitational field. (Note, all fields are of long-range interaction). David Bohm's idea on the implicit order [14] suggests more fields that may be part of the structure forming process as e.g. the constraint long-range field of a 5th basic force (which is not yet completely verified). Also, one might think of additional fields like the collective acting morphogenetic fields discussed in biochemistry [15].

All external influences on water are caused by fields, which are of the same nature as the internal fields. The fields mentioned here are very weak, and as will be explained later on, exactly in this property the origin of their influence on collective spatial action may be found. One dramatic experiment on information transfer worth mentioning is the 'gelatinized water' [16]:

it is formed when only one molecule of pseudo-iso-cyanin pigment is mixed in about 2000 molecules of water. The information 'presence of pseudo-iso-cyanin' is carried through by all hydrogen bonds, and it remains permanent, i.e. the water has gelatinized, suggesting a strong information carrying property of water.

From the hydrogen bonds an incredible number of meta stable structural states rise, where each state represents a certain structure. For a clearer picture of this, consider the hydrogen bonds as a system of logical switches, where the logical rules relate to the rules of the physical interaction. Stimulating those switches with suitable fields of suitable energy, the switches react according to the rules of interaction, like the magnetic and electric stimulated rearranging of molecules in a liquid crystal, or like a micro chip reacts with a successful memory change when it gets a proper change of voltage in a certain range of voltage. With too little voltage it will not respond, with too high voltage it will be destroyed in its structure completely *(Annotation: The concept of "proper range" may be a hint where to find the key to different effects of different homoeopathic potencies of the same tincture.)* Of course, it is a simplified picture of a possible property of water of storing information. Whether storing of information in water works this way, cannot be scientifically decided yet (note that scientific opinion may also be influenced by nonscientific motivation).

DISCUSSION
Some theories on the structure formation and information transfer

The picture of information storage and transport as given above might be a general picture for a variety of promising theories which have been recently developed. Recently Berezin discussed the isotopic diversity of chemical elements as a physical foundation of homoeopathy [17]. Homoeopathic remedies based on water or water-alcohol mixtures basically consist of molecules built of hydrogen (H), carbon (C) and oxygen (O) atoms. From nuclear physics it is known that for almost all elements (atoms) there are atoms of the same type, but of slightly different mass (caused by a different number of neutrons in the atomic nucleus). Some of those atoms with different mass (isotopes) are stable, most of them are unstable and suffer radioactive decay. The stable isotopes of hydrogen are ^1H (or H) and ^2H (or D, deuterium). The number to the left of the element symbol refers to the number of nuclei in the atomic nucleus. The natural abundance of H is 99.985%, that of D is 0.015%. Also, there is an unstable isotope of hydrogen, ^3H (or T), with half a lifetime of 12.26 years. The stable isotopes of oxygen are ^{16}O (or O), ^{17}O and ^{18}O, with a natural abundance of 99.759%, 0.037% and 0.204%, respectively.

The stable carbon isotopes are ^{12}C (or C) and ^{13}C, their natural abundances are 98.893% and 1.107%. Thus, besides the different chemical and physical properties of atoms in a molecule we get two more characteristics: the diversity in mass, and the diversity in abundance. With different mass the vibrational interaction among the molecules changes. From the natural abundance of the isotopes average distances may be estimated, and vibrational modes and mode differences determined. From Berezin's theory on isotopic diversity together with the simplified picture of cybernetic switches given above, we may imagine the (rare) isotopes as characteristic nodes in a pattern of information, and the diversity of vibrational modes as dynamical carrier (amplifier, multiplier) of information. Thus, a drug molecule may change the pattern of nodes, which causes a different coupling of vibrational modes. The pattern of isotopic nodes may be stabilized (amplified, multiplied) by the coupling of vibrational modes, even when the drug molecule has been extracted from the system. Figure 6 shows the isotopic diversity of the water molecule, and the diversity of free ions found in pure water. This theory is asking for a new type of experiments in homoeopathy, the high resolution LASER spectroscopy.

This type of spectroscopy has been developed very well during the past decade, it is highly sensitive to vibrational mode coupling, and specific molecular structures. It may provide some information about basic questions in homoeopathy (information transfer, potentisation, periodicity in potentisation - see Glossary, p. 221 -).

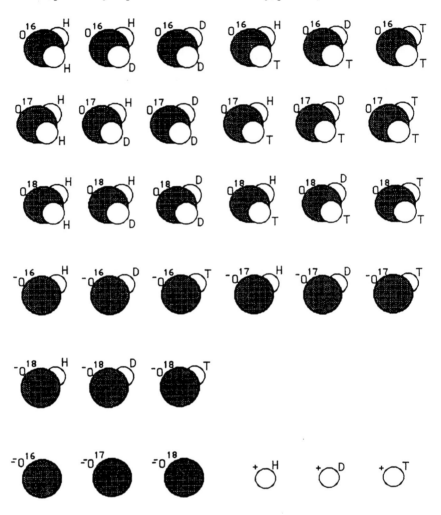

Fig. 6: Pure water is a mixture of isotopes and ions that can be tallied as 33 different substances. The 18 isotopes of the water molecule are represented in the top three rows of the diagram, the 15 kinds of free ions in the lower three rows. The number to the right of the element symbol refers to the number of nuclei in the atomic nucleus, which basically is the mass of the atom given in atomic mass units.

The imprint theory developed by Del Guidice [18] is based on a coherence assumption similar to that of Berezin. Instead of isotopes keeping coherence long-time stable, Del Guidice concentrates on coherent states induced by electromagnetic polarization fields.

It has been proven that polarization fields are extremely non-local, and may stabilize certain structural states [13]. Although Del Guidice's model seemed to be very different from Berezin's isotope model, in a dynamical (thermal) theory both models work hand in hand through the electron-phonon coupling, i.e. the coupling of electron interaction and vibrational motion of the atom core. Thermal electrodynamics couples the amplitude of vibrational modes (phonons) and the polarizability of the atomic shell through a non-local integral over vibrational modes and frequency-dependent polarizabilities.

Another promising theory on information storage and transfer in high diluted solutions has been developed by Popp [19]. In Popp's model the homoeopathic drug induces a resonance transfer of disregulatory energy from the patient's body (biological cell) to the absorbing homoeopathic dilution which then is excreted. Emission and absorption are described, through coherent states of the "drug" field and the "disregulatory" field in the patient's body, respectively. The natural weak coupling of both fields keeps the states coherent. Popp also proposes that the long-time stability of homoeopathic remedies might be due to coherent phonon states induced by the succussion process. (In this volume Auerbach discusses the fluid dynamical part of the succussion process in detail.) With regard to Popp's model, it seems necessary to give specific model fields and Hamiltonian in order to develop it further.

Mechanism of UHD and homoeopathic drug production and the limits of current theories

The homoeopathic drug and their production are subject of the following paragraphs. The major question regarding homoeopathy is how the health bringing information is transferred to the carrier substance water, or how information can be stored in water permanently, and how this information is transferred to the patient to be cured (plant, animal, human). The answer will not be solved within this book, but we shall present some pieces of that puzzle that need to be turned around to decide if they really fit.

Another piece of the puzzle is the production of drugs, where three main procedures of their production can be distinguished:

- The substance (e.g. of a herb with healing properties) is soluted in water and diluted by potentisation (homoeopathy).
- The herb or leaves of herbs make contact with the water only (recipes by Bach [20]).
- The information carrying water does not have physical contact with the donator of the healing properties (Agnihotra [21]).

With the first procedure mentioned, the interaction of the essence and water is through direct contact. Exchange of information (property) with water molecules occurs on the same level. Here we still might be able to search for physical methods and explanations within the frame of the currently accepted knowledge of science. Maybe with simple models, like the switch-model given above, it is possible to solve this problem.

The second class of recipes mentioned above in its interpretation is very much near the edge of the current horizon of science. The active substance is not soluted in water any more, it touches the surface of the water only. This is a huge restriction when considering electromagnetic interaction. There is only one layer of herb or leaf molecules and one layer of water surface molecules in contact. The distance of subsequent layers is electromagnetically large, therefore the interaction is considerably weaker and the effect of those drugs is a much more subtle one. Indeed Scheffer speaks of a transfer of the essence of the plant to the essence of the patient [20].

With Agnihotra, much more subtle levels are touched [22]. The distance is in the macroscopic range, and molecular electromagnetic interaction can not be measured any more.

For explaining the last two procedures of recipe production, the accepted paths of physics have to be extended. The collective interaction models and modern wholesome studies of dynamics involved in complex systems (dynamical phase space analysis [23]), might be able to give some insight into homoeopathy. For the more subtle recipes, currently we do not have any detail concept in physics that might be adopted. This does not mean that a real investigation of the first, second and third generation of recipes is out of reach. As we are not sure about all the types of interaction that are important, some might be unknown. The information-carrying capacity of liquid water suggests that indirect paths of investigation might be taken. Investigating problems by means of indirect paths are well-known in physics. Frequently, a physical effect is modelled within the concept of "hidden parameter" [14]. Only by studying all relations of the hidden parameter to all known parameters and interactions, some insight into the nature behind the hidden parameter might be discovered. With "hidden parameter" we do not mean adjustable parameter only, but also new introduced fields, i.e. interactions, which are implied by the complexity of underlying dynamics. *(Annotation: A hidden parameter field model in high energy physics predicted W^+, W^- and Z bosons which have recently been discovered by high energy physicists.)* With regard to homoeopathy we may think of morphogentic fields mentioned by Sheldrake [15], or the primitive reciprocal field [24]. New concepts like the quantum mechanical model of an interacting community [25] imply a further field. Theoretical investigations and model simulations on homoeopathy are far away from providing the empirical and experimental research with a necessary foundation. Independent funding of qualified research on homoeopathy may give homoeopathy a theoretical base that not only can provide its experimental and empirical research with important hints, but may provide valuable contributions to biogenetics, chemistry and nuclear engineering, too.

ACKNOWLEDGEMENTS

The author is grateful to his wife M.A. Kniseley for her patience, for discussions and for reading the manuscript.

REFERENCES

1. Knight D, Clemenger K, de Haar WA, Saunders WA. Phys Rev Lett 1985; B 31: 25-39.
   Echt O, Sattler K, Recknagel E. Phys Rev Lett 1981; 47: 11-21.
   Proc 3rd Int Meeting on Small Particles and Inorganic Clusters. Surface Science 1985; 156.
   Farges J,de Feraudy F, Torchet G. J Chem Phys 1983; 78 (8): 50-67.
2. Rao CNR in: Water, A Comprehensive Treatise Vol.1, Franks F ed., London: Plenum Press 1972.
3. Franks F. Polywater. The Massachusetts Institute of Technology 1981.
4. Dreyfuss D, Wachmann H. J Chem Physics 1981; 76: 20-31.
5. Alexandrov NL et al. X. International Symposium on Molecular Beams, Cannes 1985.
6. Frank HS, Wen W. Disc Farrad Soc 1957; 24: 133.
7. Walrafen GE. J Chem Phys 1968; 47: 20-79.
   F. Kohler. Monographs in Modern Physics 1, The Liquid State. Weinheim: Verlag Chemie 1972.
8. Anagnostatos GS, Vithoulkas G, Garzonis P, Tavouxoglou C. Proc 43rd Congress LMHI, Athens 1988, pp. 11-21. See also Anagnostaos, this volume.
9. See the respective contributions in this volume
10. Luck WAP. Structure of Water and Aqueous Solutions Weinheim: Verlag Chemie 1974.
11. Crick FCH, Watson JD. Proc Royal Soc 1954; A233: 80.
12. Schröder H. J Chem Physics 1980; 72 (5): 32-71.
13. Schulte J. Z Phys D 1991; 19: 147.
14. Bohm D. Quantum Implications, Essays in Honour of David Bohm. In: Hiley J ed. London: Routledge Kegan Paul Inc 1987.

15. Sheldrake R. Das schöpferische Universum. München: Meyster Verlag 1982.
    Fischer N, Scheibe N. J prakt Chemie 1928: (2) 100: 89.
16. Berezin AA. Berlin J Res Hom 1991; 1 (2); see also Berezin, this volume..
17. Del Guidice E, Preparata G, Vitiello G. Phys Rev Letters 1988; 61; 1085-1088. See also Del Guidice, this
    volume.
18. Popp FA. Br Hom J 1990; 79: 161-166. See also Popp, this volume.
19. Scheffer M, Bach-Blütentherapie. München: Hugendubel 1981.
20. Schulte J. Acta Medica Empirica 1990; 3:, 763.
21. Pranjpe VV. Homa Farming. Madison: Fivefold Path Inc 1986.
22. Venkatesh R, Marlow WH, Lucchese R, Schulte J. Proc Material Research Society 1991; 206: 185.
23. Schulte J. Acta Medica Empirica 1990; 39: 418.
24. Jahn RG. On The Quantum Mechanics of Consciousness, with Application to Anomalous Phenomena,
    Princeton University 1984.
25. Martin TP. In: Beneke G, Martin TP, Pacchioni G, eds. Elemental and Molecular Clusters, Springer Series
    in Materials Science. New York: Springer 1987.

IS THE "MEMORY OF WATER"
A PHYSICAL IMPOSSIBILITY ?

E. Del Giudice

SUMMARY

The retaining traces of molecules in polar liquids are discussed by means of the electro-magnetic superradiance phenomenon occurring in a dense phase of matter. The super-radiance model as proposed by Preparata is applied to polar liquids such as water. Some recent calculable results will be discussed, and summarized to a complete theory.

ZUSAMMENFASSUNG

Das 'Erinnerungsvermögen' von Wasser wird an Hand einer ursprünglich von Preparata entwickelten Theorie erklärt. Dieses, auf dem Phänomen der elektromagnetischen Superradianz basierende Modell vermag verbleibende Information von Molekülen in be-stimmten Phasen polarer Flüssigkeiten zu erklären. Einige neuere Ergebnisse werden dis-kutiert und zu einer geschlossenen Theorie zusammengefaßt.

INTRODUCTION

Recently, experiments [1,2] have been reported that should provide a laboratory test of the claim of homoeopathy that polar liquids such as water and ethanol can retain a trace of molecules (or molecular aggregates) once present in them.

Actually the very high degree of dilution (10^{-N}-molar, where N is of the order of hundreds or thousands) excludes that in the homoeopathic remedies molecules of the original solute can still be present. In the reported experiments, a molecular species is diluted in water and thoroughly shaken (succussed) so many times that reasonably at last only pure water should be present. In spite of this circumstance, the dilution, when aimed at specific biological targets, produced at least some of the effects of the original molecular species.

Is it possible to understand these reports? The possibility of storing information in a li-quid seems, at a first glance, completely out of the physical mechanisms known hitherto; thus one can understand the negative, sometimes sarcastic, reaction of the physics com-munity to the above reports.

So far it seemed that the only way of storing information was to require a solid-state array of atoms or molecules or, anyhow, a space ordered pattern, as, for instance, in nucleic acids. In this scheme, elementary components (atoms or molecules) are the "alphabet let-ters" and, when kept in spatial well-defined positions, they are able to produce the "words" of a possible message.

How is it possible to have the same mechanism in a liquid where each space configuration lasts no more than a tiny fraction of a microsecond and molecules are sliding and rolling on each other without rest? One could imagine a clustering of molecules, but the number of cluster components (at most some tens) cannot be large enough to accommodate a sizable chunk of information and, moreover, clusters are not that stable to account for the very long time (hours and days) of validity of the prepared dilution.

However, how is it possible to understand such a selective memory, where the trace of an original molecular species is protected against the dissolving influences of the numerous stray molecules and clusters passing by through the same liquid?

117

P.C. Endler and J. Schulte (eds.), Ultra High Dilution, 117–119.
© 1994 *Kluwer Academic Publishers. Printed in the Netherlands.*

All the above questions cannot find any reasonable answer within the framework of the generally accepted physics of liquids. It is not a surprise that it denies the green light to the very concept of "memory of water".

"Eppur si muove" ("still it moves") - this is what Galileo mumbled while getting out of the court of Inquisition which had just outlawed his physical claims. "Eppur ricorda" ("still it remembers") the researchers may mumble by comparing their results to the generally accepted theories.

MAIN PART

As a matter of fact, the generally accepted theory of liquid state is still unable to explain many basic features of real liquids. The very property of the onset of cohesion at the liquid-vapour transition point is still elusive and has been dealt with so far in terms of empirical concepts as nucleation, but without any detailed understanding of the mechanism through which, at 100°C and 1 atm., water molecules in the vapour phase, mutually distant about 36 Å, suddenly keep rushing one towards another and increase the density by a factor 1600; and that occurs at a sharp temperature without any gradual evolution.

Furthermore, the physics community cannot yet claim to understand liquids satisfactorily. There is, however, a different approach to condensed matter which has just taken its first steps. This approach emerges from the quantum theory of superradiance, proposed on general theoretic grounds by Giuliano Preparata [3] and has been successfully applied by the Milano group to a number of condensed systems [4,5,6,7,8].

Superradiance is the phenomenon occurring in a dense set of particles interacting through the electromagnetic (e.m.) radiative field. In classical physics, the system configuration with minimum energy would be the array of particles fixed at the equilibrium positions with respect to the static short range forces, whereas the e.m. field would vanish. In quantum physics this configuration field must fluctuate (zero-point fluctuations) as a consequence of Heisenberg's uncertainty principle. But conventional wisdom in condensed matter physics [9] maintains that zero-point fluctuations of the different components of the system are uncorrelated and incoherent, so that they don't change very much the fundamental classical picture of matter, as made up of small balls kept together by strings.

But this conventional outlook is not true. It has been proven in Ref. 3 and in the other papers [4-8] that, beyond a threshold of density, the particle and e.m. fluctuations couple, and give rise to a configuration, the energy of which is lesser than the energy of the "classical" configuration. In this new "coherent" configuration particles and field oscillate with a constant phase relationship within regions ("coherence domains"), the size of which is the wavelength of the e.m. radiation field. In order to have the above interplay, it is essential that particles have a discrete spectrum so that only the e.m. frequencies matching the differences between the discrete levels would be enhanced. This mechanism automatically provides a mode selection.

When superradiance starts up, the frequency of the e.m. field is shifted downwards, so that it finds itself in the conditions of "total reflection" well-known in optics. The e.m. field is kept trapped within the coherence domain and cannot be irradiated. In this way the basic objections of conventional wisdom [9] to the possibility of having a non-vanishing e.m. field in the minimum energy configuration are overcome.

Different coherence domains are kept together by typical quantum mechanism such as the Josephson ones. But the basic unit, the coherence domain, the size of which spans in the range of microns, gets condensed through the choral oscillation of matter and field. The gained energy is proportional to density, so that the fundamental principle of minimization of energy forces the system to increase its own density as much as possible. The upper limit to the density increase is provided by the strongly repulsive hard cores which prevent molecules from penetrating each other.

The existence of condensed matter then requires the existence of a coherent e.m. field tuned with the matter field and tied to it. Such a coherence regime starts up spontaneously when density exceeds a sharply defined threshold and that accounts for the sharpness of phase transitions.

DISCUSSION

By means of the above ideas, a first rough investigation of water has been done [10,11]. It has been proven in Ref. 10 that the collective motion of water molecules makes the system much more polarizable than in an incoherent configuration. Externally applied electric fields are able to produce sizeable static or quasi-static polarization fields over large regions, which last a long time.

Then it would be an intriguing possibility that very low frequency polarization fields could be imprinted in water by an external source (the dilute molecule clusters?). These frequencies could modulate the fundamental frequency of the e.m. field in the minimum energy configuration.

The possibility of such a modulation has not been proven yet, but it is not obviously wrong regarding today's knowledge. The mechanism of modulation could provide just the possibility of "writing in water" through a time-ordered pattern of signals (polarization field) that water is able to conserve indefinitely. As in radio communications two different modulation patterns should not interfere, conserving then the selective aspects of the message. Finally, these very low frequency fields, the source of which must be a giant macro-molecule or a cluster of small molecules, should not be disturbed by the higher frequency fields produced by the single molecules present in the liquid. The aqueous solvent does not require then to be exceedingly pure.

In conclusion, there is still a long way to the understanding of "memory of water", but the road is not obviously closed.

REFERENCES

1. Benveniste J, Arnoux B, Hadji L: Highly dilute Antigen increases Coronary Flow of Isolated Heart from Immunized Guinea-Pigs. FASEB J 1992; 6: A1610 (abs.)
2. Endler PC. This volume.
3. Preparata G. Quantum Field Theory of Superradiance. In "Problems of Fundamental Modern Physics". Cherubini R, Dal Piaz P, Minetti B eds., Singapore: World Scientific 1990.
4. Del Giudice E, Giuffrida M, Mele R, Preparata G. Europhysics Letters 1991; 14: 463. Physical Review 1991; B43 5381.
5. Del Giudice E, Enz CP, Mele R, Preparata G. Solid ^{4}He as an ensemble of superradiating nuclei. Preprint of Milano Physics Department MITH 1990; 90: 11.
6. Bressani T, Del Giudice E, Preparata G. What makes a crystal stiff enough for the Moessbauer effect? Preprint of Milano Physics Department MITH 1990; 90: 4.
7. Bressani T, Del Giudice E, Preparata G. Il Nuovo Cimento 1989; 101A: 845-849.
8. Del Giudice E, Giunta G, Preparata G. Superradiance and the mystery of ferromagnetism. Preprint of Milano Physics Department MITH 1991; 91: 16.
9. Anderson PW. Basic Notions of Condensed Matter Physics. Menlo Park, CA - USA: Benjamin Cummings Publishing 1984.
10. Del Giudice E, Preparata G, Vitiello G. Physical Review Letters 1988; 61: 1085-1088.
11. Del Giudice E, Preparata G. A Collective Approach to the Dynamics of Water in Hydrogen Bonded Liquids. In: Dore J, Teixeira J eds. NATO ASI Series. Dordrecht: Kluwer Academic Publishers 1991.

SMALL WATER CLUSTERS (CLATHRATES) IN THE PREPARATION PROCESS OF HOMOEOPATHY

G. S. Anagnostatos

SUMMARY

In a previous publication, a three-step hypothesis was introduced to help the comprehension of the technique followed in homeopathic preparation of remedies in the case that the dilution of the pharmaceutical substance does not reach the molecular level of the substance, but small aggregates of molecules of the substance are first formed and then eventually diluted into the solvent. Here, this hypothesis is extended to cover the cases where the dilution reaches the molecular level and eventually simple molecules of the substance are found among molecules of the solvent.

ZUSAMMENFASSUNG

In einer früheren Publikation wurde für die Vorgänge bei der Herstellung von homoeopathischen Präparaten im konzentrierten molekularen Verdünnungsbereich eine "drei Stufen- Hypothese" vorgeschlagen. Es wurde dabei davon ausgegangen, daß die pharmazeutische Substanz zunächst kleine Aggregate von Molekülen bildet, die dann in dem Lösungsmittel verdünnt werden. In der hier vorgestellten Arbeit wird die vorgeschlagene Hypothese erweitert, um auch mittlere Verdünnungsbereiche zu beschreiben.

INTRODUCTION

In a recent publication [1], a three-step hypothesis was introduced to explain the specific organization of the molecules of the solvent in homeopathic microdilutions able to maintain the properties of the initial substance which is not effectively present. In that publication, the cases of pharmaceutical substances which need grinding before their dilution (or, more general, the cases where the dilution does not reach the molecular level of the substance) are discussed, while in the present work substances not coming from grinding, but coming, e.g., from extracts of substances, are treated where the dilution reaches the molecular level and eventually simple molecules of the substance are found among molecules of the solvent. A summary of that previous publication is given below.

MAIN PART

1. Dilutions not reaching the molecular level

The three-step hypothesis of reference [1] briefly includes the following. *First step:* During the grinding and the initial dilutions, characteristic small clusters (aggregates of a small number of molecules) of the pharmaceutical substance are formed, surrounded by shells of organized hydrogen-bonded molecules of the solvent (clathrates). *Second step:* Because of the applied forceful successions (during the preparation of homeopathic remedies) and the different inertial properties the small clusters move out of their clathrates. Then a new clathrate is formed round each relocated small cluster, and in addition another clathrate (called mantle-clathrate) is formed round the initial clathrate (now called core-clathrate), which has shrunk and become hollow, having lost its small cluster. *Third step:* At this stage effectively no molecules of the substance are present and the role of the small cluster for the dilutions and successions to follow is totally taken on by the compact structure of the core clathrate. That is, due to successions and the different inertial properties between core and mantle clathrate, the core clathrate separates from the mantle clathrate and new clathrates are formed round each of them. In brief, through our hypothe-

121

P.C. Endler and J. Schulte (eds.), Ultra High Dilution, 121–128.

sis the properties of the initial pharmaceutical substance (which for homeopathic microdilutions is not effectively present) are preserved via the properties of the shaped void (interior to the core clathrate) characteristic of the properties of the initial substance. Figures 1 and 2 schematically show all the above stages involved in preparing a homeopathic remedy when there is no molecular dilution of the pharmaceutical substance into the solvent, but initially small clusters from molecules of the active substance are formed.

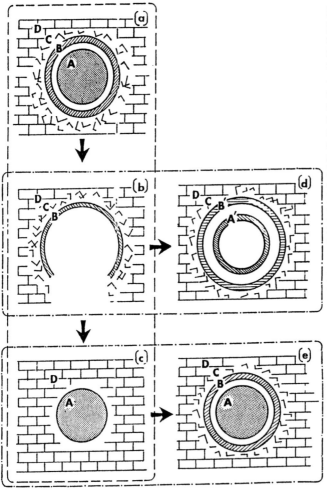

Fig. 1: Stages of step 2 in preparing a homoeopathic remedy in the case where the active substance initially forms a small cluster. A and B schematically denote the small cluster and the clathrate of Fig. 6, respectively. A' and B' denote the shrunk-compact core-clathrate and the mantle clathrate, respectively. Letter C denotes the semi-organized incompatibility layer of water in between the organized layer of water B and the unperturbed water D.

2. Dilutions reaching the molecular level

Now, our hypothesis of ref. [1] is adjusted into the following three steps. *First step:* This step refers to the formation of clathrates. If the properties of the substance permit, clathrates (from molecules of the solvent) are formed surrounding the stereo chemical structure of each isolated molecule of the substance.

A clathrate hydrate is a crystalline compound that can be obtained by the formation of a hydrogen-bonded water "host" lattice around one or more species of "guest" molecules or ions [2]. The clathrate hydrates can be classified according to two criteria (i) the chemical nature and stereo chemistry of the guest species and (ii) the structure of the water host lattice. In other words independent of the details of the cohesive forces, the size and the shape of the guest species and the ability of the water molecules to form appropriate hydrogen-bonded lattices are the primary criteria for clathrate formation. It is also necessary for the guest species to be able to exist in an aqueous environment without chemical reactions, i.e. hydrolysis. On the other hand, certain cyclic ethers and amines, which are water soluble form crystalline hydrates.

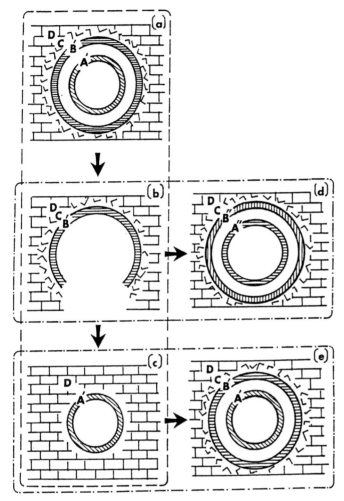

Fig. 2: Stages of step 3 in preparing a homoeopathic remedy in the case where the active substance initially forms a small cluster. All letters have the same meaning as in Fig. 1. Double primed letters A", B" denote structures derived from the structures denoted by letters B'; and C of Fig. 3(b), respectively.

With a few exceptions, the structural feature common to all clathrate hydrates hitherto studied is the pentagonal dodecahedron [2] of water molecules, i.e. $H_{40}O_{20}$. The formation of clathrates when certain substances (e.g. hydrophobic gases and liquids, and soluble accidogenic gases) are mixed with water is definite and the specific structure of these clathrates is well- known for a rather long list of substances. For example, in Fig. 3 (next page) the case of tetra-iso-anyl-ammonium Fluoride in water [3] is shown, where the stereo chemical arrangement of the substance appears at the center (bold circles) surrounded by the clathrate (open circles) in the form of a 2T.2P void. It is characteristic that this void resembles, in a general sense, the stereo chemical shape of the substance. Two more examples are shown in Figs. 4 and 5 for the cases [1,4] of $(n-C_4H_9)_3$- S+F-•$23H_2O$ and of $(n-C_4H_9)_4{}^+N$ cation, respectively. The three examples above and all other cases tabulated in the literature [2] are the results of an x-ray analysis and everything is well-specified. For other substances not included in the list of already studied ones, the same method can be applied for an unambiguous and precise specification of the relevant clathrates.

It is a matter of research to examine which of the homeopathic substances belong to the category of those forming clathrates when mixed with water under proper conditions of temperature and pressure. For the cases where the clathrate formation of substances mixed with water is documented, the present first step of our hypothesis is well-verified.

For the remaining cases, the formation of clathrates should be tested. Of course, a clathrate hydrate usually is an extended crystalline structure and what is further assumed through the present work is that the same structure of solvent molecules is reached as a host round a clathrategenic guest substance even when the molecules of the guest constitute isolated structures inside the solvent.

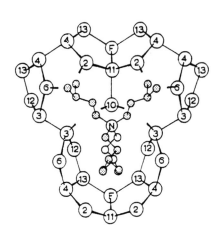

Fig. 3 (right): Orientation of a tetra-iso-amyl-amonium cation in a 2T.2P void.

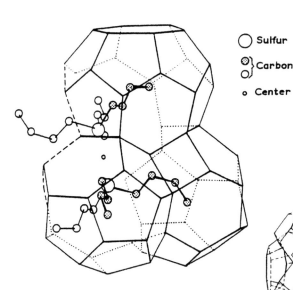

○ Sulfur

◎ } Carbon
○)

o Center

Fig. 4 (left): One half of the centro-symmetrical 60-hedron found in (n-$C_4H_9)_3$-S^+F^-•$23H_2O$. The center is at 0,1,0, and the two halves share the dotted edges. The half polyhedron shown is occupied by the shaded alkyl groups.

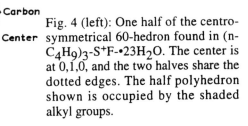

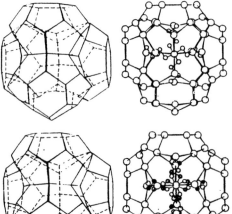

Fig. 5 (right): Clathration of a (n-$C_4H_9)_4N$ cation within the complex polyhedra (left) in a 3T.1P void, showing one orientation of the twofold disorder of the cation, and (right) in a 4T void, showing both orientations of the twofold disorder of the cation.

Second step: The second step of our hypothesis refers to the separation of the molecule of the substance from its clathrate. The molecule of the clathrate at its position at rest is at equilibrium inside the clathrate due, e.g. to repulsive forces between hydrophobic substance and water molecules. When a forceful succussion is applied, however, on the solution there is a different inertial kinematics behaviour between the relatively large density molecule of the substance and that of its small density clathrate (large volume with interior void). This difference between the velocity of the molecule of the substance and that of its clathrate leads to their separation. As the succussion becomes more and more forceful, so the separation between substance and its clathrate becomes more and more definite. In more detail, as the substance (due to its relative velocity) moves closer to the boundaries of the clathrate, the neighbouring water molecules of the clathrate (due e.g. to hydrophobic repulsive forces between substance and water molecules) move away from the substance leaving space for its passage. When the substance is finally outside the clathrate, the same water molecules gradually return to their initial positions. Finally, we obtain a relocated molecule of the substance and an empty clathrate the structure of which is more compact than before due to the absence of the repulsive forces between substance and water molecules of the clathrate.

The molecule of the substance at its new, relocated position interacts with the surrounding water molecules, a fact that leads to the formation of a new clathrate identical to the initial one. As discussed below, the behaviour of the water molecules at and round the clathrate before and after the relocation of the active molecule of the substance is more complex.

The water molecules of the water layer that comes immediately after the clathrate are effected by the symmetry of the clathrate. In order to comprehend this influence, one should resort to Fig. 5 presenting the $H_{40}O_{20}$ pentagonal dodecahedron which (as we noticed above) is the basic building block of all clathrate hydrates and thus is taken here as an example. The oxygen atoms are located at the vertices of the dodecahedron, where at the edges hydrogen atoms are situated forming hydrogen bonds with the neighbouring oxygen atoms. That is, on each edge we have the formation of O-H...H.

Thus, out of the 40 hydrogen atoms of $H_{40}O_{20}$, 30 of them are situated along (or very close to) the 30 edges of the dodecahedron. The remaining 10 hydrogen atoms are directed outwards of the dodecahedron forming hydrogen bonds with 10 oxygen atoms of 10 water molecules belonging to the layer immediately outside the clathrate. These bonds form links between the clathrate and the surrounding water layer. Additional links (hydrogen bonds) are formed between the water molecules of this same layer and the remaining 10 oxygen atoms of the dodecahedron (both hydrogen atoms of which are along the edges of the dodecahedron). Totally up to 20 water molecules of the layer that comes immediately after the clathrate possess links (one to one) with the 20 oxygen atoms of the dodecahedral clath-rate. Thus, the distribution of the protons in the initial dodecahedron is such that each oxygen atom in the polyhedron is half donor and half acceptor relative to its four nearest neighbours (three on the dodecahedron itself and one at its immediately after water layer).

This fourfold association between the water molecules of the two neighbouring (almost spherical) layers (i.e. the clathrate itself and the water layer immediately after) is the most probable way of associating neighbouring water structures. While water is such a versatile molecule in a hydrogen-bonding ionic environment and fivefold and sixfold coordination cannot be excluded *per se*, they have not yet been observed in these clathrates. The changes in the association of water molecules, at the clathrate itself and the surrounding water layer (through hydrogen bonding) after the relocation of the active molecule outside the clathrate, will be discussed below.

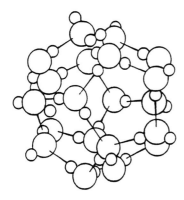

Fig. 6: Dodecahedron small cluster of water molecules interacting by hydrogen bonds, called clathrate.

After the separation of the molecule of the substance and its clathrate, the clathrate, due to its symmetric structure, survives and in addition shrinks, due to absence of the hydrophobic forces between active substance and water molecules of the clathrate. This shrinking of the initial clathrate leads to two effects. First, the clathrate itself increases its stability through the increased strength of the hydrogen bonds due to shorter distances among interacting water molecules. Second, the water molecules of the layer immediately follow the shrinking of the clathrate and thus come in shorter distances among themselves, a fact that favours the exercise of hydrogen-bonding among these molecules. Thus, this water layer immediately after the clathrate becomes more organized and finally takes the form of a pentagonal dodecahedron, i.e. the form of the initial clathrate by containing the same number of water molecules and imitating its shape vertex by vertex.

A difference exists, however, between the two similar shapes. Specifically, due to the difference in their size, the core clathrate is more compact than the mantle clathrate, a fact that makes the first to behave like solid, and the second like liquid with an increased viscosity. This difference in "phase" between the two clathrates corresponds to a difference in their inertial behaviour with respect to the force of the succussions to follow.

Third step: The third step of our hypothesis refers to the separation of the core clathrate from its mantle clathrate. The difference in kinematics between core and mantle clathrates noticed above leads to a separation of these two clathrates, when a forceful succussion is applied on the solution. Apparently, a new mantle clathrate is formed round the relocated core clathrate at its new position which is identical to the initial mantle clathrate. At the same time, the previous mantle clathrate shrinks into a smaller, more compact clathrate and finally becomes a core clathrate round which a new mantle clathrate is formed. The formation of the new mantle clathrate and the shrinking of the previous mantle clathrate mentioned above take place as described next.

The 30 hydrogen atoms of both the core and the mantle clathrates as described in the second step of our hypothesis are along the edges of these two polyhedra (pentagonal dodecahedron). Thus, between these two polyhedra repulsive forces are exercised due to repulsive forces between protons lying almost at the middle points of their parallel edges. This fact, together with the relevant hydrogen bonds, determines the size of the mantle clathrate. That is, the hydrogen bonds alone would lead to a more compact structure, identical to that of the core clathrate, the existence of repulsive forces (discussed above), however, increase the distance between protons (in the middle of parallel edges) and thus the size of the mantle clathrate leads to a rather loose structure of this clathrate. This difference in compactness between core and mantle clathrates is the reason for the difference in inertial behaviour between these two clathrates during the forceful succussions applied on the dilution finally leading to their separation. In more detail, as the core clathrate (due to its relative velocity) moves closer to the boundaries of the mantle clathrate, the neighbouring water molecules of the mantle clathrate (due to the repulsion between the relevant protons of core and mantle clathrates) move away leaving space for the passage of the core clathrate.

When the core clathrate is finally outside the mantle clathrate, the same water molecules gradually return to their initial positions reconstructing the initial shape of the mantle clathrate. The absence, now, of the repulsive forces described above between core and mantle clathrates permits the initial mantle clathrate to define its size according to its hydrogen bonds alone and thus to become identical to the previous core clathrate.

Furthermore, in order to comprehend the formation of a new mantle clathrate round the shrunk initial mantle clathrate, one should consider the hydrogen bonds between water molecules of this clathrate and water molecules of the water layer immediately following this clathrate; in a similar way we have described above the formation and evolution of the initial mantle clathrate in the previous second step of our three step hypothesis.

DISCUSSION

The basic factors in our explanation of the different stages in the preparation of homeopathic remedies are the formation of clathrates round the active molecules of the substance and the stability of these and all other clathrates involved during the procedure of the preparation. However, the formation of clathrates is well established for many substances and their stability is also supported by ref. [5] considering them as small clusters of water molecules. According to [5] clathrates with more than 8 water molecules are stable.

In brief, reviewing our three-step hypothesis we notice that the important parameters involved are (1) the guarantee that the extract leads to simple molecules diluted into the solvent, (2) the applied force in each succussion, which should be able to compensate for the cohesion between a molecule of the substance and its clathrate or between a core-clathrate and a mantle-clathrate, (3) the direction of the force, which should stay fixed in one way for all succussions so that each succussion does not partially destroy the effect of the previous ones, (4) the number of succussions per dilution in relation to the strength of the applied force and to the dimensions of the vessel used during the potentisation, (5) the frequency of succussions, that is, the time between two successive succussions, which should be large enough to permit the formation of the necessary clathrates round the molecules of the substance or mantle-clathrates round the core-clathrates, and (6) the number of sequential dilutions necessary in order to reach the desired density of holes (and their size) in the solvent and thus in the remedy.

Since the formation of a specific clathrate corresponds to a molecule of a specific substance, some of the properties of the initial substance can be traced back to the properties of its clathrate in the solvent. In other words, the specific remedy in homeopathy resulting from the microdilutions of a certain substance, has characteristic properties of the initial substance, even without its physical presence, a fact that constitutes the foundations of homeopathy.

The next stage in our effort to explain which mechanism of potentisation affects the homeopathic remedy is the performance of suitable experiments to verify all steps of our hypothesis. Finally, if our hypothesis is proved to be successful, it will contribute a great deal towards the obtaining better homeopathic remedies and towards a standardization of their preparation. As a result, all remedies derived from a certain substance with the same potentisation would have more or less fixed properties. Lack of standardization today is the major defect of our present remedies. In order to obtain this accomplishment the optimum force of succussions and its frequency of a particular case should be estimated. The necessary number of succussions of each dilution and the total number of dilutions to obtain the desired properties of the specific homeopathic remedy should also be determined.

REFERENCES

1. Anagnostatos GS, Vithoulkas G, Garzonis P, Tavouxoglou C. A Working Hypothesis on Homoeopathic Microdiluted Remedies. Proc 43rd Congress LMHI, Athens 1988; pp. 11-21.
2. Jeffrey GA, McMullan RK. "The Clathrate Hydrates" and references therein. Progr Inorg Chem 1967; 8: 43-107.
3. Feil D, Jeffrey GA. The Polyhedral Clathrate Hydrates II. Structure of the Hydrate of Tetra-iso-Amyl Ammonium Fluoride. J Chem Physics 1961; 35: 1863-1873.
4. Beurskens PT, Jeffrey GA. Polyhedral Clathrate Hydrates VII. Structure of the Monoclinic Form of the Tri-n-butyl Sulfonium Fluoride Hydrate. J Chem Physics 1964; 40: 2800-2810.
5. Rao BK, Kestner NR. Ab Initio Calculations on Negatively Charged Water Clusters . J Chem Physics 1984; 80: 1587-1592 .

MASS, FLUID AND WAVE MOTION DURING
THE PREPARATION OF ULTRA HIGH DILUTIONS

D. Auerbach

SUMMARY

The succussion phase in the preparation phase of UHDs ("potentisation") is described within the frame of fluid dynamics. The mechanical mixing of the process is discussed in terms of saddle flow, vortex flow, shear flow, and diffusion flow. With respect to the drug dissolving process the relevances of the different flows are discussed. By analyzing the flows it may be implicated that the diffusion flow gives rise to the most active force of preparing UHDs.

ZUSAMMENFASSUNG

Der Präparationsprozess von Homöopathika wird im Rahmen der Dynamik von Flüssigkeiten dargestellt. Der Mischungsprozeß von ursprünglicher Droge und Lösungsmittel wird an Hand des Sattelpunkt-, Vortex-, Scher- und Diffusionsflusses beschrieben. Die theoretische Analyse der Flußmöglichkeiten läßt darauf schließen, daß der Diffusionsfluß den Hauptanteil beim Prozeß der Zubereitung der Lösung einnimmt.

INTRODUCTION

Hahnemann describes the "dynamization" (potentisation) process in the following way [1]:

First a gran of this powder is dissolved in 500 drops of a mixture of one part brandy and four parts of distilled water; thereof one single drop is put into a little bottle. One adds an additional 100 drops of a good alcohol (the dynamization glass should be two-thirds full) and then slams the stoppered (capped) bottle 100 times against a hard, but elastic body (perhaps a leather-bound book). This is the medicine in the first dynamization level, whereby one first saturates small sugar globuli, then rapidly spreads them on blotting paper, finally dries them and stores them in a sealed bottle marked with the sign of the first (1) degree of dynamization. But a single globule is used for further dynamization. This is put in a new bottle (with a drop of water to dissolve it) and then dynamized in the same way by mixing it with 100 drops of good alcohol and with 100 slams.

A bottle containing a water/alcohol mixture (some methods substitute the alcohol for something else) and the base-substance are slammed 100 times on an elastic surface. Sugar globuli are saturated in the solution and then dried. This is the first dynamization step. For each further step one of the globuli is taken and dissolved in alcohol in a new bottle: dissolving, shaking and drying are repeated until the desired potency has been reached. Finally, the medicine is administered either in globule form or in solution.

Those who have either dynamized themselves or supervised the process will have been fascinated by the various – generally fast-changing – phenomena. The moment that the powder or tincture has been let into the quiescent water/alcohol mixture it dissolves and *diffuses* into it. The mixture first really begins to *flow* some time during the upward braking motion of the bottle. The mixture then impinges onto the top, taking on a so - called *saddle* form (so-called because of the saddle-like flow curvature), flows partly down the sides in a *shearing* motion, to impinge on the bottom as the bottle is slammed. During the short rest period it transforms from an impinging saddle flow to a large-scale *vortex flow*.

129

P.C. Endler and J. Schulte (eds.), Ultra High Dilution, 129–135.

By then the entire flow has become turbulent: small-scale vortices, saddles, waves, bubbles, drops and foam form. The bottle emits a tone and the fluid may sometimes become briefly opaque. The slam causes a new motion, its very noise indicating that we are now calling the *acoustic* properties – the compressibility – of the water into play. Details of these characteristic motions - the diffusive, the flow and the (acoustic) wave motions - depend to a more or less extent on the form of the bottle, on the quantity and nature of the fluid as well as details of the shaking motion. In the following we shall characterize the above forms of motion in the hope that they may help us understand the dynamization process better. None of them ever occurs in isolation.

MAIN PART

1. Flow

Instability and flow beginning. Let us first assume that the bottle is raised directly against gravity, that it is only partly filled and that the deceleration of the fluid is largely compared to the acceleration due to earth's gravity. Insofar as the free surface of the fluid is flat the fluid tries to rise everywhere at the same time. It must either tear it off forming a hole ('cavitation') or its flow just flows through the slightest perturbation of the surface. Fig. 1 shows three of the many possible forms which the flow may take (the full three dimensional form is more complicated). Of course, the slightest asymmetry in the motion assures that the bottom empties as the water flows upward, as shown in Fig. 1c.

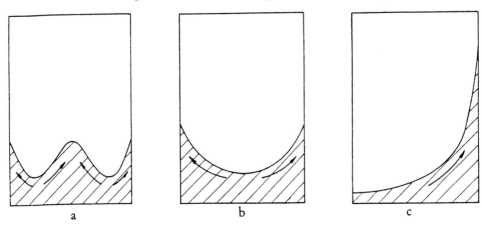

a   b   c

Fig. 1: The upward motion of a container with initially quiescent fluid is braked, causing the fluid to become unstable and begin flowing. 1a and b are two possible forms the flow after the instability can take. The flow in Figure 1c would result if there were a slight asymmetry in the motion.

Saddle flow. On reaching the top, the mixture impinges on the roof, giving rise to the *saddles* shown in Fig. 2. Here we find particular points or lines (Fig. 2, large dots) – the stagnation points or lines – where the mixture is coming to a rest. A local pressure maximum reigns at those locations. Everywhere else where the flow is deflected the velocity increases from zero and the corresponding pressure decreases from its maximum. Figure 2 shows three possible forms which the saddle flow may take resulting from the flow shown in Fig. 1. The small points and circles represent drops, foam and bubbles. The mixture flows through each saddle, taking part in it for but a moment. This is not the case with the next flow structure we are going to consider, the large scale vortex structure.

Large-scale vortex flow. Now the fluid descends and impinges onto the bottom. It rapidly turns on itself to become a continuum, and a circulation begins: Large scale vortices of large varieties may occur, three idealized ones are shown in Figure 3. Figure 3a shows a vortex ring the ring shaped axis of which is aligned parallel to the bottom. In Fig. 3b we see the sort of flow which would be generated from the flow in 1a and 2b after the fluid reaches the bottom. Its axis too, is parallel to the bottom. At this time it is not a ring but a fairly more or less straight line. The axis (represented as a point) may also be stretched out into a surface (broken line). Figure 3c shows the bathtub-type vortex which would generate if, e.g. the container were twisted while shaking. Vortices often become noticeable only when bubbles gather in (or move under centripetal force to) their axis. At the vortex axis they become larger, not only because they coalesce, but also because each bubble expands due to the suction (not pressure, as is the case with the stagnation points of the impinging flow) reigning along the vortex cores. Once particles are caught in a vortex they generally move along it. Unlike the impinging flow, particles do not flow through the vortex , but they may move along with the vortex motion. Let us now turn to the shear flow, which always comes along with a fluid in motion.

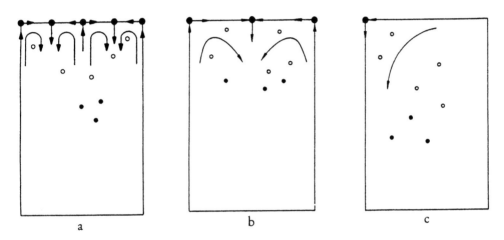

a   b   c

Fig. 2: Three possible saddle forms of the impinging flow resulting from those in Figure 1. The large dots are the stagnation points (which may be 3-dimensional lines) where the flow locally comes to a halt and has a local pressure maximum. The flow is deflected from these points along the wall (the "saddle" form). The small dots and open circles represent the bubbles and foam of the surface.

Shear. Figure 4 shows what happens when the mixture flows down the walls of the container. The closer the fluid is to the wall, the slower it moves. At the wall it rests due to viscosity. A marked blob (filled rectangle in Fig. 4) deforms as it is sheared by the decelerating forces at wall flow. Wherever the velocity changes from point to point in the fluid - in both saddles as well as vortices - we find a shearing flow. At the wall, however, we find it within a laminar flow. It causes no bubbles that may be found formed by suction in vortices, or formed by pressure in impinging saddle structures.

During further shaking cycles the shear at saddles merges into the vortex flow. With the succussion process the flow motion is repeated periodically. The turbulent vortex flow consists of myriads of vortices and saddles in random motion that effectively mix the drug with the water alcohol mixture.

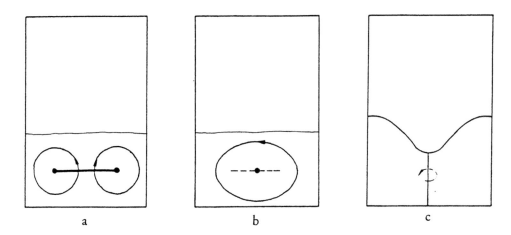

Fig. 3: In Figure 3a a vortex ring is shown, the ring shaped axis of which is parallel to the bottom. This sort of flow could result from motions shown in figures 2a or 2b after the fluid reaches the bottom. The flow in 3c, perhaps from motion 2b, turns about the straight line (the point) axis. The axis may be spread out into a surface (dashed line). The flow in 3c has its axis going from top to bottom and might result if the bottle were shaken with an additional twist. The surface is symbolic only.

Flow forms compared. We have already seen two differences between saddles and vortices: Vortices, with their centers having a suction maximum, carry essentially the same fluid with them as they move. On the other hand saddles, with their stagnation points, are associated with a high local pressure and particles constantly pass through them, giving them a more ephemeral existence. An interesting mathematical property, the Galilean invariance, emphasizes this last difference. It can be shown that saddles are not Galilean invariants, whereas the vorticity, a measure of the local rotation, is. Saddles will generally change their forms or even disappear, depending on whether the observer is at rest or in motion. Vortices yet depend on the motion of an observer. The instability, mentioned at the outset, has an even more ephemeral form. It contains the potential of flow without being such. Its sensitivity to the environment is shown by the fact that a completely different flow results from it, depending on the smallest accidental motion.

It is interesting that the most common way of sensing the speed of a flow – the so-called Pitot tube used in all airborne and waterborne vehicles – uses the saddle form as its basis. On the other hand vortices are not used to sense, but to mix, to lift and to carry: the vortex force drives ships, carries aircrafts, turns turbines and windmills. Vortices are the sinews, the muscles and even the voice of the flow: there is even a theorem on the effect that a frictionless flow can neither bring generate sound nor force. We thus see that saddles and vortices are as different from each other as the Tango is from the Waltz.

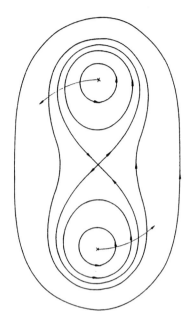

Fig. 4: The square transforms to a parallelogram under shear at a wall. The arrows of decreasing size represent the change in the velocity of the fluid. The flow is quickest distant from the wall, slowing down the closer it is due to the effect of viscosity. Fluid at the wall is at rest. The square, which deforms as it moves, is nothing more than an idealized cubic region of fluid.

2. Sound

Slamming the container onto a surface not only calls forth the flow properties of the fluid but also its inner elastic acoustic properties. If the flow impinges fast enough shock waves are generated in the fluid and these may be heard. When the fluid hits the bottom during the slam a shock-wave is generated and moves upward as shown in Fig. 5. Its speed is the speed of sound in the mixture, 1483 m/sec in pure water at 20°C (air dissolved in water reduces the speed dramatically, in extreme cases up to 20 m/sec). The wave begins as a compression wave (figure 5a), then, being reflected at the free surface, returns as an expansion wave (Fig. 5b), and reduces the pressure behind it to the original value. First on continuously being reflected at the bottom it leaves a suction behind it as it moves upwards as an expansion wave (Fig. 5c). If the suction reaches a critical value, the fluid is torn apart and cavitation bubbles are formed. The cavitation bubbles grow, vibrate and collapse as the shock wave propagates. The entire process lasts but a few milliseconds only.

An extreme situation is found during bubble implosion: gases and vapors inside the bubbles are compressed and heated. The typical temperature and pressure within the cavity may reach a thousand degrees and hundred atmospheres, respectively. Under such conditions liquid water transforms to H_2 and H_2O_2. If other substances are present, the reactions become more complicated and electromagnetic transitions may occur.

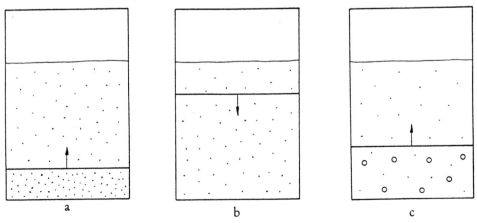

Fig. 5. For legend, see next page.

Legend to Fig. 5: Three phases in the motion of a shock wave generated by slamming the container on the bottom: 5a, the rising compression wave; 5b, the expansion wave after being reflected at the free surface and 5c, the expansion wave after being reflected at the bottom. Cavitation bubbles (circles) generate and collapse behind this last wave due to the the suction left behind it.

3. Diffusion

At the outset we mentioned the force of diffusion as a dissolving force. For even if the water seemed at rest, there is no such thing as absolutely quiescent water. At a microscopic scale it is in restless chaotic Brownian motion. Figure 6 shows the position of a particle in a colloid at intervals of 30 sec. This motion becomes the more intense the smaller the scale we look at is. The Brownian motion brings the substance to an equilibrium distribution. The average kinetic energy of the particles is a measure of the physical temperature of the system: Heat itself – controller of density, viscosity, conductivity, *inter alia* – lives in this chaotic motion.

Thus, although flow carries and stirs substances, the most dissolving mixing is due to diffusion. In the natural world the force of diffusion not only mixes, but is responsible for the stiffness of green plants (turgor pressure). It is also the force which brings plant-sap from the roots to the leaves of the tallest trees. A further interesting quality of diffusion is its speed, which seems to be so slow. Yet, theoretically, (Fick's law) if a concentrated substance is put into water then a trace of this substance will be found arbitrarily far away *immediately*: a trace of the substance moves with high velocity from a region of high concentration to a region of low concentration.

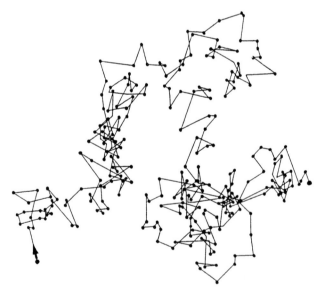

Fig. 6: Even quiescent water isn't really so. The jagged line represents the path of a single particle in a colloid recorded at 30s. intervals. The distance between the first two points is about 8mm (Perrin). The same force which drives Brownian motion also drives fluid in diffusion and is one with the heat of the fluid.

DISCUSSION
Comparison of the three motions

We have been able to identify three major forms of particle motion which occur during the dynamization process: shock wave motion, flow, and diffusion. Shock waves are propagating with sound velocity so that the fluid effectively stands still compared to the propagation speed of the wave. Note, in a shock wave there is no mass transport. The shock wave may cause electromagnetic transition, heat and chemical reactions in cavitation. Diffusion, on the other hand *is* mass transport, thus it may change with the specific soluted drug. Traces of the drug move very fast from high to low regions of concentration. Chaotic motion reigns at a microscopic level where the drugs characteristica can be seen Since diffusion is a mass transport it serves molecular reactions. Saddle flow and complicated surface instabilities are more ephemoral and associated with sensitivity and sensing. Vortices may be considered as the driving force of the flow. How close the process of diffusion is related to the vortex motion can be seen in the similarity of their mathematical formulation. Sound can be considered as an impinging motion which is that extreme that it squeezes the neighbouring fluid layers before they get a chance to mass transport.

Implications for dynamizing (potentising)

Looking first at the shock waves which may arise when the glass is slammed hard enough, it seems that the tendency to create new substances at each slam runs counter to the dynamization idea of taking a given substance through a dilution process. Whether substances generated during cavitation help healing or not is not the object of this article, but shocks certainly do not contribute to mixing or diluting. Diffusion is the force which really activates the all-penetrating quality of water. Although there is no flow the motion is effectively infinitely quick and almost down to the submicroscopic scale which defines substantiality itself. Yet here the chaos of diffusion meets with the order of substance: ordering substantiality battles against chaotic diffusion during potentisation in order to maintain its sovereignty (e.g. Pulsatilla must fight against the diffusion to maintain its Pulsatilla identity). And flow, with its myriad of saddles, shears and vortices, is diffusion's handmaiden, speed - up bulk mixture without distinguishing between concentration differences, without causing contraction or expansion.

REFERENCES

1. Hahnemann S. Organon of Medicine: English Translation of Organon der rationellen Heilkunde. Los Angeles: Tarcher 1982.
2. Suslick KS ed. Ultrasound. Its Chemical, Physical and Biological Effects. Temperatures and pressures in aerated water are substantially lower. Weinheim, BRD: VCH Verlag. 1988.

ULTRA HIGH DILUTION EFFECT AND ISOTOPIC SELF-ORGANISATION

A. A. Berezin

SUMMARY

This paper discusses the possible relationship between the homeopathic effect as a physical phenomenon and the fact that most chemical elements are mixtures of several stable isotopes. Isotopic diversity (isotopicity) of major elements provides an additional level of high-density information storage manifestable both in solid and liquid phases of matter. Informational aspects of isotopicity are discussed on the basis of classical and quantum physics. Along the classical line I review spontaneous pattern formation in systems far from thermodynamical equilibrium, application of chaos theory and catastrophe theory. On the quantum side I discuss localization and quantum coherency phenomena in which isotopicity serves as a pattern-forming and triggering agent for a quasi-random chain of reductions (collapses) of a macroscopic wave function. Furthermore, vistas with special importance for the subjective factors in a homeopathic protocol are discussed.

ZUSAMMENFASSUNG

In diesem Beitrag werden mögliche Zusammenhänge von homöopathischen Effekten und physikalischen Phänomenen beschrieben, deren Grundlage das gleichzeitige Vorkommen verschiedener Isotope in chemischen Elementen ist. Diese isotope Diversität von chemischen Elementen liefert eine bedeutsame Variante der Informationsspeicherung in fester und flüssiger Materie. Der Bezug von Isotopizität und Information wird auf der Basis klassischer und quantenmechanischer Physik diskutiert. Im klassischen Bild wird die spontane Musterbildung in Systemen abseits des thermischen Gleichgewichtes beschrieben. In diesem Zusammenhang werden hier auch Anwendungen von Chaos-Theorie und Katastrophen-Theorie diskutiert. Im quantenmechanischen Bild werden die Lokalisierung und das Koherenz-Phänomen diskutiert, wobei besonderer Bezug auf die durch Isotopizität hervorgerufene Musterbildung und Stimulierung von quasi-Zufallsketten von Zusammenbrüchen in makroskopischen Wellenfunktionen genommen wird. Weiters werden Bereiche mit spezieller Bedeutung für die subjektiven Faktoren in homoeopathischen Protokollen diskutiert.

P.C. Endler and J. Schulte (eds.), Ultra High Dilution, 137–169.
© 1994 *Kluwer Academic Publishers. Printed in the Netherlands.*

INTRODUCTION

This paper discusses possible links between the homeopathic effect and the phenomenon of isotopicity. The latter term is used to designate the fact that the majority of chemical elements are actually mixtures of two or more stable isotopes. Here we discuss at length several arguments on possible connections of isotopic diversity and the homeopathic action. In this context the latter is treated largely (but not exclusively !) as a memory effect in water sustainable in a limit of infinite dilution.

[...] It is not my agenda here to discuss [issues like the] Benveniste saga (Davenas et al., 1988) as they have appeared from the published material over the last few years. Furthermore, regardless of the outcome of Benveniste's claims, one likely runs into a serious scientific risk if the issue of medical validity of the entire lore of homeopathy is put into an one-to-one correspondence with a possible confirmation or refutation of experimental results reported by a particular research group. After all, the established scholarship of homeopathy has already existed for many generations of its practitioners and clients.

Instead, my aim here is to discuss a possible model of a homeopathic effect in terms of what can generally happen within the boundaries of the present-day physical paradigm. Of course, it can be said that even if the suggested scheme of a certain effect is functional, self-consistent and does not violate any accepted principle of physics, this still provides no guarantee that it describes an actual physical reality. However, the very fact of the existence of an *operational model* for the alleged effect might be a stimulating factor for more focused experimental studies. A quote ascribed to the Nobel prize physicist Murray Gell-Mann says: "*Anything which is not prohibited, is compulsory*" (cited by Comorosan, 1974). This implies that under the proper circumstances it is very likely that Nature will find a way to utilize almost any conceivable physical scenario. In what follows I will demonstrate that a combination of isotopicity and homeopathy does indeed provide a viable candidate to exemplify Gell-Mann's dictum mentioned above.

The present article further develops a hypothesis suggesting possible links between the mechanism of homeopathic action and isotopicity (Berezin, 1990a, 1991a). The reader is advised to keep in mind that the use of such words as "mechanism" or "model" in the present text should not be taken as my adherence to an ultimate mechanistic (or reductionistic) explanatory model for homeopathy. On the contrary, the following presentation pays significant attention to a possible participation of mind-matter interactions in homeopathic effects. Some physical aspects of these interactions were discussed in recent literature (Berezin, 1990b, 1990c, 1991a, 1992a; Berezin and Nakhmanson, 1990; Burgers, 1975; Germine, 1991; Miller, 1991; Jahn and Dunne, 1986; Radin and Nelson, 1989; Penrose,

1989a; Stevens, 1990; Wolf, 1989). As I will describe below, isotopicity is capable of providing a unifying framework to connect an organism level to atomic-molecular and, perhaps, even to nuclear and sub-nuclear levels. Therefore, in terms of transcending restrictions of a narrow reductionism, the present article emphasizes a working dichotomic alliance between the reductionistic and holistic approaches to the phenomenology of a homeopathic effect.

In brief, the essence of my hypothesis of relating the homeopathic effect and isotopicity is that the distribution of stable isotopes in water or other substances can play the role of an information-carrying pattern. Such patterns are capable of preserving and perhaps even amplifying the information on chemical nature and medical functioning of the originally presented substance ("drug") upon subsequent dilution steps. Our proposed scenario also includes several additional aspects discussed below. The point to emphasize at start, however, is that I do not offer my model of isotopicity as "the only true physical model" of a homeopathic effect which would deny all other alternative hypotheses focused on factors other than isotopic diversity of chemical elements. On the contrary, my model is quite "open-ended" and allows for a large degree of interactive compatibility with a number or earlier theoretical ideas some of which have been recently reviewed by Beverly Rubik (Rubik, 1989, 1990).

As such, the following presentation emphasizes a "multidimensional" approach to the homeopathic effect and looks at it from several viewpoints based on concepts of classical and quantum physics. This article should not be seen as an authoritative "final synthesis" of all seemingly relevant physical ideas into a full-fledged universal theory of the homeopathic effect. Without claiming such a monumental task at the present stage, I will nevertheless demonstrate that a single physical phenomenon (isotopicity) turns out to be capable of serving as a viable connecting factor between several different facets of classical and quantum physics and the marvel of the homeopathic action as the latter is attested by the personal experience of many thousands of people.

MAIN PART

1. The Phenomenon of Isotopicity

Isotopes are atoms of the same chemical element having a fixed number of protons but different number of neutrons in their nuclei. Correspondingly, their atomic weights (A) are distinctly different already in a crude sense - even for quite heavy stable elements with A > 200 the addition of just one extra neutron adds up about 0.5 % of a mass - a clearly "macroscopic" increase by all quantitative standards. For lighter elements, central to Earth's biology (carbon, oxygen, nitrogen, sulfur) isotopic mass variations amount to several percent, while for a hydrogen-deuterium case it reaches 100 %. Besides weight differences, isotopes of the same element may differ in the spin and magnetic moments of their nuclei, have different interactions with other elementary particles (e.g. neutrons or gamma-rays), etc. Similarly, in terms of macroscopic characteristics, compounds which are identical chemically but vary in isotopic composition, have slight but usually noticeable differences in their properties. Isotopic variations are known for melting temperatures, atomic diffusion, chemical reaction rates, and many other physical and chemical properties (Jancso and van Hook, 1974; Klein, 1975; Willi, 1983; Berezin, 1991b, 1992b). For example, freezing points of "ordinary" and "heavy" (deuterated) water differ by as much as 4°C. Similar isotopic variations exist for viscosity, molecular relaxation times, atomic hopping rates, etc. It is quite likely that some of these characteristics may, in principle, affect the homeopathic protocol in a number of conceivable ways (e.g. through the alteration of optimal dilution ratios, influencing the dynamics of successions, etc.). This argument alone is sufficient to warrant the bringing of isotopicity to the attention of theoretical homeopathy.

Another remark on isotopes may also be of importance in terms of linking isotopes and Earth's biology (Berezin, 1990b). With an exception of phosphorus (P) and sodium (Na), all elements critical of the phenomenon of biological life are poly-isotopic elements. For instance, carbon and nitrogen have 2 stable isotopes each (^{12}C, ^{13}C and ^{14}N, ^{15}N), oxygen has 3 (^{16}O, ^{17}O, ^{18}O), etc. It is interesting to note that by some sort of a "coincidence", the main elements of organic life on Earth and their isotopes (H, C, N, O) are the same elements which, together with several of their isotopes, are essential for the Bethe-von Weizsäcker "carbon cycle" in the Sun which generates the life-vital Sun energy in the first place! Thus, chemical and energetic aspects of life may indeed have common sources pointing to the direction of isotopic diversity of underlying elements.

2. Isotopic Individuality and Isotopic Freedom

In this section we will discuss the classical distinguishability of isotopes and relate this issue to "isotopic individualization" of atomic-scale structures due to variations in isotopic distribution. This concept leads to a notion of "isotopic freedom" which, in turn, will be related in the next section to a propensity of isotopes to form correlated or "informationally-loaded" patterns.

Among the most important and commonly held assumptions of modern physics is the principle of quantum indistinguishability of elementary particles. Within the conceptual framework of the present-day quantum physics, identical elementary particles (electrons, protons, neutrons, etc.) are supposed to be "exactly" the same. They are not allowed to carry any "personal labels" and, correspondingly, the interchanging of any two particles of the same category (e.g. two protons) does not lead to any observable physical changes. Therefore, physically speaking, such permutation is a "non-event". Similar indistinguishability applies to identical atoms and molecules.

This "sameness" of elementary particles is usually taken for granted and is supported by authoritative philosophical arguments (e.g. Popper and Eccles, 1977, p.71). Nevertheless, at its very foundations the principle of quantum indistinguishability is far from trivial (Berezin and Nakhmanson, 1990). Although at present, there seems to be no strong experimental or theoretical motivation for a radical rejection of this principle, there are some continuing on-going attempts to challenge it or at least to suggest some amendments to it. At the present moment, any admission of individual characteristics of elementary particles is likely to require a major revision of basic principles of quantum physics. Nevertheless, such a seemingly odd and fancy trend as to look for possible "individual" features of microscopic particles culturally (and even to some extent scientifically !) is quite understandable. Indeed, almost all "serial" entities which we know from our direct experience have individual variations. People or animals of the same biological species are never exactly the same. So are all other tangible and untangible, natural or man-made objects, from snowflakes or hamburgers to stars and galaxies. Not even two of them are ever exactly identical. As a result, it might be even quite human-natural to feel some kind of antipathy towards the theories which deprive elementary particles of their "individual rights".

An example of such attempts to attribute some sort of personal labels to atomic particles is a long-standing issue of "hidden parameters" initiated by David Bohm many years ago (Bohm, 1952). In a rather crude and largely metaphorical sense hidden parameters can be portrayed as changeable individual features of particles. In the last decades the issue of the existence (or not) of hidden parameters was put into a relationship with the so-called Bell's theorem. Numerous recent experiments on quantum correlations seem to suggest that hidden variables could not be consistently kept in a theory as local (particle specific) labels and the widespread trend now is to interpret these experiments as confirming the existence of quantum non-localities. This change of emphasis, however, does not discard the issue of the "individuality" of particles, but rather moves it into another domain related to non-local inter-particle connectivity.

At present, the whole matter of quantum non-localities remains a hot scientific topic with a high likelihood of unforeseen twists in forthcoming experiments and interpretations. Fortunately, however, in a context of isotopic diversity and isotopic patterning, an issue of proper interpretation of hidden variables and non-local quantum correlations is relatively peripheral. Isotopes as classically distinguishable particles are quite robust ("insensitive") to these issues. Isotopicity, to some extent, breaks down the above described indistinguishability of identical atoms. Contrary to a quantum indistinguishability of identical micro-species, nuclei of different isotopes are classically distinguishable particles (the word "classically" means here that they are different entities even from the point of view of classical physics). Therefore, structurally identical micro objects (molecules, atomic clusters, etc.) which, however, have different isotopic configurations do indeed attain some degree of "individuality". Isotopes can be freely moved and rearranged within chemically fixed structures and this leads to microscopically distinguishable patterns. Furthermore, such isotopic pattern can be informationally loaded, i.e. it can store some externally input information which will be coded in a specificity of isotopic distribution (Berezin, 1992b). It is important to note that "isotopic freedom" (the possibility of isotopic permutations between the sites of the same chemical element) exists within the constraints of a chemically fixed (in terms of given chemical bonds) structure. The latter remark not only applies to solid crystal lattices but also, to a certain degree, to quasi-crystalline structures of liquid systems which are relevant to homeopathy.

3. Self-organizational Propensity of Isotopicity

The idea of spontaneous pattern formation in non-linear systems was promoted by a milestone paper by Alan Turing (Turing, 1952). He showed that an originally uniform system of identical cells, coupled by a non-linear interaction, may undergo a *spontaneous symmetry breaking* (Peierls, 1991) and, as a consequence, develop a non-trivial, spatially non-uniform, pattern. The ideology of this phenomenon, also known as "Synergetics", was later extensively elaborated by schools of Prigogine (Nicolis and Prigogine, 1977) and Haken (Haken, 1978) and numerous other explorers. These developments are closely related to the modern theory of chaos. Another important recent addition to the arsenal of these ideas is the theory of self-organized criticality suggested by Per Bak and co-workers (Bak et al., 1988). Expressing it in a simple way, the latter theory demonstrates that in some circumstances dynamical systems spontaneously evolve to the least (not to the most!) stable configurations. All these concepts may bear some (perhaps variable) relevance to a dynamics of the homeopathic effect.

One specific example of the spontaneous pattern formation especially worth mentioning in a homeopathic context is the so-called oscillatory patterning in minerals (Allègre et al., 1981). Characteristic features of oscillatory patterning in natural crystals may vary over several orders of magnitude - from macro scales (i.e. centimeters or even meters) to micrometers (and, probably, down to nanometers, i.e. to atomic scales). A similar patterning sustainable over a wide range of scales might be a characteristic feature of a robustness of a homeopathic efficacy upon dilution. We will return later to the reported concentrational oscillations in the homeopathic effect (Davenas *et al.*, 1988). The point of importance for us here is that (contrary to a popular belief) an oscillatory zoning in minerals is not always caused by varying sedimentation conditions, but may also (at least in some instances) be interpreted as a spontaneous pattern formation in a system governed by non-linear kinetic equations. In the context of the present article it is especially interesting to note that similar spontaneous microscopic patterning has been recently reported for isotopic distributions (Wada, 1988). Figure 1 illustrates a potential richness of microscopic isotopic ordering even for simple one-dimensional structures with just two isotopes. It is clear that two or three- dimensional structures, especially those which contain more than one poly-isotopic element, have a much greater degree of isotopic positional flexibility and as a result may generate an enormous number of distinguishable isotopic patterns.

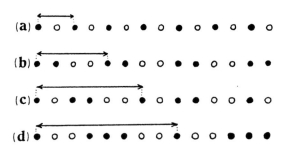

Fig. 1: Several examples of isotopic ordering for a one-dimensional linear chain for two isotopes with equal abundances. The lattice pe-riod for cases (a) to (c) is equal to 2, 4, 6, and 8 units, respectively.

Generally, positional isotopic correlations in various condensed matter systems may originate by several alternative (or complimentary) scenarios. Additionally to a spontaneous patterning in non-linear non-equilibrium systems of distinguishable species due to mass differences, isotopic variations in diffusion rates can also result as a consequence of the Pauli exclusion principle. This principle governs the structure of electronic shells in atoms and is one of the foundations of the periodical law of chemical elements. Quantum statistics divide all particles into two categories - fermions and bosons. Bosons have an integer spin, obey the Bose-Einstein statistics and disrespect the Pauli principle. The fermions have a half-integer spin, obey the Fermi-Dirac statistics and follow the restrictions on the occupation of quantum states imposed by the Pauli principle. It is peculiar that different isotopes of the same element may obey different quantum statistics. For example, oxygen isotopes ^{16}O and ^{18}O have even a number of nucleons and therefore behave as bosons, whereas isotope ^{17}O with an odd number of nucleons is a fermion. This discrimination by statistics among the atoms of the same element may result in further (additionally to the mass-effect) isotopic variations of tunnelling rates and lead to a build-up of spatial isotopic non-uniformities of a varying size.

A ubiquitous presence of self-organization phenomena in non-linear systems far from equilibrium (Nicolis and Prigogine, 1977; Haken, 1978; Yates, 1987) can be interpreted as an illustration of a general propensity of Nature for pattern formation. Of course, it is noticeable that attributing to the Nature of some sort of inherently built-in quest for patterns (Berezin, 1989) can be suspected as an approach to a panpsychistic philosophical platform. This is indeed likely so, but I do not apologize for emphasizing these links. Just a few years ago even a very carefully veiled panpsychistic tilt would almost unanimously have been considered as a highly weakening (if not outright defeating) feature of any physical theory pretending to a serious scientific attention. In my opinion this situation is beginning to change gradually now, especially in those areas of theoretical physics which deal with quantum foundations of consciousness, mind-matter interactions, universality of consciousness, general theory of adaptive systems etc. Despite a wide variety of opinions and the existence of radical disagreements even within this (still quite narrow) group of researchers, some of these recent developments may be potentially far too important for a theoretical basis of homeopathy to be ignored at this time. Therefore, I will give later (in Section 13) a brief account of these ideas, as some of them can be fitted into the context of the isotopicity scheme.

4. Informational Aspect of Isotopicity

Another aspect of isotopicity with a possible importance for the homeopathic effect is a very high density of isotopic information storage. This density is, in principle, of the same order of magnitude as atomic density (save, perhaps, one or two orders of magnitude to assure the robustness and redundancy). This amounts to about 10^{20} bits per cubic cm (Berezin, 1992b). Biologically speaking, informational potential of isotopicity can provide an alternative source of randomness, irreproducibility and individuality in a living matter (Berezin, 1988, 1990c). This source of randomness acts as a freedom-enhancing factor due to random and non-repeatable combinations of various isotopes of the same chemical element within the functionally critical biological structures carrying the genetic information. As was mentioned above, there is a significant freedom of isotopic combinations within structural limitations of ordinary hetero-atomic chemistry.

In view of this, isotopicity can be seen as the most obvious (in terms of its ubiquitous availability) natural challenger to a chemical determinism. If seen as a kind of "shadow chemistry", isotopicity is capable of greatly enhancing the informational content of matter per atom or a unit volume. Using an anthropomorphic metaphor and reapplying the sentiments of Section 4 to a common organic biology, one can say that the living Nature should likely be "tempted" to try to take advantage of an ample informational resource provided by the diversity of stable isotopes within the limits of chemical specificity. In this connection, it is worth noticing that even in the framework of ordinary hetero-chemical biology, Nature was smart enough to find some use for almost all members of the periodical Table - additionally to major bioelements (H, C, N, O, S, Na, K, etc) most other elements also have some specific functions in living organisms. Many of these elements have very low relative concentrations and are known as "microelements" (e.g. chromium, manganese, nickel, copper, etc.). Furthermore, in terms of its capacity to store and transmit the information at atomic level, isotopicity to some extent breaks a barrier between a highly organized living matter and (allegedly much lower organized) so-called non-living matter, be it non-organic crystal or liquid water.

As far as information is concerned, isotopically mixed substances can be quantified in a number of ways. A straightforward way will be to use the Shannon index originally introduced in a theory of information capacity of communication channels. Its use for a quantification of isotopic diversity was suggested earlier (Berezin, 1991b) in a manner similar to Shannon entropy. Here we re-define Shannon indices in a somewhat different way than it was done in the paper just quoted. Here we introduce Shannon index, H, for a given chemical element in the form:

$$H(z) = -\sum_i p_z(i) \ln p_z(i)$$

$$(1)$$

Here Z is the atomic number to designate the element, the sum is taken over all stable isotopes of a given element and $p_z(i)$ is the percentage abundance of the i-th isotope in a given substance; "ln" means a natural logarithm. For example, for an oxygen $Z=8$ and the natural abundance of its 3 isotopes is $p_8(i) = 0.99756, 0.00039$, and 0.00205 ($i= 1,2,3$).

The equation (1) then produces $H(8) = 0.01819$. It is clear that for all mono-isotopic elements $H = 0$ (as for them $p = 1$ and $\ln 1 = 0$).

At first sight isotopic information storage appears to be positionally inflexible and non-dynamic. In order to argue otherwise, let us consider two different isotopes occupying neighbouring sites in a solid or molecular structure. Take both sites to be energetically equivalent. Permutational exchange of two isotopes between both sites is possible as a quantum mechanical tunneling process (quantum ring diffusion) with a characteristic time T. The rate of such processes $1/T$ can be written in a way similar to the case of a radioactive decay. In common symbols it is a product of an exponentially small tunnelling factor and an "attempt frequency" f of the quantum tunnelling:

$$\frac{1}{T} = f \exp(-S) \tag{2}$$

where S is some proper combination of the height and the width of the potential barrier.

The meaning commonly ascribed to Equation (2) is statistical: at every attempt the system has an equal probability, $\exp(-S)$ to experience a quantum transition (here: a tunneling passage of the potential barrier). In the case of tunnel exchange of sites by two isotopes, an attempt frequency is defined by the Heisenberg delocalization rate. Figure 2 gives a simple explanation of this latter notion. It is a consequence of the uncertainty (Heisenberg) principle which asserts that any particle confined in a region of a size x has a velocity spread of an order of $\hbar / Mx$ ($\hbar$ is the Planck's constant, M is the mass of the particle). This means that any confined particle has some non-zero positive kinetic energy.

Therefore, like a prisoner in a cell, the restrained particle tries to escape (diffuse away) from the confinement region.

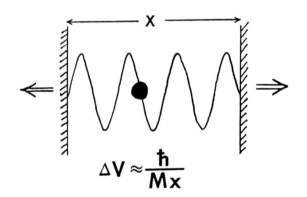

Fig. 2: Heisenberg delocalization effect of a confined quantum particle. If the walls of the potential well have a finite height and thickness, the particle "tries" to escape using a finite smearing (spread) of its velocity. These "escape attempts" are shown by left and right arrows.

For an atom of a mass M (typically, M is about 10^{-25} kg) Heisenberg delocalization rate is of an order of $x^2 M/h$ (x is a characteristic interatomic distance of, say, 3 Å). This leads to an attempt rate comparable to typical frequencies of phonons (atomic vibrations) which are about 10^{12} sec^{-1}. Such values mean that isotopic rearrangements can proceed very fast provided that a certain physical condition can be imposed on a system to enhance the tunnelling factor (even for a short time) from exponentially small values to values comparable to unity. Such a condition can be seen as an analog to the establishing of a coherent resonance tunnelling activity in the entire system (overall quantum transparency). In "ordinary" (non-coherent) conditions such enhancement normally does not happen and exceedingly small tunnelling factors drastically suppress the observable isotopic exchange rates at low temperatures despite a quite high Heisenberg delocalization rate. The model entertained in the present article suggests that a build-up of a coherency and the presence of quantum non-localities can affect the rate of tunnel rearrangements of isotopes and this mechanism, in turn, can be linked to the informational patterning in the material constituents of the homeopathic protocol.

5. Isotopic Structuring and Homeopathic Templating

When looking at the homeopathic effect as a physical phenomenon it is desirable to explore the following issues:

(a) how a specific information on a chemical nature and biological action of a particular chemical substance (drug) is "imprinted" (or "templated") on a diluting matrix (e.g., water), and

(b) how this information can withstand a thermal disordering and, moreover, how it can proliferate into subsequent dilution stages without a proportional loss of its efficiency (such loss should be expected on the basis of a common-sense chemistry).

To address both these issues it is important to demarcate the energetic and informational aspects of a physical situation. The following remarks concerning the notion of information may serve as an introductory basis for our further discussion.

The concept of information does not comfortably fit into the family of physical categories. Despite a relatively advanced stage of the modern information theory based on the work of Leon Brillouin, Leo Szilard, Claude Shannon, Rolf Landauer and many others (see e.g. Leff and Rex, 1990; Landauer, 1991; Atmanspacher and Scheingraber, 1991), the use of information as a physical term often remains somewhat ambiguous. It has a multiplicity of meanings and interpretations. Despite a common notion of information as a negentropy (Wheeler, 1988), not in all instances the information can be easily quantified or put into a one-to-one correspondence with an entropy of a system. One of the reasons for the present difficulties of the physics-based approaches to a homeopathic effect may likely be attributed to this unspecifity of terms in the physics of information. This situation is further exacerbated by an over-emphasis of energetic aspects of systems, typical of the contemporary mainstream physics, as, for instance, the trend of over-universalization of a minimum energy principle.

I certainly do not attempt to resolve all the above issues regarding the physics of information in the present article. Nevertheless, the following confronting of energetic and informational sides of the physical reality may provide some useful guidelines for further thinking about the informational aspects of the homeopathic effect. In a somewhat simplified form the following pair of equations:

$$E + E \rightarrow 2E \quad (3)$$

$$I + I \rightarrow I \quad (4)$$

illustrates the "dichotomic split" between energy E and information I in quite a general way.

Equation (3) symbolizes the *additivity* of energy as a physical quantity (the law of energy conservation). There is, however, no similar "conservation law" for the information. Instead of additivity, the information has a property of *redundancy* symbolized by the Equation (4). Two city maps, even edited differently, are essentially no more useful than just one, as they contain the same (or largely overlapping) information.

Even more noteworthy is to re-read these equations in the opposite direction:

$$E \quad \rightarrow \quad \frac{E}{2} + \frac{E}{2} \qquad (5)$$

$$I \quad \rightarrow \quad I + I \qquad (6)$$

In the last pair of equations the additivity of energy [Eq.(3)] translates into its divisibility [Eq.(5)] while the redundancy of information [Eq.(4)] translates into its duplicability [Eq.(6)]. This ability of information for the unbounded proliferation ("cloning") constitutes the basis of almost all biological activity (DNA duplication) as well as all informational technology (printing, xeroxing, TV broadcasting, etc.). A specific (but important for the purpose of this article) chemical process with informational features is a catalysis. A classical catalyzer changes the reaction rate of a chemical reaction but is not wasted itself in the course of the reaction. In a sense, a catalyser provides an algorithm ("blueprint") for the chemical reaction instructing it how to proceed faster. This is a typical informational action and, in the case of homeopathy, its likely possible parallel is the process of informational duplication in the course of homeopathic dilution (potentisation, s. Glossary, p. 221).

From the reported oscillatory behaviour of the drug efficiency as a function of dilution (Davenas et al., 1988) and the critical role of "succussions", which are normally seen as an essential part of the homeopathic protocol (Rubik, 1989), one can further deduce that this information has even some capability of "self-amplification" under a particular set of external conditions. This self-iterative amplification can apparently be related to the fact that a strong shaking action drives the system into a non-equilibrium state with an excess of free energy. As we have already mentioned, states far from equilibrium are especially prone to pattern formation.

An isotopic model of a homeopathic templating can be outlined in the following way. Let us assume, for specificity, that an information-carrying matrix is represented by pure water (addition of ethanol does not change the principles of our model). In the case of a H_2O-matrix there are 3 isotopic degrees of freedom (H to D and ^{17}O or ^{18}O to ^{16}O) and the concentration of minority isotopes is not at all negligible: e.g., for ^{18}O it is 0.2 % (1 part per 500) which corresponds to only about 8 inter-atomic spacings ($500^{1/3} = 8$) for the average separation between two neighbouring ^{18}O atoms. Atomically speaking, this is a very short distance.

The very possibility of different positional organizations of minority isotopes (D, ^{17}O, ^{18}O) within the main H_2O-matrix leads to an enormous degree of "isotopic redundancy" of potentially available isotopic patterns. For the sake of a vivid metaphor I earlier named such information-carrying patterns "isotopic lattice ghosts" (Berezin, 1991a). The first link in our chain of hypothesizing is to postulate that the presence of certain molecules (e.g. antibodies) invokes some specific readjustments in a positional distribution of minority isotopes in a vicinity of this "seed" molecule. In the spirit of quantum mechanical analogy, we can describe this process as a "focusing down" of a random (and highly degenerate or redundant) isotopic distribution into a particular isotopic lattice ghost. An analogy will be a spread of a specific contagious disease (e.g. through a virus, which is essentially an information-carrying DNA) during the epidemics. Spontaneous symmetry breaking paradigm (Section 4) provides a unifying methodology to describe such effect of "contagious" information proliferation in physical terms. It is known that the spontaneous symmetry breaking in a given system results in a (generally unpredictable and probabilistic) choice of a particular low-symmetry state out of a highly degenerate manifold of possibilities. A known textbook example of this is a spontaneous magnetization of a ferromagnetic material below the Curie temperature. The original (non-magnetic) state has a spherical symmetry (isotropic) as the directions of individual atomic magnetic moments are distributed randomly and have no preferential direction. Below the Curie temperature atomic magnetic moments align along particular direction, i.e. the system loses ("breaks down") its original spherical symmetry to a lower axial symmetry.

6. Isotopic Patterning and Quantum Transitions

The discussion of the previous section was based on the understanding of information within the framework of classical physics. In this section we will look at some quantum analogies of information cloning. Let us recall the features of the measurement process in quantum mechanics. According to a widely held (mainstream) interpretation of quantum mechanics, the measurement of a particular physical variable in a state with the wave function ψ of a quantum system

$$\psi = \sum_k c_k \varphi_k \quad ; \quad \sum_k |c_k|^2 = 1 \tag{7}$$

produces "the reduction" (focusing down) of the superposition ψ to one particular term, which is φ_k. This term is an eigenstate of the operator of the variable which is measured. The choice of a particular eigenstate is a probabilistic selection with an amplitude c_k. But once the reduction (measurement) is done the system is "locked" into the eigenstate φ_k and all subsequent measurements will bring the same result (i.e., once it has been prepared, the state will not change further at the repetitive measurements of the same variable).

There are several possible ways of how a similar state selection could work for a dynamics of the process of isotopic ordering, i.e. how the information on a dissolved molecule can be templated into a positional arrangement of isotopes. A clear conclusion would be rather premature at this stage. Yet, as a possible key to the mechanism of the formation of isotopic ghosts I will mention here the differences in bonding energies for different isotopic combinations for atomic pairs like $^{16}O-^{16}O$, $^{18}O-^{18}O$, etc. Another factor possibly contributing to an isotopic arrangement comes from isotopic variations of collisional cross sections. In any case, the result will be the appearance of some variations in the statistics of actual isotopic combinations in comparison with purely isotopically-random arrangement. It should be noted at this point that the functioning of a dynamical isotopically correlated cluster ("isotopic ghost") as an information-bearing structure does not necessarily assume the full-fledged isotopic patterns. Due to an enormous information storing density (redundancy) of isotopic combinations even just slightly traceable, immature, isotopic correlations (indeed, "ghosts" !) could already be capable of doing the job - to provide the channelling (proliferation) of information into the next stage of a dilution process.

7. Symmetry Breaking and Informational Self-Stabilizing

Spontaneous symmetry breaking in a physical system results in a decrease of its entropy and is equivalent to a generation of information within this system. The next important link in this chain of arguing is to indicate some mechanism(s) by which isotopically coded information could be protected from thermal disordering and, moreover, multiplied ("xeroxed") in the course of the dilution process. In this regard I would like to indicate two possible options.

The first option is that the imposition of a weak perturbation on a dynamically degenerate system prone to a spontaneous symmetry breaking often "dictates" the choice of a pattern to which the system will eventually arrive ("degenerate" means that a system has a variety of physically distinguishable configurations with the same total energy). For example, if a system which has several phase modifications is in the process of crystallization then (under some specific conditions) the placement of a micro crystalline seed of a particular modification could be "contagious". The entire crystal will grow in this particular crystal-lographic phase. This has some analogy with the "informational" chemical action of a catalyzer mentioned above (in Section 6).

The second option for a mechanism of a "locking-in" of isotopic patterns is a possible participation of *polarization effects*. It is known that the ionic polarizability is mass-dependent (the vibrational frequency is proportional to $M^{-1/2}$, M atomic mass) and, therefore, it is a rather sensitive effect in terms of local isotopic variations. Some positional combinations of isotopes could enhance local values of the polarizability of the medium. The deepening of the polarization potential wells can work as a stabilizing factor similarly to the polaronic self-stabilization in ionic crystals. One illustration of such an effect is a spontaneous symmetry breaking in a system of two neutrals in a polarizable medium (Berezin, 1983). For example, a system of two originally neutral gold centers in a silicon crystal may convert itself spontaneously into an asymmetric state:

$$Au^{(o)} + Au^{(o)} \rightarrow Au^- + Au^+ \tag{8}$$

and, once selected, such ionic configuration remains stabilized by the polarization interaction.

This reaction is an example of a spontaneous symmetry breaking by a mechanism of electronic charge transfer. Here $Au^{(o)}$, Au^- and Au^+ indicate a neutral atom, negative ion and a positive ion, respectively. The meaning of this process is that, contrary to a common intuition, the most symmetric state (both atoms are neutral) is not always the one which has the lowest energy. In more involved cases of isotopic clusters, a high level of degeneracy of possible isotopic configurations renders the energy minimization criteria almost irrelevant in defining the final ("target") isotopic pattern.

The pattern selection primarily proceeds on informational grounds as considered above. However, once the "choice" of a target isotopic pattern is made (by a route of either classical or quantum dynamics), the polarization effects can *energetically* stabilize the isotopic configuration and assist it in withstanding the thermal disordering.

For the purpose of the following discussion it is important to note that this polarization stabilization can be quantum mechanically accounted for by a non-linear Schrödinger equation (Berezin and Evarestov, 1971). Using one-electron approximation, the corresponding Hamiltonian in atomic (Hartree) units has the form (Berezin, 1973):

$$\hat{H} = -\frac{\nabla^2}{2} + V(\vec{r}) - \sum_k \frac{p_k}{2} \left| \vec{\nabla}_k \int \frac{\varphi^2(\vec{r}) dV}{\varepsilon |\vec{r} - \vec{R}_k|} \right|^2 \tag{9}$$

where the first two terms are ordinary kinetic and potential energy terms and the last term is the *polarization potential* describing the "back action" of a given ("designated") electron with an orbital φ on electronic clouds of the surrounding atoms. In Equation (9) p_k is the polarizability of k-th atom located at a position with a radius-vector $\vec{R}_k$ is a dielectric permitivity of a medium and ∇_k is a gradient operator in respect to the radius-vector $\vec{R_k}$. The squared expression in the last term of Eq.(9) is, actually, the strength of an electric field of a designated electron at the location of the k-th atom. A possible importance of the homeopathic information processing of the above polarization term is that it features both quantum non-linearity and quantum non-locality. Non-linearity enters through the dependence of the potential field itself (pseudopotential) on the state of an electron, the integral over the electron's coordinate reflects the non-local character of this action. The latter property (non-locality) is often expressed by the use of the term "pseudopotential" to refer to such self-consistent (Hartree-Fock) potential fields.

What could be the possible homeopathic implications of these non-linear and non-local effects reflected in the polarization term of Eq.(9) ? The presence of the integral of φ generally implies some sort of informational connectivity of all points in a system. This provides one (but not necessarily the only !) possible mechanism of a holographic-type information storage. As the atomic polarization is sensitive to the mass variations, the effects of isotopic diversity may naturally contribute to the dynamics of informationally mediated microscopic patterning. Subsequent sections will discuss further aspects of non-linearity and non-localities of isotopic dynamics in the homeopathic effect. As a last remark in this section, I will bring up some kind of similarity between the above notion of a polarization pseudopotential and the notion of *quantum potential* introduced by David Bohm (Bohm, 1952; Bohm and Hiley, 1975). Although quite different conceptually, both these constructions directly appeal to the effects of quantum non-localities and suggest the possibility of an overall informational linkage in quantum systems.

8. Catastrophe Theory and Isotopic Informational Dynamics

Catastrophe theory (CT) invented by René Thom has later been successfully applied to a wide range of problems in fundamental and applied physics and other areas of science and technology (Zeeman, 1977; Stewart, 1982; Berezin, 1991c). It was used for areas so drastically distant from each other as, e.g., light focusing in crystals, behaviour of financial markets and theoretical linguistics. The reason for such a universal attractiveness of CT for researchers in so different areas lies in its easy-to-grasp ability of a pictorial representation of processes in systems of various degree of complexity. Models based on CT combine strong sides of qualitative analysis (retaining only the major features of a complex system or a process) with a capability to offer some quantitative relationship between the governing parameters. The latter is especially important for identifying and describing critical regimes, e.g. regions of parameters near the thresholds of phase transitions and other jumpwise changes of a symmetry or a dynamical behaviour of a system. Historically, CT preceded the modern theory of chaos (Gutzwiller, 1991) but it also has mutual grounds with the latter in a communality of their ideological basis (the importance of non-linearity) and in emphasizing the discontinuity aspects of a dynamical evolution.

This and the next sections will outline some relevances of CT-based description of a homeopathic effect. In the present article I am going to discuss almost exclusively isotopicity-related aspects of homeopathic processing. The following comments can be related to the use of CT for the featuring of (a) dynamics of isotopic patterning and (b) the effect of polarization stabilization of isotopic patterns discussed in Section 8. However, apart from our isotopicity hypothesis, CT may also provide a useful illustrative tool for a number of non-isotopic explanatory theories of homeopathy, e.g. for models based on cavitation effects in water etc. Figure 3 (next page) shows the most frequently used cusp catastrophe surface of CT. Skipping some details of CT-parametrization (for more extended discussion on CT see, e.g., Zeeman, 1977; Stewart, 1982; Berezin, 1991c), the general meaning ascribed to this surface can be explained in the following way.

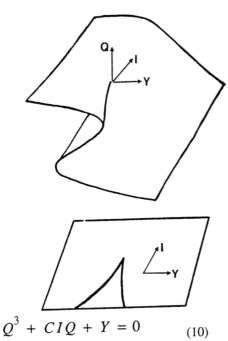

Fig. 3: Cusp catastrophe surface of the Catastrophe Theory (top) and its projection on a horizontal plane (bottom). For every point inside the cusp there are three values of Q for any fixed combination of a normal factor Y and a splitting factor I.

For example, let parameter Q represent a configurational state of isotopic pattern (e.g. averaged correlation coefficient for a selected isotope), parameter I describe the level of "informational impact" and Y be a third independent parameter describing the intensity of coupling between isotopic distribution and organizing the informational field. The latter could be, e.g., the rate of a change of the isotopic tunnelling coefficient upon the change of the local value of an electric and/or magnetic field. Let us now describe the linkage of these parameters by a cubic approximation of the type:

$$Q^3 + CIQ + Y = 0 \qquad (10)$$

where C is some fitting parameter. The upper part of Figure 3 shows the surface corresponding to Equation (10). Turning it to the isotopic case, this surface allows to picture isotopic rearrangements caused by external signals as cusp catastrophes leading to symmetry changes. Such approach could bring in an analogy between structural phase transitions in systems of confined ions (Berezin, 1991c) and informationally-stimulated isotopic tunnelling jumps. In our isotopic model the CT allows for a vivid description of the "switching" of a symmetry of isotopic pattern at different points of corresponding control parameters, e.g. some concentrational isotopic variables.

9. Isotopic Bifurcation Cascades in Homeopathic Patterning

For a CT-description of a polarizational stabilization of isotopic patterns let us reintroduce the notion of the so-called *isotopic Arnold's tongues* (Berezin, 1989) illustrated in Figure 4. In order to comprehend the concept of Arnold's tongues, also known as phase-locking tongues (Cumming and Linsay, 1987), a helpful preliminary step is to look at the bifurcation diagram shown in Figure 5. Suppose that a response of a certain system to an external variable X can be approximated by a function F(X,P) of the form:

$$F(X,P) = X^4 - PX^2$$

(11)

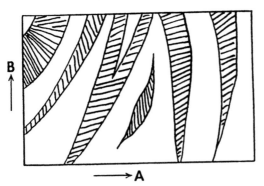

Fig. 4: Isotopic Arnold's tongues. For legend, see next page.

Legend to Fig. 4: Parameters A and B refer to two independent parameters of isotopic concentrational variability. Each shaded area refers to a particular type of isotopic ordering. White areas (between tongues) refer to an isotopically random structure.

Note the principle existence of disconnected "islands" of isotopically ordered phases within the randomness. where P is a fixed (in a given run) control parameter. The specific nature of F,X, and P is unimportant for the argument. However, to keep our thoughts along the isotopicity-homeopathy link, let us presume that in some quantifiable way the function F reflects the efficiency of a homeopathic information transfer at a single potentisation step and X and P are some isotope-related parameters, e.g. X and P could represent the deviations of some chosen isotopic ratios from their "original" values. For various P's Equation (11) produces a family of curves, 3 of them are shown in the upper part of Figure 5. For $P < 0$ all curves have a single minimum (stable point) with $X=0$. However, for $P > 0$ there are two minima at $x = \pm\sqrt{P/2}$. The latter means that at $P=0$ a single stable point splits (bifurcates) into two, and for all $P > 0$ the system has two possible stable states.

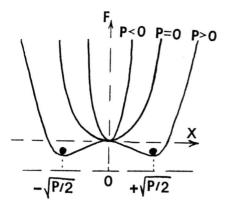

This is shown in the bottom part of Figure 5. In our example with the above stated isotopic assignments for the participating "playactors" $(F,\ X,\ P)$ this bifurcation means that the "same" level of the homeopathic efficacy, F, can be achieved with two different values of isotopic ratio X, symmetric in respect to its value symbolized by $X=0$.

Fig. 5: Bifurcation of a stability point for a simple symmetric system described by the equation $F(X,P) = X^4 - PX^2$. For all $P>0$ there are two equivalent symmetrically positioned stable states.

Suppose now that the dynamics of a system is such that the continuous increase of P leads to new and new bifurcations. Such bifurcation cascade is shown in Figure 6. A vertical line corresponding to a particular value of P produces a range of stable points, each one corresponds to a different isotopic state (symmetry) achieved for the same combination of isotopic ratios. The meaning of Figure 4 can now be explained as follows. A point on an AB-plane corresponds to a given combination of two isotopic parameters (e.g. some isotopic ratios). All points inside a given tongue have the same type of isotopic symmetry or the same type of isotopic ordering. ·

The finite width of Arnold's tongues indicates the relative stability (robustness) of a particular type of ordering. Such robustness is an important factor in producing a certain degree of stability (self-protection) of isotopic patterns against the disordering effects. Recall a possible role of the polarization (Section 8) as a likely self-amplifying factor in enhancing the degree of this robustness. An interesting analogy to this process can be drawn from the physics of corrosion. To some extent polarizational stabilization is similar to the phenomenon of spontaneous passivation ("self protection") of a metallic surface in the presence of a corrosive agent. Here the "reflex" of a metal surface is to form a thin oxide layer to inhibit the corrosion propagation.

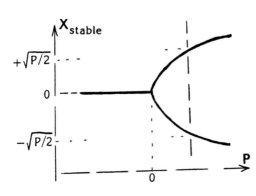

X_{stable}

$+\sqrt{P/2}$

0

$-\sqrt{P/2}$

0

P

Fig. 6: Bifurcation of stability point for $P > 0$ for the potential function shown in Fig. 5.

In a homeopathic scenario the main medium for such "reflex" reactions is a quasi-crystalline matrix of water or other fluid. In our model, the polarizational self-amplification of isotopic patterning is important for the stability of isotopic information storage and for a finite thermal tolerance of a homeopathic patterning in liquids.

10. Isotopic Neural Network

This section discusses some aspects of the homeopathic effect which may have common elements with the alleged health-monitoring properties of the so-called "healing crystals" (Berezin, 1991d). At a physical level such effects may be related to holographic-type memory effects in quartz (and other) crystals. For instance, some "crystal therapies" presently offered by alternative medicine practitioners could, in fact, have their physical foundation in complexes involving minority isotopes of oxygen and/or silicon in a crystalline quartz (SiO_2). For example, sublattices of magnetically active isotopes ^{17}O and ^{29}Si in quartz can form a connected entity which should be properly called an *isotopic neural network*, similar to model neural network suggested earlier on the basis of spin glass (Ising) models by Hopfield (Hopfield, 1982).

The fact that quartz and most other crystals have more than one isotopic sublattice provides a basis for a "natural division" of functions between one isotopic sublattice (e.g., ^{17}O) playing the role of *neural sites* (two possible spin states) while the other isotopic sublattice (e.g. ^{29}Si) serves as a so-called *synaptic network*. In a quartz crystal there are about 60 atoms of ^{29}Si per one atom of ^{17}O. One can visualize the whole "community" of those ^{29}Si-atoms (say, 20 or 30 of them) which are located between any two neighbouring ^{17}O-atoms providing a synaptic link between them. Such ^{29}Si-cluster has many inner spin states, i.e. each of 20 or 30 atoms may have two directions of nuclear spins, so for each ^{29}Si-cluster there are many hundreds possible spin combinations. As a result, the so designated ^{29}Si sub-network acquires a possibility for almost continuous synaptic adjustments. As far as virtually boundless informational capacity of the isotopically diversified crystal is postulated, the problem of interaction between the crystal and the host (e.g. human body) becomes relatively secondary since such informational exchange could be attained by a whole range of means, e.g., by resonances at infra-sound frequencies, magnetic or electrostatic effects etc. Recalling the quasi crystalline structure of water, the whole idea of isotopic information storing network can be translated to a case of homeopathic solutions. Hopfield's neural network (NN) can be described by the following spin Hamiltonian (Sompolinsky, 1988):

$$E = -\sum_{i<j} J_{ij} S_i S_j \quad (12)$$

where the N-component "spin" variables S_i, S_j mimic the activity of neurons (real or imitating) and the coefficients J_{ij} measure the efficacy of synaptic links. The latter themselves are the functions of "spin variables" and are defined by the so-called Hebb's rule, an example of which is:

$$J_{ij} = \frac{1}{N} \sum_{\mu=1}^{p} S_i^\mu S_j^\mu . \quad (13)$$

Here the summation runs over spin variables referring to all p memory patterns previously "learned" by a NN. This self-adjustability of synaptic strengths provides a bridge between energetic description based on the Hamiltonian (12) and the dynamics of *information* fluxes; the latter enter through the coefficients J_{ij}.

A recent popularity of spin-glass models of NNs is based on two advantages. The first is their capability of learning and self-directed ("unguided") information processing. The second is a high degree of physical universality making NN modelling an efficient heuristic tool for a wide variety of physical systems. The above described polarizational non-locality and isotopic connectivity could be fitted into a kind of "soft bonds" NN suitable for a featuring of the dynamics of the homeopathic effect. One possibility is to treat local values of isotopic Shannon indexes given by Equation (1) to define continuum analogies of spin variables and synaptic links, i.e. to treat S's and J's as spatial-time continuous variables H(x,t) and to apply methods of quantum dynamics to forecast the system's evolution. One possibility, e.g., is to use Fokker-Planck equations for H(x,t) to account for the informational modulation of the corresponding "world lines" of quantum particles (Cufaro Petroni, 1991). Originally inputted externally (see the next section), homeopathic information can be translated into the local values of the corresponding H(x,t)-functions; different H(x,t)-functions could "split" their participatory roles as neuronal and synaptic variables. Subsequently, the evolved patterns of H(x,t)-functions can be converted back to the informational variables relevant to the homeopathic protocol.

An important and still poorly understood issue common to crystal healing and homeopathy is the role of *placebo effects*. Although the reality of placebo actions seems to be well-confirmed in a wide variety of ways, their interpretation in terms of hard science still remains quite open. It is quite likely that any progress in our understanding of placebos will not only have an academic value but also may lead to a more efficient use of a placebo action in actual medical practice. At the same time, it is almost certain that any further progress in the understanding of the placebos and other imagery-induced effects is subordinated to our general level of comprehension of mind-body and mind-matter interactions. The next section will discuss some aspects of the quantum theory of consciousness as it may be relevant to a homeopathic effect.

11. Possible Connection of Homeopathy to Quantum Theory of Consciousness

Since after the invent of quantum mechanics there have been numerous attempts to include the phenomenon of consciousness into the realm of physical description. Few references are listed in my earlier paper (Berezin, 1992a). Here we will discuss a possible relationship between the isotopic informational patterning and the homeopathic effect in the light of a general role of consciousness in physical systems.

The common theme for the majority of suggested speculations that links consciousness to a quantum level is a drawing of a parallel between the spontaneous localization of mental patterns and (spontaneous or induced) reduction of the wave function in quantum mechanics. Most of the developments in this regard are along the "role-of-an-observer" line of arguing, i.e. the consideration of an induced collapse of a wave function of a system during the act of measurement (Penrose, 1989a; Berezin, 1992a).

A somewhat separate thread of ideas in this direction stemmed from the attempts to find a natural "built-in" mechanism for repetitive spontaneous (as opposed to the observer-induced) reductions of the wave functions in order to account for an apparent lack of quantum delocalization of meso- and macro-objects. This ("Heisenberg") delocalization was mentioned earlier in connection with Figure 2. Károlyházy and others (Károlyházy, 1966;

Károlyházy *et al.*, 1986; Frenkel, 1990; Rosales and Sánchez-Gómez, 1992) discussed the effects of gravity as an inherent route which Nature uses to achieve an "automatic" spontaneous collapse of the wave functions of extended objects. In his book Penrose (Penrose, 1989a) looks for the connections between consciousness, quantum mechanical reduction and gravity on the basis of these suggestions. In short, the essence of these ideas is that subtle *gravitational effects* at the *microscopic level* force the brain (or any other "thinking" system) into a particular mental state. This leads to a "materialization" of a potential (virtual) mental image by a route similar to the reduction of the wave function in quantum mechanics [see Eq.(7) and the accompanying discussion]. The latter essentially means that conscious processing in a quantum system can be identified with a chain of fast reductions of the manifold of all potential "possibilia" to a single given sequence of realizations. The combining of Penrose's ideas relating consciousness and gravity with a notion of isotopicity can form an informationally efficient "trio", largely because of the basic and straightforward fact that, due to their mass differences, isotopes differ in terms of the above mentioned gravitational action.

The difficulty of placing consciousness among the physical categories lies in the dichotomic nature of consciousness as a process with both ascending and descending aspects. While the former aspect requires a reductionistically describable mechanism ("from-matter-to-mind" extreme), the latter calls for some kind of immediate and coherent involvement of material processes into a consciential realm. In terms of possible microscopic observables this second (holistic) aspect may phenomenologically appear as some sort of "violation" of commonly expected microscopic behaviour (e.g. as an apparent violation of the second law of thermodynamics or as a change in a factual rate of tunnelling processes against the rate expected from the standard quantum mechanical model, etc.).

In order to specify a physical model of consciousness two routes, in turn, can be suggested as two opposing extremes. The first route is to relegate to all specific information on the anatomy, physico-chemistry and physiology of the brain and its function-ing. The second route is to suggest a model which is relatively unspecific to the brain structure and as such might be applicable outside the realm of standard organic biology. Such biology is commonly seen as the only legitimate stage of the emergence of con-sciousness. As I argued earlier (Berezin, 1990b, 1992a), it is possible to suggest a more extended definition of consciousness as a connected material process with a high degree of inner coherency which primarily (but not exclusively) refers to a specifically identifiable system (e.g. brain) and exercises non-local excursions in both space and time in a seemingly selective way. By the latter we mean that the "priority of external interactions" may end in a seeming violation of a common physical expectancy, i.e., the interactions which are decisive for the defining of a particular state of a conscious system are not necessarily the strongest in terms of energy exchange, spatial closeness, etc.

Let us recall what was said earlier (in Section 5) on the polarity of "energy" and "information" as two (almost alternative !) underlying principles of system dynamics. As for the state of the conscious system itself, its inner coherence at the macroscopic holistic level can be seen as "mirrored down" to the microscopic quantum level. This proceeds (at least in the framework of my present arguing) by a route of *seeming* violations of statistical and probabilistic rules of a "common law" physics. The model discussed in the present article gives a specific line for such *macroscopic-down-to-microscopic* physical "violations". As I proposed earlier (Berezin, 1992a), the correlated tunnel jumps of stable isotopes in consciousness-relevant material structures may proceed with a significantly higher "speed" than the speed expected from the corresponding standard tunnelling probabilities. In other words, a conscientious process works as a kind of catalyser or enhancer of an isotopic tunnelling exchange rate. In a more formalizable physical approach the effects of conscientual action can be seen as a symmetry breaking trigger at a microscopic level. The latter refers to a breaking of a symmetry of tunnelling probabilities in comparison with their ad hoc symmetric tunnelling rates.

A common feature of a number of versions of the quantum theory of consciousness is that they assume a two-way linkage between consciousness and the physical world. In other words, consciousness as a quantum phenomenon and the surrounding physical reality are jointly locked into an interactive loop. Some recent experiments seem to confirm the direct action of consciousness on physical systems, such as electronic random number generators (Jahn and Dunne, 1986; Radin and Nelson, 1989). The tentative conclusion from these experiments is that there may exist some kind of "cooperation" (resonance) between the physical system and the consciousness of an observer. Extending these ideas to a case of homeopathic protocol, one can suggest that the conscious intention of the observer (in this case - a doctor) forms an interactive informational loop with a homeopathic remedy under preparation, provided it is prepared within the restrictions of a properly applied protocol. In some sense, a patient also could be included into this informational interaction. Similar ideas could perhaps be applicable to a quantum-level discussion of various types of placebo effects.

Furthermore, a very high density of states (high degree of positional degeneracy) of isotopic permutations opens a possibility for an extremely high sensitivity of the evolving isotopic patterns to these energetically very subtle mind-matter effects. In a physics of chaos such high sensitivity of a non-linear dynamics to subtle changes in the initial conditions is known as a *Butterfly effect* (one extra wingbeat of a butterfly in Brazil today can trigger a tornado in Texas a few days later).

Therefore, chaotic features of spontaneous pattern formation give an additional argument for a possibility that energetically subtle (almost purely informational) mind-matter interactions can indeed affect the dynamics of isotopic patterning in a homeopathic procedure. A number of other interesting ideas related to the possibility of purely informational (rather than energetic) actions have recently been discussed in connection with Bohm's quantum potentials, possible detection of "empty waves", experiments to test Bell's theorem, faster-than-light information transfer and other quantum non-local effects (Selleri, 1990; Sarfatti, 1991). Many of these effects can, in principle, also have some relevance to the homeopathic effect. However, for the purpose of economy, only a few options related to quantum non-localities will be persued further in the present article.

12. Quantum Panpsychism and Homeopathic Effect

Connections between homeopathy and the quantum theory of consciousness can be traced even deeper by looking into some alleged-volitional or quasi volitional features in the behaviour of micro particles. Here we are going to mention some works developing the line of "quantum panpsychism", as such ideas may appear relevant to mind-matter connections in homeopathy.

It should be noted, though, that there is a significant degree of philosophical freedom in interpreting some features of quantum behaviour as "personal" or "volitional" properties of micro particles and microsystems. Thus the following cited references are largely a reflection of my personal tastes. The following citations are not very crucial for the overall understanding of the presentation offered in this article. They are given here mainly as an additional possible resource of ideas for further thinking on the issues reviewed in the present article.

Surveys of panpsychistic ideas in a pre-modern science and philosophy can be found elsewhere (e.g. Berezin and Nakhmanson, 1990; Popper and Eccles, 1977). In a vein of theoretical homeopathy volitional aspects of particles are related to an issue of the individualization of physical systems (Section 2). Interesting ideas suggesting some sort of purposeful or quasi-intelligent activity at a level of microworld were entertained earlier by, e.g., Cochran (1971), Wigner (1982), Margenau (1984), Jahn and Dunne (1987), and Miller (1991). These sources also contain many other references.Many of these vistas are also somehow relevant to physical foundations of self-awareness and personal identity problems (Margenau, 1984; Shalom, 1985), treated at both microscopic (particle) and organism levels. Cosmologist John Wheeler (Wheeler, 1988) discusses microworld and megaworld as two opposite "asymptotes" for the manifestation of a universal intelligence. Not going further into these far reaching speculations, I will mention a recent paper by Westerlund (1991) where the author deduces the presence of a memory effect in a polarizable matter on the basis of classical electrodynamics. This may fall in line with the above discussed role of polarization stabilization of isotopic patterns. Yet another facet of individualization at a microlevel may stem from a recently formulated paradigm of quantum computation (see Section 16). Another recent article (Cufaro Petroni, 1991) suggests an interesting idea of informational "conditioning" of particles in quantum mechanics.

13. Quantum Localization due to Isotopic Diversity

As shown earlier (Berezin, 1984), isotopic disorder in a periodical crystal lattice can under certain conditions lead to the formation of localized states. This phenomenon falls into a class of effects commonly known under the term *quantum localization* (also known as *Anderson localization*). The interference of electronic waves on isotopic non-uniformities may lead to a "locking" of propagating waves in localized spatial regions similar to quasistationary and metastable states known in ordinary quantum mechanics.

Such localized or quasilocalized states can maintain quantum coherency for much longer times than travelling (Bloch) waves and, according to what was said above, these states should be beneficial for consciential processing.

On the other hand, the formation of localized electronic states means some sort of temporal-spatial patterning of electrical charge density. The latter can affect the neural system through Coulomb effects, e.g. acting on ions Na^+ and K^+ critical for the transmission of signals along neurons, or through the monitoring of membranes' penetrability for a charge transfer, etc. The reverse communication (from neural system back to an isotopic pattern) can be based on the isotopic selectivity of through-membrane diffusion, differences in magnetic moments of isotopes responding differently to weak local changes of magnetic fields, etc.

I have mentioned earlier (Berezin, 1990c) a possible relevance of the Anderson localization model to the establishing of patterns of mental activity. In a vein of the present article this model can form a useful descriptor for the interaction of isotopic and neuronal levels. Namely, isotopicity can perhaps work as a monitoring factor in a two-way connection between the informational content of consciousness and neuronal system. Further down, the neuronal system links the consciousness to the rest of an organism (with all its activities) in a "usual" neuro-chemical way. This latter aspect is beyond the scope of the present article. Concluding, one can extrapolate our model to the extreme and postulate (provisionally) that all the content of consciousness is stored in isotopic diversity only. Whether this is this the case or not is (at least in principle) an experimentally verifiable question, e.g. by advancing some experiments on isotopically shifted and isotopically purified organisms (Berezin, 1988).

14. Localization, Coherency and Isotopicity

We now turn to possible links between quantum localization and quantum coherency. The above exposure emphasizes two important aspects of isotopicity in terms of its potential relevance to the operation of consciousness. The first is the informational storage capacity of isotopic permutations. The second is the ability of isotopicity to affect the dynamics (e.g. time scale and frequency) of a gravitationally-induced quantum reduction process through the isotopic mass variations due to the fact that the effects of gravity are mass sensitive. In this and next sections we will take a closer look on how, possibly, isotopicity can provide a viable means of controlling the frequency of quantum reductions. My suggestion is that isotopicity can provide (in a self-regulated way) a kind of "reduction-rate-control-mechanism". The latter means that the very structure of the spacio-temporal pattern of quantum reductions can be linked in a feedbacked way to a content of a flowing mental process in such a way as to maintain quantum coherency of the brain (or other "thinking system") for a period of time sufficient to allow the evolution of a cohesive mental image during this time interval of uninterrupted quantum coherency. The latter is a time interval between two successive quantum reductions. In this scenario the mental activity of a quantum system can be illustrated by a cinematography: each frame is an analogue of a coherent (between the two consecutive reductions) period, while each quantum reduction simulates the transition to the next frame which (in most cases) has some sort of cause-effect relationship with a previous frame. This leads to a piece-wise dynamics of consciousness of intermittent coherency periods (preservation of linear quantum superposition) abrupted by quantum reductions which are, in essence, non-linear "intrusions" into the otherwise linear flow of quantum dynamical evolution.

Let us now describe a possible scenario of how this piece-wise dynamics of quantum coherency can be "assisted" by isotopic diversity. In main bio-elements (H, C, N, O) isotopic diversity is omnipresent and the absolute concentration of minority isotopes is quite high. For example, 1 % of all carbon atoms are isotope ^{13}C. Even a tiny micro cell of, say, 1000 Ångströms in size has an enormous number of isotopic permutations in its carbon chains. I now argue that in isotopically diversified micro objects the frequency of occurrence of quantum reductions can be quite different from the mentioned frequency of the structurally similar but isotopically homogeneous objects.

As an example we can consider a fragment of a DNA chain. Let us suppose, for simplicity, that we can at will prepare identical replicas of this fragment: one with and one without the isotopic diversity. The latter means mono isotopic form. From the point of view of gravity-induced reductions isotopic diversity can be seen as random mass density fluctuations on an atomic scale. For an isotope with a volume concentration n the characteristic length scale of such mass fluctuation is of the order of

$$b = \frac{1}{\sqrt[3]{n}} \qquad (14)$$

Several elements, together with their respective isotopes, produce several values of b. For example, the trio of carbon, nitrogen and oxygen (which have 2,2,3 stable isotopes, respectively) produces 7 different values of b for a segment of a uniform chemical composition. The gravitational quantum reduction rate generally has non-linear dependence upon the mass of an object (Károlyházy et al., 1986). For extended objects it can likewise also depend on local values of density gradients. The majority of the existing schemes of quantum collapses (reductions) suggests that the space-time point of the "next" quantum reduction is a random variable, defined in itself by some kind of "uncertainty principle"

(Ghirardi *et al.*, 1986; 1988). If so, an intricate mass-density wavy pattern arising from the isotopic diversity can (due to the interference of various length scales *b*) affect the local probabilities of quantum reductions *suppressing* or *enhancing* them at various regions of a given object.

Note that, at first sight, the same argument can apply to atomic-scale mass variations due to a regular chemical diversity. However, the important difference is that while the isotopic diversity manifests itself "within" the chemistry, it is not subjected to chemical bond restrictions common to a regular hetero-atomic chemistry. As such, isotopic mass diversity exists "over-and-above" chemical diversity. It is probably appropriate to say that isotopicity as having a high degree of inner freedom within chemistry (see above) furnishes such an independent mechanism to control local probabilities of gravity-induced quantum reductions without any need to re-arrange chemical order or modify "regular" chemical activities. On the contrary, microscopic mass-density profile caused by chemical diversity is much more deterministic and does not possess informational flexibility which is characteristic of the isotopic degree of freedom.

A matter of special interest in the above scheme would be a possibility of such isotopic arrangements which, due to some kind of "negative interference" are capable of suppressing the rate of quantum reductions. This will lead to greater survival times for macroscopic quantum superpositions (coherent states) allowing both for a proper mental focusing (similar to a minimal exposure time for a photo-shot) and for the engagement/disengagement of consciential information with outer-to-the-consciousness neural mechanisms. As a further possible specification of such reduction-suppressing mechanism based on isotopes, I will point out quasistationary vibrational configurations which can be implemented by specific isotopic arrangements within a given chemical structure. For quasistationary configurations the overlap of localized vibrational levels with the surrounding continuum is very small (a total cancellation interference of all outgoing waves) and, consequently, they have an unusually long lifetime. This is similar to metastable excited states of atoms when the interaction with an electromagnetic continuum is suppressed due to selection rules, symmetry restrictions, etc. In other words, some specific isotopic patterns may account for an especially efficient quantum entanglement (understood here primarily as an informational richness of a corresponding quantum superposition). This leads to longer-than-average survival times of coherent states of a conscious system. The possibility of similar slowly decaying isotopic configurations has been indicated earlier (Berezin, 1989) on the basis of the mechanism of polaronic self-stabilization (Austin and Mott, 1969; Berezin, 1983).

15. More Information on Isotopicity and Quantum Collapse

We sketched earlier (in Section 7) the process of quantum reduction which happens in the course of quantum measurement (the alleged "observer effect"). In Section 12 we outlined a "competing" mechanism of spontaneous quantum reductions under the action of gravitationally induced space-time distortions. It should be pointed out that the term "reduction" in this context should not be associated with the term "reductionism" in a philosophical meaning. Quantum reduction (sometimes also called "*quantum collapse*") is a process, be it induced or spontaneous, in which the wave function of a system experiences a sudden increase of its degree of localization.

No matter that it is microscopically very weak, a gravitational field brings some non-linear effects in quantum dynamics. For instance, the originally linear Schrödinger equation for a center of mass of N-particle system acquires an additional non-linear term (Rosales and Sánchez-Gómez, 1992):

$$ i\hbar \frac{\partial \psi}{\partial t} = -\frac{\hbar}{2Nm} \nabla^2 \psi - \frac{4}{3} \frac{N^2 \pi G \hbar^2}{c^2} |\psi|^2 \psi , \tag{15} $$

where for simplicity each particle has the same mass m; G is the gravitational constant and c is the speed of light.

It is notable that the relative role of such gravitational non-linearity grows faster than the number of particles N (in the above equation it is proportional to N^3, other theoretical treatments may lead to somewhat different dependencies). Comparison of the second (non-linear) term in the right-hand side of Eq.(15) with the first term (kinetic energy) shows that the gravitationally-induced quantum non-linearities become important for:

$$ N^3 \geq \frac{3}{8\pi} \frac{M_p}{m} \frac{W}{L_p} , \tag{16} $$

where M_p and L_p are the Planck's mass ($2.2 \cdot 10^{-8}$ kg) and Planck's length ($1.6 \cdot 10^{-35}$ m), respectively. Here the characteristic quantum dispersion of the center of mass, W, is identified with the Gaussian quantum collapse width. The meaning of the latter is explained after Eq.(17); we will take here W = 1000 Å. Taking for m the proton mass ($1.67 \cdot 10^{-27}$ kg), the Eq.(16) gives $N > 2 \cdot 10^{15}$ which corresponds to a very tiny particle of matter (e.g. a micro droplet of water). Correspondingly, gravitationally induced quantum non-linearties can indeed be shown up at the scale of importance for homeopathic solutions.

One of the consequences of these quantum non-linearities is the alleged presence of a fast sequence of quantum reductions (quantum collapses) mentioned in earlier sections. The act of an elementary quantum collapse (EQC) in a single particle system is a seemingly spontaneous (triggered stochastically) process at which the wave function of a particle experiences a sudden shrinkage of its size. If prior to EQC the particle has a wave function, at the act of EQC it instantly gets multiplied by a squared exponential (Gaussian) factor of the form (Frenkel, 1990):

$$ \varphi(\vec{r}) \rightarrow \exp\left[-\frac{|\vec{r} - \vec{r}_o|^2}{2W^2} \right] \varphi(\vec{r}) \quad . \tag{17} $$

In Eq.(17) r_0 is a random (stochastic) variable indicating the center about which a given quantum collapse occurs. The width of the Gaussian W is of an order of 10^{-5} cm (i.e. about 1000 Å) and the probability of such a jumpwise EQC per particle per second is about 10^{-16}. The latter means that for a single particle EQC happens only once every 300 million years (!) on average and, consequently, the process of EQC for a *single particle system* for all practical purposes can be neglected. This is, however, not the case for a system of "normal" (microscopic) size. A peculiar feature of quantum collapse is that for a system of interacting particles (such systems have so-called nonfactorizable wave functions) EQC for any particle results in the multiplication of the entire wave function of the whole system by the above written Gaussian factor. This goes, so to say, in analogy with an automobile engine operation when the local (pointwise) ignition of an air-fuel mixture leads to a practically instant chemical explosion of the whole combustible mass in the interior of a given cylinder.

Therefore, the rate of quantum reductions of macroscopic bodies is quite significant. For instance, one cubic cm of water contains about 10^{23} atoms and for it the resulting frequency of quantum collapse is about 10^7 reductions per second. In the context of a homeopathic effect we can look at isotopicity as a connecting "transducer" between unformationallyloaded patterning and quantum reductions. The kinematics of the latter is sensitive to isotopic mass variations through the gravitational triggering mechanism outlined above. Gravitational reduction can proceed at different time and mass scales and due to isotopic mass variations the dynamics of local reductions can be sensitive to isotopic patterning. Because of this, gravitational quantum reduction can serve as a non-local connector enhancing some ("favourable") isotopic permutations (possibly at the expense of some other, "unfavorable" permutations to keep the overall statistics unchanged) towards the establishing of the holistic instantaneous pattern of isotopic arrangements. We reiterate that isotopic patterning, contrary to chemical spatial (positional) permutations, is not subjected to chemical valence constrains and proceeds within a variety of energetically (almost) degenerate states. Global isotopic patterns can be a result of iterative reductions to the conscious attractors and the very dynamics of constant isotopic rearrangements can be seen as a goal-driven (teleological) process. This allows for a dichotomy of a reductionistically describable microscopic mechanism (local isotopic hoppings) and a holistically conditioned overall isotopic patterning caused by non-local attractors.

The discussion above was focused on an alleged mechanism of the wave function collapse triggered by gravitational effects. This preference of mine naturally follows the fact that isotopes differ in their mass and hence variation of their gravitational action is a first-order effect (proportional to the mass). It should be noted, however, that similar (or, perhaps, alternative) schemes can be devised on the basis of other mechanisms of quantum reduction, e.g. the stochastic quantum reduction scheme (Ghirardi et al., 1986, 1988; Pearle, 1989), reductions induced by zero-point fluctuations of the electromagnetic field, or even using such versions of quantum mechanics which avoid the concept of a wave function reduction altogether (Cini and Lévi-Leblond, 1990). We will not review all these alternatives in the present article. However, isotopically-based tunnelling mechanism of informational linkage (Berezin, 1992a) can likely be matched to other interpretational quantum schemes as well. Another promising alternative is discussed in the next section.

16. Quantum Computing Paradigm in Homeopathy

Quantum computing or the theory of *quantum automata* is a new segment in theoretical physics (Albert, 1983; Feynman, 1985; Deutsch, 1985, 1989, 1991; Penrose, 1989a). Its emerging importance is quite interdisciplinary and may far transcend the physics of computational systems in a proper sense. It relates such apparently quite distant things as cosmology, nature of time, evolution, emergence of information in non-equilibrium nonlinear systems and the physics of consciousness. Here I will give some brief remarks on possible implications of this emerging paradigm of quantum computing for the physics of homeopathic effect in the context of our general discussion of informational aspects of isotopicity and homeopathy.

Although still *gedankensytems*, quantum computers differ conceptually from their classical counterparts. They are *not* advanced digital systems based on atomically-fast microscopic processes. They are the systems which make the use of the superposition principle of quantum mechanics [see Eq.(7)], i.e. the notion that in some sense a quantum system can be simultaneously in a number of mutually alternative ("virtual") states. This leads to a concept that different terms of a superposition evolve *independently* along different "quantum paths". The latter can be hypothesized as world-lines (or rather world-tubes) of the "copies" of the same system (quantum computer) existing in alternative universes. This notion comes from the so-called *Many World Interpretation of Quantum Mechanics*

(MWIQM) put forward in 1957 by Hugh Everett. According to MWIQM, each elementary quantum event with many possible results leads to a multiplication of the whole (!) Universe into many copies of itself, in each so-branched universe the outcome of this quantum event is different. For a vivid (but a somewhat superficial) illustration of this process one can "reuse" Figure 7 which was discussed in quite a different connotation.

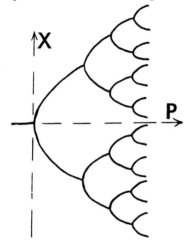

Fig. 7: Bifurcation cascade of stability states for a dynamical system with a non-linear feedback [this is *not* the system described by the Equation (11)]. The continuous change of a control parameter P leads to a breeding multiplicity of stable values of a configurational parameter X.

Such "chain breeding" leads to a horrendous number of parallel universes forming a kind of a tree-like "Mega Universe" with all thinkable evolutions of elementary (as well as global) events happening "somewhere" in it. Despite its seeming weirdness in front, MWIQM has recently enjoyed a new boost of popularity in connection with ideas of quantum computing (Deutsch, 1991; Penrose, 1989a) and inflationary cosmology. Conceptually, the ideas of quantum computing and MWIQM scenario are closely related (Deutsch, 1991), at least at the present stage of theoretical development. In what way can these ideas be relevant to homeopathic mechanisms? Some authors see MWIQM-scheme and EQC-scheme as mutually excluding interpretational alternatives (Cini and Lévi-Leblond, 1990). It is not clear for me at this time to what extent these two theories can be reconciled or perhaps blended into some kind of a synthesis. Correspondingly, instead of opting between MWIQM and EQC, I shall indicate in the rest of this section some clues on how the ideas of quantum computing (and MWIQM) can possibly turn out to be handy for the advancement of a conceptual framework of homeopathy.

Generally, a physical computation is a process which produces outputs that depend in some desired way on given inputs (Deutsch, 1989). The nature of inputs and outputs should normally have some informational interpretation. This generalized vision admits under the notion of computing such processes as, e.g., ontogenesis, bio-evolution, operation of the immune system, and many other bio-processes, including consciousness. Taking into account the fundaments of informational aspects of homeopathy, the process of preparation and administration of a homeopathic remedy can also fit the above definition of computing. The incorporation of the superposition principle in a quantum version of a computing process allows, so to say, to "delegate" blocks of information processing to various terms in quantum superposition (in MWIQM - to replicas of a "computer" in various branches of a Mega-Universe). This principle possibility of processing the information "somewhere else" is one of the prime reasons which makes the ideas of quantum computing so interesting. The homeopathic protocol (especially at a deep potentisation level, i.e. at a very high dilution stage) is a tempting candidate for the applicability of quantum computing paradigm. This idea invites some analogy of a homeopathic remedy with a TV-set: the originally "input" information (e.g. by a prime "seed" drug) is processed by a distributed non-local quantum network and then (after the potentisation) returned back to the point of origin. In this scheme a homeopathic remedy acts as a receptor ("TV-set") of an outwardly processed (or stored) information. The non-local quantum network mentioned above involves all terms in a quantum superposition corresponding to various virtual states of a homeopathic solution. In our model the latter can be related to various virtual isotopic patterns.

A somewhat unusual frame of discussion of quantum computing can be provided by the physics of formation of the so-called quasicrystals. The latter are recently discovered (in 1984) metallic structures which do not follow the principle of the translational invariance (long-range periodicity) of a standard solid-state physics. Again, the notion of a quasicrystal in this connotation should not be confused with a quasi-crystalline structure of water and other liquids which was mentioned above. The symmetries of metallic quasicrystals seemingly contradict the rules of crystal growth known from the traditional crystallography.

One of the explanations for the existence of such peculiar crystalline forms is a participation of the effects of quantum non-localites. Such possibility has recently been suggested by Penrose (Penrose, 1989a, 1989b). Quasicrystals known so far are predominantly metallic alloys having a crystallographically forbidden five-fold (pentagonal) symmetry. They have a local order (local symmetry) while lacking the long-range translational symmetry. In some ways, these structures exemplify a dichotomy of periodicity and aperiodicity. If non-local quantum interactions are indeed essential for the formation of quasicrystals, it may deliver an additional analog to a possible emergence of correlated isotopic patterns in a homeopathic effect. In both cases the process is driven by a non-local (holographic) algorithm. Similarly to quasicrystals, informationally-driven isotopic patterning in the course of homeopathic dilutions and successions can be seen as a specific case of a quantum computing.

To summarize, the combination of isotopicity and quantum computing proposes to view on isotopically patterned molecular structure in a homeopathic solution as a kind of a quantum mechanical measuring system or a "quantum computer" in the sense explained above. The fundamental capability of isotopicity to perform such a task lies in its information potential. This was explained above using a "freedom-within-fixed-structure" paradigm.

17. Isotopic Rearrangement and Nuclear Process

Informational entanglement in an isotopic scenario can be further supported by a conceptual unification of a whole system through a single neutron field. This means the following: from the point of view of quantum field theory our description of isotopic tunnelling is equivalent to jumps of "pure" neutrons in neutronic narrow bands spanned on the atoms of a given chemical element. For instance, tunnelling exchange ^{17}O and ^{18}O can be seen as a neutron diffusion from the original ^{18}O-site to an ^{17}O-site. The total number of neutron bands is equal to the total number of different poly-isotopic elements housed in a given crystalline or liquid structure.

Representation of isotopic rearrangements in terms of narrow neutron bands allows to visualize these processes a similar to small polaron hopping. In this representation the inter-band transitions (jumps of neutrons between the bands originating from different elements) are equivalent to nuclear transmutations. Such picture establishes some similarity of these inter-band neutron hoppings (resulting in nuclear transmutations) to *umklapp-* (inter-band) processes for phonons in solid state physics. It also raises the possibility of some linkages between our isotopic model of a homeopathic effect with recent theories of cold fusion (Preparata, 1991) and are an interesting but now almost forgotten earlier piece of work by Louis Kervran on biologically induced nuclear transmutations (Kervran, 1972).

18. Quantum Coherency and Thermal Effects

One of the most obvious objections to the idea of isotopic patterning in homeopathy may come from the effects of thermal disordering. It is known that the average thermal energy per atom is roughly equal to kT, where k is the Boltzmann constant and T is the absolute (Kelvin) temperature. The value kT (26 meV at room temperature) is several times bigger than typical isotopic fluctuation of a site energy which is normally a few meV (Berezin, 1989). At first glance, it appears that such an excess of thermal ("noise") energy over the claimed effect (isotopic self-organization) will fully block any coherent manifestations of isotopic effects in comparison with thermal disordering. This, however, is not a necessary conclusion.

It is known that in situations of spontaneous pattern formation a very weak *but spatially coherent* perturbation can often serve as a pattern-forming factor even at the presence of a much stronger *but stochastic* (non-coherent) thermal randomness. It is not so much a matter of energetic scales but rather entropic considerations. A low entropy coherent signal can thus take a precedence over a high entropic thermal bath smearing action even if the latter has a much higher energy content. We mentioned earlier (in Section 3) the phenomenon of self-organized criticality. In some scenarios noise can produce a *stabilizing* rather than randomizing effect on a non-linear dynamical system (Herzel and Pompe, 1987). Similarly, a *removal* of some conducting paths or mechanical supports can sometimes lead to an *increase* (!) of a circuit resistance or mechanical strength (Cohen and Horowitz, 1991). This might provide an additional by-analogy argument that the homeopathic dilution not necessarily destroys the healing information, but on the contrary, may even lead to its amplification (perhaps in an oscillatory way, see next section).

Numerous examples of self-stabilization are provided by spontaneous symmetry breaking effects in non-linear systems possessing a plethora of configurations which are quasi-degenerate energetically. Transitions between such close configurations, e.g., Arnold's tongues related to different positional phases (Berezin, 1989) can be induced by subtle (i.e. energetically weak and primarily informational) interactions.

Quantum effects (e.g. the wave function collapse and coherency emergence in many-particle systems) can provide an additional facet to pattern stabilization against the thermal disordering action. It is known, for instance, that quantum effects can "suppress" chaos in non-linear systems (Gutzwiller, 1991). This feature of a quantum chaos is the result of the discreteness of quantized energy levels. Finite energy intervals between quantum levels lead to some degree of robustness of quantum states against the external perturbations.

In summary, "kT-argument" is not a detractor from the possibility that isotopic distribution can affect homeopathic informational patterning. In fact, a possibility of strong pattern-forming actions in the effects that have a characteristic per-particle energy far below kT level is a relatively common feature in non-equilibrium thermodynamics (Nicolis and Prigogine, 1977; Haken, 1978).

19. Quantum Interference and Oscillations of Homeopathic Activity

Oscillations of homeopathic activity (Davenas *et al.*, 1988) can, in principle, be addressed in a number of ways. It is not uncommon for a physical system, especially non-linear, to show an oscillatory behaviour of a given parameter upon a continuous increase of another parameter. Many phenomena based on resonance and/or interference effects may exhibit various types of oscillatory behavior.

We mention oscillatory effects in spectra of disordered systems in the regime of Anderson localization in a line with our isotopic hypothesis (Nieuwenhuizen, 1990). A connection to a possible role of quantum localization in the homeopathic effect (Sections 13, 14) can be established through the model of isotopic triggering of quantum localiza-tion (Berezin, 1984).

The list of possible oscillatory keys for homeopathy can include a few other options, both isotopic and not. Here I mention: (a) alternating regions of periodicity and aperiodicity in patterns of isotopic correlations, (b) the possibility of Wigner crystallization of ions and other effects of electrostatic ordering including clusters with "magic numbers" (Berezin, 1991c), and (c) formation of almost delocalized Rydberg states for chemically unbound (solvated) electrons. All these effects are, in principle, prone to a rich dynamics upon homeopathic succussions.

20. On a Hierarchy of Structural Levels in Homeopathy and Homeopathic Synergetism

Holographic information storage has some degree of scale invariance. This essentially means a self-similarity (self-identity) of information recorded at different scales of spatial resolution. This feature of a redundancy of information was discussed above in connection with Equations (4) and (6). Here we note some further quantum aspects of this informational redundancy which may be important for the understanding of informational role of homeopathic dilutions and succussions. We also give here a remark on a possible synergism of isotopic and non-isotopic effects in homeopathy.

Self-similarity at different spatial levels is known as fractal symmetry. Several possible scenarios for fractalization of isotopic distributions can be suggested. One is the formation of isotopic density waves with a wide range of wavelengths. In the simplest ("monochromatic") case isotopic density wave could be defined as a sinusoidal varying deviation of the local value of isotopic abundancies from their basis level (e.g. from the standard natural abundance). Another way is to treat a local value of isotopic correlation function $C(r)$ as an "order parameter". The function $C(r)$ was discussed earlier (Berezin, 1990c) in connection with isotopic ordering.

It is interesting to note that in a poly-isotopic system these functions can be defined in quite a large number of ways. Even for such a "simple" system as pure water 3 isotopes of oxygen and 2 isotopes of hydrogen provide several different combinations to define isotopic ratios and $C(r)$-functions. It is also possible, in principle, that different isotopes have different local degrees of ordering.

The remark above on isotopic density waves assumes time-independent (static) isotopic distributions. A non-static case can be even more intriguing in terms of its informational and pattern-forming richness. Isotopic distributions can "breathe" and this process, again, can have a whole spectrum of possible frequencies. Furthermore, both spatial (wavelength) and temporal (frequency) components of isotopic distribution can be informationally-loaded with either overlapping (redundant) or non-overlapping (complimentary) information carried by each of these components. These scenarios can proceed in pure or mixed versions. The latter means that both wavelength and frequency spectra of multicomponent isotopic correlation functions $C(r,t)$ can carry partially overlapping information with some qualified (e.g. preferential) information sharing their time and spatial functional dependencies.

As for a possible synergism (co-action) of isotopicity with other non-isotopic factors which may contribute to the homeopathic effect (some of them are mentioned in Section 19) I restrict myself to the following comment: on the basis of presently available data it is hardly possible to rule out *a priori* the relevance to homeopathy of any other information-carrying mechanism, any combination of such mechanisms or their combination with isotopicity. Most of these mechanisms are related to various microscopic and mesoscopic structures in water, like turbulent vertices, patterns of hydrogen bonds, etc. All these possible options will not be discussed in the present article. Nevertheless I will mention just one - the effects of local energy cummulation. Recently these effects have been discussed in connection with sonoluminescence which is a generation of light by the action of sound (acoustic) waves in liquid (Johnston, 1991). The effects of energy focusing may be of special importance for homeopathy as they allow for the formation of highly non-equilibrium microscopic centers of activity where informational and energetic fluxes can interact in a pattern-forming way.

DISCUSSION
Some Lines of Suggested Experiments

Exploratory-wise one of the advantages of the isotopic model for homeopathy lies in a relatively promising technical availability (and perhaps practical affordability) of some simple experiments. Isotopes of most chemical elements are commercially available and, therefore, homeopathic experiments with isotopically shifted liquids are relatively easy to design and perform.

One of the most straightforward lines of experimentation could be to study the efficacy of a given homeopathic protocol (e.g. a fixed scheme of potentisation ratios) using samples of water with a variable isotopic composition. Various isotopic compositions can be conveniently labelled by Shannon indexes introduced in Section 5. Several participating isotopes provide a variety of possible parametrizations for such experiments. For example, for a given experimental run one particular isotopic ratio can vary, while some others may remain fixed. There are more elaborated sampling techniques for multi-parametric systems aimed towards more efficient optimization of sample testing (e.g. the quickest descent method of a simultaneous variation of several parameters, Monte-Carlo-based techniques, etc.).

A range of methods of the efficient analysis of experimental data have recently been suggested on the basis of the chaos theory. For example, suppose that the homeopathic efficacy (HE) could be measured for N experimental points i ($i = 1,2,...,N$), each point i is labelled by corresponding values of several experimental variables. In the context of our present article variables refer to various isotopic ratios (IR) available in a given system, e.g. ratios of oxygen and hydrogen isotopes in the case of pure water. Generally speaking, any other experimental variables, not necessarily isotopically-related, can be used as labels. If now HE can be plotted as a function of several variables, the result will be N "isotopic points" (Berezin, 1991b) which can be treated informationally by methods of chaology (Grassberger and Procaccia, 1983).

In such a case a natural way to analyze isotopic points is to calculate the correlation sum $U(R)$ as:

$$U(R) = \frac{1}{N^2} \sum_{1 \le i < j \le N} \theta(R - \left| \vec{R}_i - \vec{R}_j \right|)$$

(18)

Here $\left| \vec{R}_i - \vec{R}_j \right|$ is the distance between two isotopic points in a corresponding parametric space, $\theta(u) = (1+sgn(u))/2$ is the Heaviside step-function, and R is a floating length in a given D-dimensional parametric space (Berezin, 1991b). Function $U(R)$ should not be confused with the microscopic positional correlation function $C(r)$ mentioned above, as in the Eq.(18) the distance R is a distance in a parametric space spanned on parameters HE and experimental variables; it is *not* a regular 3-dimensional spatial distance. In the case when $U(R)$ can be approximated by a power function:

$$U(R) \propto R^n \qquad (19)$$

the slope of $\log U(R)$ vs. $\log R$ plot gives an estimate of power n which, in analogy to the theory of chaos, can be called the dimensionality of an *isotopic attractor*. In the chaos theory equations (18) and (19) are usually understood in a limiting sense, i.e. in the limit $N \to \infty$.

However, in practice, even a relatively small number of isotopic points (perhaps, as small as, say, N = 5 or 7) can be sufficient for a meaningful estimate of the dimensionality of a corresponding isotopic attractor (Berezin, 1991b). This whole approacn is certainly not restricted to isotopic variability and any non-isotopic homeopathic variables can be used as well.

Not speculating here any further about such *gedankenexperiments* with homeopathic remedies of a variable isotopic composition, I will nevertheless express my guess that the homeopathic efficacy, HE, in such experiments will somehow easily tend to show a plateau-type behaviour as a function of a particular isotopic ratio (R) chosen as a variable (Figure 8). The reason for this may lie in a high redundancy of isotopic information storage. Only an extreme "depletion" of a particular isotope at the edges of the plateau, when isotopic ratio IR approaches 0 or 1, will noticeably affect the efficiency of isotopic distribution to store homeopathic information. In particular, "a negative" experiment could be based on isotopically purified water with D, ^{17}O and ^{18}O removed below the expected threshold of the self-amplification of isotopic non-linearties.

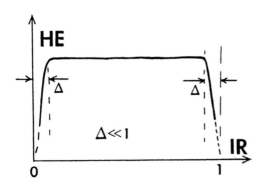

Fig. 8: Homeopathic efficacy, HE, as an expected function of a particular isotopic ratio, IR. The existence of a long "plateau" is a likely consequence of a high redundancy of isotopic information storage.

Epilogue

There is a plethora of other possible connections between physics and homeopathy which is left outside of the scope of this text. Along with several items mentioned throughout this article I additionally note two more promising lines here:

(a) application of a formalism based on *quantum logic* can possibly lead to a reveal of some unnoticed connections hidden in statistical data of numerous homeopathic trails performed so far (quantum logic should not be confused with quantum computing; for some explanations of it see, e.g., Oi, 1988), and

(b) some possible physical parallelism of homeopathy with a class of phenomena known as *hormesis*. Hormesis is an effect of a beneficial action of a toxic agent (chemical, radiative, etc.) when the latter is administered in very low ("homeopathic") doses. A resemblance between hormesis and homeopathy is almost obvious (Luckey, 1980), however, the effects of isotopic diversity may uncover some additional aspects of this communality.

To put the major theme of this article in a nutshell: I propose a physical relationship between the homeopathic effect and the informational structuring in water utilizing a mechanism of correlated information-bearing isotopic patterns. This hypothesis does not dismiss the possibility of other alternative explanations, furthermore, it could turn out to be complimentary to them.

What, then, could be an actual "proportion" (or relative "weight") of isotopicity in a homeopathic effect? Applying a Hegelian rule of the "negation of negation" we can suggest the following dichotomic argument. Let us use the analogy with an immune system of a living organism. The stronger the viral attack is, the more vigorously the immune system will react in producing antibodies as a defense reaction. This is a negative feedback loop. Perhaps, in some cases isotopicity (isotopic information storage) may not be strong enough to bear the full load of the homeopathic protocol. Then other complimentary mechanisms (e.g. structures of hydrogen bonds) can be attracted "to help" isotopicity to do its part of the job. The relative role of isotopicity and other competing and/or cooperating mechanisms may vary from case to case. At the present time all these possibilities, of course, remain fully open for experimental endeavours.

REFERENCES

Albert DZ. On Quantum-Mechanical Automata. Physics Letters 1983; A98: 249-252.

Allegre CJ, Provost A and Jaupart C. Oscillatory zoning: a pathological case of crystal growth. Nature 1989; 294: 223-228.

Atmanspacher H and Scheingraber H. Information Dynamics. New York and London: Plenum Press, NATO ASI Series, vol. 256, 1991.

Austin IG and Mott NF. Polarons in Crystalline and Non-Crystalline Materials. Advances in Physics 1969; 18: 41-102.

Bak P, Tang C and Wiesenfeld K. Self-Organized Criticality. Physical Review 1988; 38A: 364-374.

Berezin AA. Theory of the Polaron Effect in Boron. Soviet Physics - Solid State 1973; 15: 1298-1299.

Berezin AA. Spontaneous Tunnel Transitions Induced by Redistribution of Trapped Electrons over Impurity Centers. Zeitschrift für Naturforschung 1983; 38a: 959-962.

Berezin AA. An Isotopic Disorder as a Possible Cause of the Intrinsic Electronic Localization in Some Materials with Narrow Electronic Bands. Journal of Chemical Physics 1984; 80: 1241-1245.

Berezin AA. Isotopic Randomness as a Biological Factor. Biological Journal of the Linnean Society 1988; 35: 199-203.

Berezin AA. Some Effects of Positional Correlations of Stable Isotopes. Physics Letters 1989; A138: 447-450.

Berezin AA. Isotopical Positional Correlations as a Possible Model for Benveniste Experiments. Medical Hypotheses 1990a; 31: 43-45.

Berezin AA. Isotopic Diversity in Nature and Quantum Theory of Consciousness. Raum & Zeit 1990b; 2: 2, 66-71.

Berezin AA. Isotopic Diversity as an Unexplored Mind-Matter Dimension. Science Progress, Oxford 1990c: 74: 495-512.

Berezin AA. Diversity of Stable Isotopes and Physical Foundations of Homeopathic Effect. Berlin J Res Hom 1991a; 1: 85-92.

Berezin AA. Isotopic Relatives of Strange Attractors. Physics Letters 1991b; A161: 295-300.

Berezin AA. Application of Catastrophe Theory to Phase Transitions of Trapped Particles. Physica Scripta 1991c; 43: 111-115.

Berezin AA. On the Possible Physical Foundations of Health Related Effects of Crystals. Medical Hypotheses 1991d; 36: 213-215.

Berezin AA. Correlated Isotopic Tunnelling as a Possible Model for Consciousness. J Theoretical Biology 1992a; 154: 415-420.

Berezin AA. Isotopicity - Implications and Applications. Interdisciplinary Science Review 1992b; 17: 74-80.

Berezin AA and Evarestov RA. Electron-Positron Colour Centre in Alkali Halides. Physica Status Solidi 1971b; 48: 133-140.

Berezin AA and Nakhmanson RS. Quantum Mechanical Indeterminism as a Possible Manifestation of Microparticles Intelligence. Physics Essays (Toronto) 1990; 3: 331-339.

Bohm D. A Suggested Interpretation of the Quantum Theory in Terms of "Hidden Variables". Physical Review 1952; 85: 166-193.

Bohm DJ and Hiley BJ. On the Intuitive Understanding of Nonlocality as Implied by Quantum Theory. Foundations of Physics 1975; 5: 93-109.

Burgers JM. Causality and Anticipation. Science 1975; 189: 194-198.

Cini M and Levi-Leblond J-M eds. Quantum Theory without Reduction. Bristol and New York 1990: Adam Hilger.

Cochran AA. Relationship Between Quantum Physics and Biology. Foundations of Physics 1971; 1: 235-250.

Cohen JE and Horowitz P. Paradoxical Behaviour of Mechanical and Electrical Networks. Nature 1991; 352: 699-701.

Comorosan S. The Measurement Problem in Biology. Int J Quantum Chemistry: Quantum Biology Symposium 1974; 1: 221-228.

Cufaropetroni N. Conditioning in Quantum Mechanics. Physics Letters 1991; A160: 107-115.

Cumming A and Linsay PS. Deviations from Universality in the Transition from Quasiperiodicity to Chaos. Physical Review Letters 1987; 59: 1633-1636.

Davenas E, Beauvais F, Amara J et al., Human Basophil Degranulation Triggered by very Dilute Antiserum Against IgE. Nature 1988; 333: 816-818.

Deutsch D. Quantum Theory, the Church-Turing Principle and the Universal Quantum Computer. Proceedings of the Royal Society of London 1985; A400: 97-117.

Deutsch D. Quantum Computational Networks. Proc Royal Society of London 1989; A425: 73-90.

Deutsch D. Quantum Mechanics near Closed Timelike Lines. Physical Review 1991; D44: 3197-3217.

Feynman RP. Quantum Mechanical Computers. Optics News 1985; 11: 11-20.

Frenkel A. Spontaneous Localizations of the Wave Function and Classical Behaviour. Foundations of Physics 1990; 20: 159-188.

Germine M. Consciousness and Synchronicity. Medical Hypotheses 1991; 36: 277-283.

Ghirardi GC, Rimini A and Weber T. Unified Dynamics for Microscopic and Macroscopic Systems. Physical Review 1986; D34: 470-491.

Ghirardi GC, Rimini A and Weber T. The Puzzling Entanglement of Schrödinger's Wave Function. Foundations of Physics 1988; 18: 1-27.

Grassberger P and Procaccia I. Characterization of Strange Attractors. Physical Review Letters 1983; 50: 346-349.

Gutzwiller MC. Chaos in Classical and Quantum Mechanics. New York: Springer-Verlag 1991.

Haken H. Synergetics - Nonequilibrium Phase Transitions and Self-Organization in Physics, Chemistry and Biology. 2nd Edition, Berlin, Heidelberg, New York: Springer Verlag 1978.

Herzel H-P and Pompe B. Effects of Noise on a Nonuniform Chaotic Map. Physics Letters 1987; A122: 121-125.

Hopfield JJ. Neural Networks and Physical Systems with Emergent Collective Computational Abilities. Proceedings of the National Academy of Sciences of the USA 1982; 79: 2554-2558.

Jahn RG and Dunne BJ. On the Quantum Mechanics of Consciousness, with Application to Anomalous Phenomena. Foundations of Physics 1986; 16: 721-772.

Jahn Robert G and Dunne Brenda J. Margins of Reality - The Role of Consciousness in the Physical World. Harcourt Brace Jovanovich, Publishers 1987, First Harvest/HBJ Edition San Diego, New York, London: 1988.

Jancso G, van Hook WA. Condensed Phase Isotope Effects.Chemical Reviews 1974; 74: 689-750.

Johnston AC. Light from Seismic Waves. Nature 1991; 354: 361.

Karolyhazy F. Gravitation and Quantum Mechanics of Macroscopic Objects. Nuovo Cimento 1966; 42A: 390-402.

Karolyhazy F, Frenkel A and Lukacs B. On the possible role of Gravity in the Reduction of the Wave Function. In: Quantum Concepts in Space and Time. R Penrose, CJ Isham eds., Oxford 1986: Clarendon Press, 109-128.

Kervran CL. Biological Transmutations. Crosby Lockwood, Bristol 1972.

Klein FS. Isotope Effects in Chemical Kinetics. Annual Review of Physical Chemistry 1975; 26: 191-210.

Landauer R. Information is Physical. Physics Today 1991; 44: 23-29 (May issue).

Leff HS and Rex AF eds. Maxwell's Demon: Entropy, Information, Computing. Princeton, New Jersey 1990: Princeton University Press.

Luckey TD. Hormesis with Ionizing Radiation. Boca Raton, Florida: CRC Press Inc 1980.

Margenau Henri. The Miracle of Existence. Woodbridge, Connecticut: Ox Bow Press 1984.

Miller DA. Useful Perspective on the Relation Between Biological and Physical Description of Phenomena. J Theoretical Biology 1991; 152: 341-355.

Nicolis G and Prigogine I. Self-Organization in Nonequilibrium Systems. From Dissipative Structures to Order Through Fluctuations. A Wiley-Interscience Publication, New York, London, Sydney,Toronto: John Wiley & Sons 1977.

Nieuwenhuizen Th M. Singularities in Spectra of Disordered Systems. Physica 1990; 167A: 43-65.

Oi T. Biological Information Processing Requires Quantum Logic. Zeitschrift für Naturforschung 1988; 43c: 777-781.

Pearle P. Combining Stochastic Dynamical State-Vector Reduction with Spontaneous Localization. Physical Review 1989; 39A: 2277-2289.

Peierls R. Spontaneously Broken Symmetries. J Physics A: Mathematical and General. 24: 5273-5281.

Penrose Roger. The Emperor's New Mind. Oxford University Press, New York 1989a.

Penrose R . Tilings and Quasi-Crystals; a Non-Local Growth Problem? In: Introduction to the Mathematics of Quasicrystals, MV Jaric ed, 1989b: Academic Press, Inc, pp. 53-79.

Popper K and Eccles JC. The Self and Its Brain. Berlin 1977: Springer Verlag.

Preparata G. Some Theories of "Cold" Nuclear Fusion: A Review. Fusion Technology 1991; 20: 82-92.

Radin DI and Nelson RD. Evidence for Consciousness - Related Anomalies in Random Physical Systems. Foundations of Physics 1989; 19: 1499-1514.

Rosales JL and Sanchez-Gomez JL. Non-linear Schrödinger Equation coming from the Action of the Particle's Gravitational Field on the Quantum Potential. Physics Letters 1992; A166: 111-115.

Rousseau DL. Case Studies in Pathological Science. American Scientist 1992; 80: 54-63.

Rubik B. Report on the Status of Research on Homeopathy with Recommendations for Future Research. Br Hom J 1989; 78: 86-96.

Rubik B. Homeopathy and Coherent Excitation in Living Systems. Berlin J Res Hom 1990; 1: 24-27.

Sarfatti J. Design for a Superluminal Signalling Device. Physics Essays, Toronto 1991; 4: 315-336.

Selleri F. Quantum Paradoxes and Physical Reality. Kluwer Academic Publishers, Dordrecht 1990.

Shalom A. The Body / Mind Conceptual Framework and the Problem of Personal Identity. Atlantic Highlands, New Jersey 1985: Humanities Press International, Inc.

Sompolinsky H. Statistical Mechanics of Neural Networks. Physics Today 1988; 41: 70-80 (December issue).

Stevens HH Jr . Evolution of Minds and Quantum Mechanical Fields. Physics Essays, Toronto 1990; 3: 126-132.

Stewart I. Catastrophe Theory in Physics. Reports on Progress in Physics 1982; 45: 185-221.

Turing AM. The Chemical Basis of Morphogenesis. Philosophical Transactions Royal Society of London 1952; B237: 37-72.

Wada H. Microscale Isotopic Zoning in Calcite and Graphite Crystals in Marble. Nature 1988; 331: 61-63.

Westerlund S. Dead Matter Has Memory! Physica Scripta 1991; 43: 174-179.

Wheeler JA. World as System Self - Organized by Quantum Networking. IBM Journal of Research and Development 1988; 32: 4-15.

Wigner E. The Limitations of the Validity of Present-Day Physics. In: Mind in Nature (Nobel Conference XVII), Elvee RQ ed, San Francisco 1982: Harper & Row Publ., pp.118-133.

Willy AV. Isotopeneffekte bei Chemischen Reaktionen. Stuttgart, New York 1983: Georg Thieme Verlag.

Wolf FE. On the Quantum Physical Theory of Subjective Antedating. J Theoretical Biology 1989; 136: 13-19.

Yates FE ed. Self-Organizing Systems: The Emergence of Order. New York and London 1987: Plenum Press.

Zeeman E. Catastrophe Theory. Reading, Massachusetts: Addison-Wesley Publishing Company 1977.

ABOUT QUALITY AND STANDARDS IN ULTRA HIGH DILUTION RESEARCH AND RESEARCH ON HOMOEOPATHY

J. Schulte

SUMMARY

Research on ultra-high dilution and on homoeopathy is discussed with respect to quality, standards and needs.

ZUSAMMENFASSUNG

Die Forschung über extreme Hochverdünnungen und Homöopathie werden in Bezug auf ihre Qualität, ihren Standard und notwendige Verbesserungen diskutiert.

MAIN PART

As briefly touched at the beginning of the chapter "Outline of physical methods", research in homoeopathy, as well as in ultra low dose, is not well-respected in current academic and industrial research communities. Besides some preoccupied views, there are serious critics on UHD research, and research on homoeopathy that need to be dealt with. Since this branch of research has had little (or no) space in established international academic research so far, a research subcommunity (subculture) has been formed apart from the established research. Thus, a variety of researchers from different sciences and different levels of qualification is roaming about the literature on complementary medicine. Although most of the researchers are experts in their very area only, they do not hesitate to give general interdisciplinary interpretations of the presented (mostly not verified) results. Since there is no real academic referee board that can cover the diversity of issues in UHD- and homoeopathy-research, based on academic credibility and competence with respect to a publication being proposed, almost anything can be published. As a consequence of this situation, the standard of many studies in research on UHD and homoeopathy is still poor. An evaluation of the literature on research in homoeopathy published in the English language has shown that only a few percent of the experimental research in physics and physiology can reach the desirable standard of scientific publications [1]:

a) complete documentation of the experimental set-up,

b) complete documentation of the uncertainties of the results presented,

c) independent reproducibility.

Similar results have been obtained by other authors [2-6]. Thus, for many publications, it cannot be concluded that they have a reliable scientific fundament. In spite of this fact, however, the publications are often considered as scientifically proved within the complementary science community. One can frequently find publications ending with a "to be proven" implication that somewhere else is referred to as "the author xyz has shown", i.e. feeding a cascade of interpretations and publications based on very weak foundation.

There is one issue that has to be addressed to research on homoeopathy in particular: even if research on homoeopathy passes all current standards of scientific publications, there is still one important factor left to be considered, actually to be considered first: the homoeopathic drug itself. Except in those countries where a legal, permissible, binding homoeopathic pharmacopoeia exists, i.e. where every single step in the preparation of homoeopathic remedies is prescribed, the drugs may be prepared according to homoeopathic

P.C. Endler and J. Schulte (eds.), Ultra High Dilution, 171–174.

principles, but it cannot be presupposed that they are manufactured in a standardized way [1]. These irregularities in the manufacturing method lead to several difficulties regarding the comparability of experimental data. A proof of these difficulties has already been given by some workers screening samples of different origin (same drug, same potency) and using identical assay methods for all of the samples [7,8]. Distinct results have been observed for almost each of the specimen!

Hornung and Linde [9] proposed some guidelines for the exact description of the preparation and mode of application of serial dilutions and potencies in publications on UHD effects and homoeopathic research. Although their proposal is under current discussion [10], it is worth mentioning their proposed publication requirements in detail here, especially for the journal, in which it has been published, is out of print:

Proposed Publication Requirements:

1.Mother Tincture / Primary Solution (denoted as P)

1a.The preparation of the mother tincture must be described. A reference to a homoeopathic pharmacopoeia is sufficient if the preparation is prepared in exact accordance with the instructions in that specific pharmacopoeia.

1b.A specification of the chemical composition of the mother tincture should be as detailed as possible. For agents with a clear chemical definition, both the concentration in mol/l and the molecular weight should be given. If this is not possible, e.g. with phytopharmaceutica, the concentration of the main constituents should be given in g/l; for certain agents, e.g. Interferon, in internationally determined units.

1c.The dilution (e.g. in steps of 1:10 -"D"- or 1:100 -"C"-) corresponding to the mother tincture must be specific, e.g. $P = 0C$, $P = 1C$ or $P = 2C$.

2.Dilution Procedure

2a.The solvent or medium used for dilution must be specific.

2b.The manufacturer of the drug must be specific.

2c.The dilution process must be described exactly, including information on sterility, laboratory vessels, and dilution proportions.

2d.Handling between dilution steps must be specified, including information on frequency and duration of vortexing and succussion, on mixing of any kind, and on periods of rest after succussion or vortexing.

2e.Dilution notations (e.g. 7C, 14C, etc.) should indicate a maximum of information about the preparation and concentration of the active agent.

3.Control

3a.The organization of control must be specified exactly. This applies to the control subjects as well as to the control preparations (placebo, etc.)

3bThe composition of the control substance (placebo, solvent, etc.) must be specified.

3c.The preparation of the control substances must be specified, including whether or not they underwent the same dilution or potentisation process as the probe preparations.

4.Mode of Application

4a.The dose applied must be exactly specified. The absolute number of molecules of the diluted agent should be indicated where it is possible to do so, e.g. if the active agent is a well-defined chemical substance. In cases where the absolute number of molecules cannot be given, the exact volume of the applied probe and control preparation must be at least specified.

4b. Due to possible chronobiological variations, the application of the test and control preparations must be specified with regard to time, day and year in *in vitro* as well as in *in vivo* experiments. The age and source of any animals used in the experiments should be given.

5. Dilution and Potency Notations (see Glossary, p. 221)

To express a dilution, the vast majority of publications on serial dilution effects uses the terms DH (Hahnemann decimal, 1:10) and CH (Hahnemann centesimal, 1:100), along with the number of dilution steps. In many cases, however, the preparation of a dilution does not follow the instructions in any homoeopathic pharmacopoeia. The notation CH has often been used to indicate vortexing instead of succussion. Therefore expressions such as 7CH no longer have a clear meaning. Here it is proposed to drop the H and express it with D and C, respectively, where dilutions are prepared in serial ten or hundred-fold steps with succussion or any intense mechanic manipulation. In order to identify different, already well-established (and protected) national homoeopathic pharmacopoeia, and for better differentiation, the following notation is proposed:

7CV for a vortexed 7C dilution,

7CD for a potency prepared according to the official German homoeopathic pharmacopoeia, 7CF for a potency prepared according to the official French homoeopathic pharmacopoeia etc. The letters D, F, may be the country identification letters as used for international vehicles identification.

Furthermore, guidelines for clinical research in homoeopathy are developed [6,11].

In a recent publication, Popp [12] suggested a criterion for the reliability of homoeopathic experiments that is based on the credibility of results, i.e. on the real significance level. For the order of the real significance level Popp proposed the significance relation

$$a = 1 - (1 - a')^m ,$$

where a' is the announced significance level, and m' is the number of possible dependent measurements, which have been documented in such a way that makes one believe that the measurements are definitely independent. For a series of experiments where one out of ten different dilutions is significantly different from the control, and a claimed significance of $a' = 0.05$, the real significance level would be

$$a = 1 - (0.95)^{10} = 0.4 .$$

With Popp's proposed real significance level the results of the experiments are reliable on a significance level of about 60%, i.e. far off from what the experimenter claims. Thus, as a criterion for the reliability of homoeopathic experiments, Popp proposed that the experimenter should be required to verify the efficacy of the remedy claimed in 20 blind samples (including 10 controls), while allowing for only one failure. Besides the technical publication requirements, the publications themselves need to pass the standard referee procedure well-known in academic and industrial science, i.e. proposed publications have to be referred by people that are competent with respect to the particular subject of the publication. Critically refereed publications serve the quality and reliability of all research in UHD and homoeopathy. Also, it is a base of opening pathways to academic interdisciplinary work and independent academic funding.

ACKNOWLEDGEMENTS

The author is grateful to K. Linde, Munich, and to P.C. Endler for their help.

REFERENCES

1. King G. Experimental Investigations for the Purpose of Scientific Proving of the Efficacy of Homoeopathic Preparations - A Literature Review about Publications from English-speaking Countries. Thesis, Tierärztliche Hochschule Hannover 1988.
2. Majerus M. Kritische Begutachtung der wissenschaftlichen Beweisführung in der homöopathischen Grundlagenforschung. Gesamtbetrachtung der Arbeiten aus dem frankophonen Sprachraum. Thesis, Tierärztliche Hochschule Hannover 1990.
3. Scofield AM. Experimental research in homeopathy - a critical review. Br Hom J 1984; 73 (3):161-180; 73 (4): 211-226.
4. Righetti M. Forschung in der Homöopathie - Wissenschaftliche Grundlagen, Problematik und Ergebnisse. Göttingen: Burgdorf Publisher 1988.
5. Hill C, Doyon F. Review of randomized trials of homoeopathy. Rev Epidem et Sante Publ 1990; 38: 139-147.
6. Kleijnen J, Knipschild P, ter Riet G. Clinical trials of homoeopathy. Br Med J 1991; 302: 316-323.
7. Khan MT, Saify Z. J Am Inst Hom 1975; 68: 97-104.
8. Verma, HN. Adv Hom 1981; 1: 70-71,73.
9. Hornung J, Linde K. Guidelines for the Exact Description of the Preparation and Mode of Application of Serial Dilutions and Potencies in Publications on Ultra Low Dose Effects and Homoeopathic Research - A Proposal. Berlin J Res Hom 1991; 1(2): 121-123.
10. Linde K, Melchard D, Jonas WB, Hornung J, Wagner H. Methodological guidelines for confirmatory experimental studies on serial dilutions: a proposal. Abstracts of the GIRI Meeting, Munich 1992.
11. Hornung J. An overview of formal methodological requirements for controlled clinical trials. Berlin J Res Hom 1991; 1 (4/5): 288-297
12. Popp FA. Some remarks on research in homeopathy. Berlin J Res Hom 1991; 1(3): 172-173.

Information transfer from an ultra-high dilution

PART 3: BIOPHYSICS

"Now there is evidence that it is the informational aspect of biological systems that characterizes the essential view of life. And this is less reflected by biochemical findings, than by a level beyond the domain of chemical reactivity, namely that of electromagnetic fields. Actually, all the recent high-lights of biosciences, as for instance, dissipative structures, deterministic chaos, coherent excitations, as well as solitons, excitons and other oscillations in living tissues, have their real origin in electromagnetic couplings of small up to rather large-sized sub-units of biological systems."

K.H. Li in Popp, Electromagnetic Bio-Information, 1989

"Think of a singer, her song and yourself - that's it"

J. Benveniste, annual meeting of the ECRH, spring 1993

SOME BIOPHYSICAL ELEMENTS OF HOMOEOPATHY

F.A. Popp

SUMMARY

The extremely high sensitivity of biological systems provides the key to the understanding of the "homoeopathic effects". According to the classical analogue of a pendulum, the "coherent states" described in quantum physics are able to explain the basic effects of agitated high dilutions as described in "homoeopathy" (the law of similarity as well as the potency rule, see Glossary, p. 219) as interrelated phenomena. The basic effect is always a delocalization of the energy in a resonance-like interaction between emitter and absorber. The sick body works there as a boson store where "wrong" oscillations give rise to perturbations of the regulation of the body's processes. The homoeopathic remedy acts there as a resonance absorber of wrong oscillations as soon as the correct remedy is chosen.

ZUSAMMENFASSUNG

Es ist die extrem hohe Empfindlichkeit biologischer Systeme, die den Schlüssel zum Verständnis "homöopathischer Wirkungen" darstellt. Entsprechend der klassischen Analogie eines Pendels können die aus der Quantenphysik bekannten "kohärenten Zustände" die grundlegenden Wirkungen verschütteter hochverdünnter Lösungen, wie sie in der Homöopathie beschrieben werden (Simile-Prinzip und Potenzierungsregel), als zusammengehörige Phänomene erklären. Der wesentliche Effekt ist dabei immer eine Energie-Übertragung in einer resonanzartigen Beziehung zwischen Energiequelle und Absorber. Hierbei arbeitet der kranke Körper als ein Bosonenspeicher, in dem "falsche" Schwingungen Störungen der Regulation der körpereigenen Prozesse verursachen. Das homöopathische Heilmittel wirkt hierbei als Resonanz-Absorber falscher Schwingungen, falls das passende Heilmittel gewählt wurde.

INTRODUCTION

Since the publication of the papers by Benveniste's group (1), the pressure put on scientists who are working on questions of homoeopathy (and related fields) has increased in such a way that only extremely careful work can help to overcome this critical phase of understandable scepticism.

In the present situation, it is nevertheless worth-while to focus again on the problem of high potencies (= agitated high dilutions), in order to answer the decisive momentary question as to whether high potencies have the capacity to work at all, or, in other words, whether the "unbelievable" (2) is actually believable.

Apart from Benveniste's approach, a variety of experimental investigations was already devoted to the proof of the efficacy of high potencies, starting in 1828 (3) and continuing up to the present day, and covering the fields of physio-chemical investigations (e.g. the papers of Boiron -4-), studies on plants (e.g. L.Kolisko -5-, and Pelikan and Unger -6-), animals (e.g. Stearns -7-, Wurmser -8-) and clinical investigations (e.g. Paterson -9-, Reilly -10-). Although most of the results indicate that potencies can be actually effective, none of them are reliable, generally accepted and adequate to exclude non-reproducibility (e.g. Galen Ives -11- on papers of Boiron), or to exclude hidden errors with certainty, due to the complexity of the method.

The author of the present paper has been both theoretically and experimentally involved in the question of the efficacy of high potencies for many years (12-15), in particular as

P.C. Endler and J. Schulte (eds.), Ultra High Dilution, 177–185.

supervisor of some governmental research projects on this topic (16). Although some recent new approaches to the problem exist from the side of theoretical physics (17; see this volume), postulating, and also giving some reasons, for "memory effects of water" (18,19), the most serious basis and starting point of a discussion at present seem to be the scientifically well established and generally accepted phenomena of

1. placebo effects,

2. hormesis effects and the

3. extraordinarily high sensitivity of living systems.

All three of these phenomena are directly linked to the crucial questions of homoeopathy, and are, on the other hand, subject to the investigations of modern, established science. They are not, however, understandable in terms of conventional drug research. This means that clear and undisputable effects in the field of homoeopathy force us to correct our paradigms of drug action in such a way that a more profound understanding of homoeopathic treatments may be reached. Let us confine ourselves here to some necessary conclusions which invite a more fundamental, but simple model of understanding of high potency effects within the general framework of drug research.

MAIN PART

Contradictions within the conventional approach

1. *Placebo thesis*

According to the present classical medical understanding, homoeopathy should work as a placebo therapy, where the essential factors of help are the "mutual trust relationship" between physician and patient and the "faith" in the medication but not the medicine, and hence certainly not a substance-specific component of the medicine.

Actually, this means that the conventional medical school believes in "non-substantial" interactions based on some kind of specific "information", which helps in at least 50 to 70% of "psychosomatic disorders".

The strange point in this situation is the fact that (as shown in Table 1) mainstream medical thinking postulates that there exists a gap between substantial (biochemical) and non-substantial (placebo) interactions, thus rejecting the possibility that there is a continuous transition, based on interactions with more or less impact on the biochemical as well as on the "psychological" control system of the body.

Non-substantial influences on biochemical pathways as well as substantial interactions on the psychic system are excluded in that case.

Interaction	Target	Effect
Nonsubstantial	Belief-system	Placebo
Substantial	Biochemical	Drug action

Tab. 1: Conventional model of healing.

Experience, however, contradicts this rigorous division of interactions into a completely biochemical and an entirely psychological one. Drugs can obviously influence the psychic situation as, in turn, psychological states (e.g. stress, frustration, etc.) may distinctly change the pattern of biochemical products in the body. Consequently, the model of a continuous transition from substantial interactions to non-substantial ones is not only the more general situation than the discontinuous division of the reality into a local and a non local one, but it also reflects the truth much better than the oversimplified conventional approach. It is worthwhile noting that the elementary rule of psychotherapy, i.e. to remind the patient of events that stimulate almost the same feelings as those of his "trauma", is in accord with the simple principle (see below) of homoeopathy. This invites us to extend Table 1 to a more general picture of interactions (Table 2).

Field	Distance of interactions	Substance specifity	Biological significance
Nuclear physics	10^{-12} - 10^{-8} cm	very high	low
Molecular physics, chemistry	10^{-8} - 10^{-7} cm	relatively high	moderate
Homoeopathy	10^{-7} - 10 cm	moderate	high
Psychology	10 cm - 100 cm	low	very high

Tab. 2: For explication, see Table 1 and text.

With decreasing substance specifity and at the same time increasing delocalization, the "informational" aspect of the interaction gains importance as compared to the purely chemical (substantial or local) one.

2. Hormesis effects

With increasing delocalization a further effect is brought up which is of increasing interest to modern biochemists (20). This is the hormesis (or hormoligosis) effect (21) which clearly shows biphasic dose/response relations. This means that the Arndt-Schulz law (22) (which states that weak stimuli stimulate vital activity, medium strong stimuli may support this activity, while the strongest stimuli may annul vitality) is now an experimentally proven principle of biochemistry. Such reversed biological effects in various ranges of concentrations of the same agent correspond exactly to the basic principles of the simile and the potency rule, clearly supporting these elementary, but at present not generally accepted, laws of homoeopathy. Unfortunately, this argument does not convince biochemists. It was concluded:

The Arndt-Schultz Principle which is again and again used as a possible scientific explanation for the efficacy of homoeopathic remedies might, at best, be used for lower potencies. According to Stebbig (21), who investigated the hormesis effects, it is even extremely doubtful whether the Arndt-Schultz principle is valid at all at the potency levels which are normally used in homoeopathy (20).

Homoeopaths may object that the liquid homoeopathic medication is not merely a dilution but a succussion of the agent. It might be that the simile and potency principle (see below: Elements of a model of the effects of homoeopathy) based on the Arndt-Schultz law becomes more marked on succussion. The most recent results on changes in the water structure through interactions with solutes (23) emphasize this hypothesis.

It should be realized, however, that the mechanism of the Arndt-Schultz principle is not and cannot be explained in terms of biochemistry and that biochemists will therefore be unable to make valid predictions about its validity and its connection to the law of similarity. Their usual approach is to attribute unexplained mechanisms to complex reactivities of receptors or enzymes.

However, even the most intelligent "X-ors" or "Y-ases" are unable to lead to an understanding of a non-linear allosteric reaction (feedback) at low substance concentrations within the scope of the local reciprocal action based on biochemistry. For concentration values C_j of agent j with C_j tending to 0, all biochemical models become necessarily linear, where

$$\Delta W = \sum_j \alpha_j \Delta C_j \qquad [1]$$

ΔW is the change of any effect, and ΔC_j the change in concentration of substance j. α_j is local in regard to its biochemistry and must therefore be seen here as a parameter.

One further observation can be added which cannot be explained by the above relation, but the fact is undisputed: within the scope of the usual concentrations used in homoeopathy, synergetic biological effects appear. We observed (16) them clearly in experiments with formaldehyde up to higher potencies, and with oleander up to the fifth decimal potency. This means: While biochemists can interpret α_j in the relation only locally, the observations (Arndt-Schultz principle and synergisms at low concentrations) demand that α_j is non-local, and with that the relation is non-linear.

The summarizing conclusion is that the biochemical interpretation of homoeopathy must fail as much as the placebo thesis does; however, they become consistent when non-local substance-specific effects are included in their framework.

3. Extraordinarily high sensitivity of biological systems

Besides placebo effects and hormesis, a third undisputable phenomenon, not yet integrated into the common scientific frame, points to the direction of homoeopathic efficacy, namely the extraordinarily high sensitivity of biological systems. Living systems are able to emit coherent electromagnetic waves and also show sensitive responses throughout the whole spectral range. This emission depends sensitively on all biological processes, as cell cycle phase or growth. Examples and some explanations have been presented elsewhere (24,25).

Let us confine ourselves here to the thermodynamic aspect of "electroreception", which may lead to an introduction of our model of homoeopathic efficacy.

According to our measurements in the optical region of the spectrum (26), the relative signal/noise-radio (SNR) of an ideal living system follows the law

$$\beta(v) = \frac{f}{\exp\left(\dfrac{hv}{kT}\right) - 1} \qquad [2]$$

where f is a constant. b(n) is the *relative* SNR compared to that of a therminal equilibrium system.

For the radio-wave range we simply get:

$$\beta(v) \approx f \cdot \frac{kT}{hv} \qquad [3]$$

indicating that a living system gains relatively more and more sensitivity with decreasing frequency n of the signal. C. Smith (24,25) reported a variety of examples where just this high sensitivity of humans in the radiowave-range has been demonstrated (see also the contribution of Smith in this book). For external optical signals the SNR of biological systems may, according to [2], obtain lower values than expected from an equilibrium system. This can be interpreted as the possibility of protection against relatively high quantum energies. However, since the system can work as an amplifier, too, the optimum SNR in the optical range in the interior of the system amounts to

$$SNR = -\dot{n}_v \tau_v \quad [4]$$

where τ_v is the coherence time of the signal of intensity $\dot{n}_v$ (number of photonquanta /sec).

From [2], [3] and [4] we realize that the sensitivity of biological systems cannot be described in terms of biochemistry. Rather, it is based on deviation from thermal equilibrium.

Sickness can in general be traced back to definite changes of the SNR (n), where f as a *constant* according to [2] provides health, and the pattern of characteristic deviations from f = const, e.g. the temporal behaviour of the spectral distribution f(n) of occupation of the manifold excited stages in the biological system (translations, spin states, rotations, vibrations, electronic excitations) determines the spectral sensitivity, the transparency and regulatory activity within the system, as well as its "behaviour" in terms of health and disease.

Elements of a model of the effects of homoeopathy

A theory of homoeopathy is incomplete if it does not describe the basic statements, namely

(1) the law of similarity, and

(2) the potency rule, including high potencies.

(Annot. by the editor: The following example is given to illustrate the reactional similarity -'Law of Similarity' - which is the basis of homoeopathy: A bee sting can induce an inflammatory reaction concomitant with cutaneous erythema. Homoeopathically prepared dilutions of bee and bee venom are used in order to inhibit and cure cutaneous erythema induced not only by bee stings, but also by ultraviolet rays - see Part 4, Glossary.) The 'potency-rule' says that through the homoeopathic process of stepwise agitation and dilution of a mother substance, information is transferred to the solvent. It is assumed that this information is the more marked ('sharp'), the more often the process of agitation and dilution has taken place.

Any consistent model which explains *(1)* and *(2)* is sufficient to reject the postulate that homoeopathic efficacy is 'unbelievable'. At the same time it can act as a platform on which convenient experimental investigations can (and should) be performed.

Most, if not all, hypotheses of homoeopathic efficacy do not explain *(1)* and /or *(2)*. They postulate some memory-functions of water, which, of course, are necessary, but certainly not sufficient conditions for the efficacy of high potencies. The memory function of water is postulated in all models on the basis of a possible entropy change. No model explains, however, why the thermal dissipation of water structures takes an extraordinarily long time (27). But this point is crucial, since from a theoretical point of view an infinite number of different mechanisms may provide some substance-specific memory of a fluid or solid. However, the memory *time* , which has to last for some weeks or even years depending on the thermal dissipation, is the most decisive parameter of such a model: how can thermal dissipation be avoided to such an 'unbelievable' extent?

Let us confine ourselves to some elementary questions. At least we should find a way to describe the simile principle and the potency rule in a consistent way, including possible answers to the problem of thermal dissipation. Therefore we may start with a mechanical model, e.g. a system of two weakly coupled pendula (Figure 1).

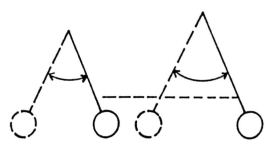

Fig. 1: Two weakly coupled pendula represent the most simple physical model of homoeopathic efficacy, since they satisfy *(1)* the simile principle in the case of the resonance of their frequencies, *(2)* the potency rule in so far as with the lowest starting amplitude the highest energy transfer can take place.

The physical state of this system can be described by the amplitudes $A_1 (t_1)$ and $A_2 (t_2)$ at any instant t_1 and t_2, and by the frequencies v_1 and v_2 of the pendula 1 and 2 respectively. The initial values may be $A_1 (0)$ and $A_2 (0)$. A_1 represents some spectral component of a wrong regulatory ('disregulatory') oscillation within the patient's body, while A_2 stands for the homoeopathic medicine, e.g. an oscillation away from thermal equilibrium. The goal of the treatment is then to transfer energy from system 1 to system 2.

It is easy to show (16) that the highest possible efficiency for a weakly coupled system can be achieved, when $v_2 \approx v_1$, thus reflecting the simile-principle, and $A_2 \rightarrow 0$, just according to the potency-rule.

The result of this coupling is $A_2 (\infty) = A_1 (\infty)$ to $A_1 (0)/2^n$, where ∞ represents an infinitite time after coupling.

In this model, the homoeopathic medicine induces a resonance transfer of disregulatory energy from the patient's body to the absorbing homoeopathic dilution which is then excreted. The n-times repeated treatment leads to a decrease of the original amplitude $A_1(0)$ to $A_1(0)/2^n$.

Just the same result is obtained if one takes into account, more generally, coherent photon states. The latter are defined as eigenstates of the annihilation operator a, where γ is the field amplitude.

$$a \, |\gamma\rangle \; = \; \gamma \, |\gamma\rangle \qquad\qquad [5]$$

Let us start with the initial state $|i\rangle \; = \; |\gamma_1\rangle + |\gamma_1'\rangle$, where $|\gamma_1\rangle$ represents the disregulatory field in the patient's body, while $|\gamma_1'\rangle$ describes the field of the medicine.

We provide a very weak coupling c which shall keep coherent photon (boson) states coherent. This means that it commutes with the Hamiltonian of the system. Secondly, the conservation of the boson numbers must be valid, since the interaction away from equilibrium provides a photochemical potential (26). The final state $|f\rangle$ can be simply calculated by diagonalization of the matrix

$\langle i|\chi|f\rangle$, where

$$|f\rangle = |\gamma_2\rangle + |\gamma_2'\rangle \exp(i\Phi),$$ exp(iΦ) is a phase-factor. After straight-forward calculations (16) we get as a result

$$|\gamma_1|^2 + |\gamma_1'|^2 = |\gamma_2|^2 + |\gamma_2'|^2.$$

Consequently, the highest efficacy of energy transfer is obtained for χ working as a delocalized operator, coherently and in resonance, just according to the law of similarity, and

$$|\gamma_1'\rangle \equiv |0\rangle,$$ according to the potency rule, where $|0\rangle$, (g=0),

represents the vacuum state.

DISCUSSION

Let us assume that, in most general terms, coherent states are responsible for the efficacy of homoeopathy in the biological system as well as in the homoeopathic medicine. This supposition is based on the fact that the simile - and potency - rule is to some extent reflected by physical properties of coherent states. However, the following questions have to be answered. If we accept that biological systems display coherent states (24),

- What is then the mechanism for the creation of coherent states in homoeopathic remedies?

- How can it be verified experimentally?

There exists a variety of proposals for the creation of coherent states in homoeopathic medicines. According to E. del Guidice (18), the Goldstone theorem alone provides a sufficient condition of coherent states, as soon as a certain symmetry breaking takes place by the process of succussion. We showed in 1978 that the introduction of a momentum operator into the Hamiltonian of the dilution describing the succussion procedure considerably favours the creation of coherent states. Their lifetime is in the order of 10^7s, or even higher. From a theoretical point of view, coherent states must not decay at all. However, this ideal situation will never occur in reality. Rather, the high number of non-thermal bosons introduced by mechanical shaking provides relatively long decay times when dropping the quasi-thermal occupation probabilities. In addition, with progressive relaxation, the decay will turn (by the change of its exponential decay law into a hyperbolic one) more and more into the stable subradiance regime, which has been postulated by Dicke, and which is now experimentally verified (28). Consequently, our model does not only describe the simile and the potency rule on the basis of coherent states. It also makes it possible to propose a clear physical mechanism (phonons of the succussion frequency), which provides coherent states of a sufficiently long lifetime, as soon as Dicke's theory and the high sensitivity of biological systems are taken into account.

As a result, we are able to cite and propose experiments which help to examine the hypothesis. For instance, the experiments of C.W. Smith et al. with allergic patients indicate that sensitive biological systems respond to coherent radiation, even when usual technical antenna systems do not show significant signals (see Smith's contribution).

Own experiments and some which were performed in our laboratory (16) point to possible significant changes in the sensitivity itself, when biological systems (in our case plant seedings) are treated with homœopathic potencies. Indications of synergetic effects and hormesis were found in these cases.

Some essential consequences of our model are the following:

- It is useless to look for physical/chemical differences of dilutions and succussions, as long as non-thermal phonons of intensities of at least 10^{-12} W cannot be registrated by the measuring equipment.

- Best suited for experimental investigations of this kind are bio-indicators, which respond according to the simile principle as well as to hormesis. The problem is to find an optimum range where the antagonism of low sensitivity and good reproducibility on the one hand (for instance by the use of cell cultures), and high sensitivity and bad reproducibility on the other hand (e.g. in animal experiments) has to be minimized. A way to solve this problem can be a systematic study on the same medicine over a wide range of species, covering different degrees of differentation without changing specific response functions.

- Further progress can be expected by a continuation of C. Smith's experiments, linking the gap between purely substantial and purely electromagnetic coupling of the biological system and the external agent. Smith's results indicate that there is no fundamental difference between electromagnetic interactions and the effects of substances, which again according to both C. Smith and our own investigations work by means of their electromagnetic pattern (see the contribution of Smith and Endler). A systematic analysis of dose-response relations of biological systems exposed to a variety of electromagnetic fields over the whole spectral range will certainly help us to understand homoeopathy more profoundly

ANNOTATION by the editor

Important ideas outlined in this paper have been described by the author in the British Homoeopathic Journal, vol. 79, n. 3, July 1990. This presentation has been reviewed, enlarged and adjusted for this book.

Researchers involved in this field are presently discussing the Arndt-Schultz observation in detail.

REFERENCES

1. Davenas E, Beauvais F, Amara J et al. Human basophil degranulation triggered by very dilute antiserum against IgE. Nature 1988; 333: 816.
2a) When to believe the unbelievable. Nature 1988; 333: 787.
 b) 'High-dilution' experiments a delusion. Nature 1988; 334: 287.
3. Hahnemann S. The Chronic Diseases. New Dehli: Jain Publishers 1981.
4. Boiron J. Etude experimentale des conditions d'efficacité de dilutions hahnemanniens. Vol des Rapports, Congress LMHI, Lyon, 1985, 89.
5. Pelikan W, Unger G. The activity of potentized substances. Br Hom J 1971; 60: 233.
6. Kolisko L. Physiologischer und physikalischer Nachweis der Wirksamkeit kleinster Entitäten, 1923-1959. Stuttgart: Arbeitsgem. antroposoph. Ärzte, 1961.
7. Stearns GB. Experimental Data on One of the Fundamental Claims of Homoeopathy. J Am Inst Hom 1925; 18: 433; 790.

8. Wurmser J. Pharmacologie des microdoses. Hom Fr 1985; 73: 113.
9. Paterson J. Report on Mustard Gas Experiments (Glasgow and London). J Am Inst Hom 1944; 37: 47; 88.
10. Reilly D, Taylor M. Is Homoeopathy a Placebo response; Controlled Trial of homoeopathic potency, with pollen in Hayfever as model. The Lancet 1986; 19: 881.
11. Ives G. Relative Permittivity as a measure of homoeopathic potency effect: Negative results on repeating previous work. Br Hom J 1983; 72: 65.
12. Popp FA. Gutachten zum Wirksamkeitsnachweis der Homöopathie, Bundesgesundheitsamt, Berlin 1979.
13. Popp FA. Deutungsversuche homöopathischer Effekte aus moderner physikalischer Sicht. Allg hom Ztg 1978; 223: 46; 93.
14. Popp FA. Homöopathie - Placeotherapie oder Leitschiene einer modernen Medizin. Erfahrungsheilkunde 1980; 29: 570.
15. Popp FA. Wirkmodelle der Homöopathie. Dt J f Hom 1985; 3: 215.
16. Popp FA. Bericht an Bonn. Essen: Verlag für Ganzheitsmedizin 1986.
17. Dicke RH. Coherence in Spontaneous Radiation Processes. Phys Rev 1953; 93: 99.
18. Del Giudice E, Preparata G. A Collective Approach to the Dynamics of water. MITH 1989 /10, Geneva: CERN-Library. See Del Giudice's contribution in this book.
19. Berezin AA. Isotopic 'Lattice Ghosts' as a Possible Key to Memory Effects in Water, Pers. Comm. 1989 from CERN (F. Caspers). See Berezin's contribution in this book.
20. Wagner H. Symposium: Biologisch pharmakologische Wirkung kleinster Dosen. Inst f Pharmaz Biologie, Universität München, 2./3. April 1987.
21. Stebbing A. Hormesis - The Stimulation of Growth by Low Levels of Inhibitors. Total Environ 1982; 22: 213.
22. Schultz, Pflügers Archiv f Physiologie 1888; 42: 635.
23. Hüttenrauch R, Fricke S. Veränderung der Wasserstruktur durch Gelbildner. Pharmazie 1987; 42: 635.
24. Popp FA, Warnke U, König HL, Pleschka W. Electromagnetic Bio-Information. Munich: Urban & Schwarzenberg 1989.
25. Smith CW. Electromagnetic Man. London: J.M. Dent 1989.
26. Popp FA, Li KHY, Mei MP et al. Physical Aspects of Biophotones. Experientia 1988; 44: 576.
27. Righetti M. Forschung in der Homöopathie. Göttingen: Burgdorf Publishers 1988.
28. Crubbellier A, Pavolini D. Superradiance and Subradiance II. J Phys B At Mol Phys 1986; 19: 2109.

ELECTROMAGNETIC AND MAGNETIC VECTOR POTENTIAL BIO-INFORMATION AND WATER

C. W. Smith

SUMMARY

The phenomena of bio-information and water are considered in relation to electromagnetic fields. This first came to the fore when treating electromagnetically hypersensitive patients for whom the use of water treated with an alternating magnetic field, or vector potential, at a frequency specific to the patient, can be therapeutic. Resonances thus written into a sample of water can be read out by a subjective method, but have now also been detected with a low-noise, narrow-band high-gain amplifier. The various threshold conditions found for magnetic potentisation (= influence on the water) are given. Resonance frequencies found in glass ampoules of water held by patients have been used for diagnosis and subsequent therapy. Resonances have also been measured in homoeopathic potencies (= agitated high dilutions) and these must relate to the bio-information imparted during their preparation and "read" by the body following clinical application. It is proposed that the frequency information is carried on the magnetic vector potential and that the magnetic field or succussion (= agitation) "formats" the water.

ZUSAMMENFASSUNG

Phänomene von Bio-Information und Wasser werden in Beziehung zu elektromagnetischen Feldern betrachtet. Anhaltspunkte dafür hatten sich in der Behandlung elektromagnetisch hypersensitiver Patienten ergeben, für die die Verwendung von mit einem magnetischen Wechselfeld oder einem Magnet-Vektor-Potential behandeltem Wasser therapeutische Wirkung haben kann. Resonanzeigenschaften, die auf diese Weise in eine Wasserprobe "eingeschrieben" wurden, können mit Hilfe einer subjektiven Methode herausgelesen werden; sie wurden nun aber auch mittels eines rauscharmen Schmalband-Hochleistungsverstärkers gefunden. Die unterschiedlichen Schwellenbedingungen, die für magnetische Potenzierung (= Beeinflussung des Wassers) gefunden wurden, werden angegeben. Resonanzfrequenzen, die von Glasampullen, welche von Patienten gehalten werden, abgeleitet wurden, wurden in Diagnose und nachfolgender Therapie verwendet. Resonanzen wurden auch in homöopathischen verschüttelten Verdünnungen gemessen; diese müssen in Zusammenhang mit der Bio-Information stehen, die während der Zubereitung eingebracht wurde und vom Körper nach der klinischen Anwendung "gelesen" werden. Es wird der Gedanke vorgeschlagen, daß die Information der Schwingungen durch das Magnet-Vektor-Potential übertragen wird und daß ein magnetisches Feld oder Verschüttelung das Wasser "formatiert".

INTRODUCTION

General remarks

The objective the writer set himself two decades ago was to understand the physics involved in the electromagnetic interactions of living systems. Energy calculations in general are only dependent on the initial and final states of the system and are independent of the physical processes by which this change of states is achieved. The fundamental energy limitations of the thermal environment (kT) lead to critical threshold electric and magnetic fields, and incident energy flux densities for effects on coherently responding living systems of given dimensions [1]. There is an upper limit to the field or energy "window", this occurs at fields strong enough to break the coherence of the system.

P.C. Endler and J. Schulte (eds.), Ultra High Dilution, 187–201.
© 1994 *Kluwer Academic Publishers. Printed in the Netherlands.*

It is on the threshold of electromagnetic interactions that the fundamentals of the physics are to be observed. This was demonstrated in measurements on simple cell systems such as bacteria and yeast [2] for which the unexpected magnetic field effects were of particular significance. The onset of these effects occurred when the magnetic flux density was sufficient to have an average of a single quantum of magnetic flux linking the measured area of a single average cell.

Synchronously dividing cells may emit radiofrequency oscillations which are too coherent to be a narrow-band noise and must be limited by quantum fluctuations [3, 4]. The implications of this finding are that living systems may be able to use both quantum sensitivity and the Josephson effect as discussed on the basis of quantum field theory by Del Giudice and co-workers [5].

The interactions of the human system with electromagnetic fields are best demonstrated through the measurement and therapy of those hypersensitive persons who have acquired a hypersensitivity to electromagnetic fields [2, 6]. The existence of this phenomenon has now been confirmed through double-blind trials [7]. This hypersensitivity appears to be the result of a failure in some patient-specific regulatory system which results in patient specific reactions to normally imperceptible electromagnetic fields at precise and patient-specific frequencies, resulting in resonances forming a non-linear harmonic series extending from hertz to gigahertz [6].

As a consequence of the work with these patients, the writer has found out that they can also respond to water in which the same frequencies have been imprinted. This imprinting may be done with an alternating magnetic field or magnetic vector potential. It may also be done by letting a patient hold a tube of water (e.g. a spring water) in the hand for a short while. The water picks up the body resonances and the water can be used for the surrogate testing of patients too ill to be exposed even to very weak coherent electromagnetic oscillations.

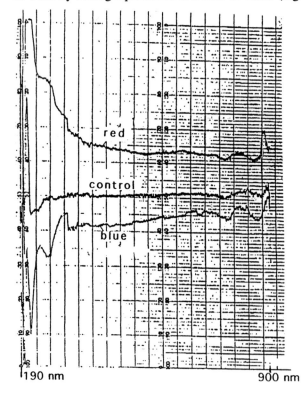

The impared resonances can be detected in water by subjective methods as described by the writer in Dallas in 1991 (33; see also the contribution by Smith et al. in this book), but further work has enabled these to be detected with a suitable low-noise, high-gain narrow-band amplifier [8] as shown in Figure 1.

Fig. 1: The upper tracing shows a magnetically imprinted resonance at 1.0 kHz and serves as a control for the lower tracing which is the same sample of water into which the frequency 1.1 kHz has additionally been imprinted.

190 nm   900 nm

Similar results have been obtained at Wekroma AG using a Fourier transform Spectrometer to signal average over a large number of measurements.

There is a close similarity between the potentisation of homoeopathic remedies and the preparation of neutralising dilutions of allergens by serial dilution as used for the treatment of allergy by provocation-neutralisation therapy. This reflects the fundamental duality between coherent frequencies and the chemical bond, which may be considered as another of the "consciousness conjugates" proposed by Jahn [9]. The fact of chemical analysis by spectroscopy demonstrates that for each molecular structure there is a corresponding and unique frequency pattern corresponding in turn to the appropriate summation of all the chemical bond frequencies. Superimposed on these are frequencies derived from the function of the living system.

The broad background to electromagnetic field phenomena and living systems, including diagnostic techniques and therapy for electromagnetic hypersensitivity, is covered in "Electromagnetic Man" [10] and the research papers cited therein.

Homoeopathy

The basic bio-information phenomena of homoeopathy requiring explanation in terms of the physics of water are:

1. Similia Similibus Curentur - Like cures like. ("Law of Similarity", see p. 221.)

2. Potentisation - Serial dilution with succession. (See p. 221.)

3. Commonly prescribed homoeopathic remedies are on a logarithmic progression of potency,

4. A wide range of molecular structures are readily potentised by dilution and succussion or trituration.

These range from single atoms such as Ag, Au, Cu, Fe, Pd, Se, Zn, to macromolecules of biological origin. All "mother tinctures" can be serially diluted with water, alcohol and many other liquids, and after succussion (solids after trituration) they give clinically effective potencies far beyond the 12th centesimal dilution, which is the serial dilution corresponding to the Avogadro (Loschmidt) Number. According to the laws of chemistry no original material from the "mother tincture" should remain beyond this point. Yet, the observed clinical effectiveness of such potencies and the laws of physics remain at all dilutions. Hahnemann [11] was well aware of the clinical effects of potencies of "electricitas", "magnetis" and "X-ray".

5. A sealed glass ampoule containing a homoeopathic remedy can produce measureable effects while placed directly in contact with the skin or connected to the skin by a metallic conductor [12] (see also the contribution of van Wijk et al. and the contribution of Endler et al. in this book).

Hypersensitivity

The basic phenomena of allergy therapy using neutralisation of the reactions with serial dilutions of the allergen require corresponding explanations similar to those listed above for homoeopathy:

1. The neutralising dilution of an allergen will alleviate the symptoms which a provoking dilution of the same allergen will trigger.

2. Serial dilution by syringe appears to provide effective succussion.

3. The neutralisations of allergic responses are logarithmic with the number of serial dilutions of the allergen. In the case of electromagnetic field-triggered reactions, the successive neutralisation frequencies are logarithmically spaced harmonics.

4. It is not possible to distinguish clinically between the effects of an allergen serially diluted by syringe and the same allergen potentised in a homoeopathic pharmacy [Jean Monroe, Breakspear Hospital, Personal Communication].

5. The list of allergens includes a very wide range of chemical and biochemical and biomaterial substances as well as electromagnetic fields. A sealed tube of clinically inactive water or saline can be given a clinical effectiveness for such sensitive persons by exposing it to a magnetic field at a specific frequency. Furthermore, its contents can subsequently be further potentised by dilutions and succussions.

Order and Coherence

The statements mentioned above clearly imply that explanations for homoeopathy, serial dilution allergy therapy and effects of electromagnetic fields on living systems must be sought from the laws of physics applied to effects imparted into water or other diluent medium as first proposed by Barnard [13].

The major theoretical consideration in respect of these bioelectromagnetic phenomena has come from the work of the late Herbert Fröhlich on coherence in active biological systems [14, 15]. The writer has been fortunate in being able to cooperate with him for over 18 years [16].

Coherence is a measure of the spectral bandwidth of a signal and is related to the "Q" of a resonant circuit and its build-up and decay time constants. For a living system the coherence must be adequate to carry a modulation of bio-information for purposes of effective biocommunication. A living cell would need a serial communication channel of optical bandwidth to be able to do all its "housekeeping" in real time. Hence, the development of parallel communication channels through the endocrine system.

The basic theoretical requirement for understanding homoeopathy and allergy therapy is that some ordering should be set up in the dilutent that is characteristic of the original remedy and can be retained after the original substance has been diluted away to zero concentration. The mechanism must be equal to all the media used for homoeopathic remedies. The bio-information must be stored in a form which can be read and acted upon by living systems.

Evidence that high potencies can give measurable effects on biological systems has been obtained by various working groups [see the respective contributions in this volume]. The idea of structured water is not new to biology. Clegg [17] has discussed this in relation to cell water structuring or "vicinal water" which exhibits properties significantly different from pure water and extents 50 nm into the water phase. This implies that the 0,9 mole fraction comprising cell water will be ordered by the remaining 0,1 mole fraction reresenting the non-aqueous constituents of the cell. Coherence in the space domain persists in the time domain throughout the life and activities of the cell or organism.

The basic cylindrical symmetry of current flows and magnetic fields immediately suggests that a helical form of structure in water could provide this memory [18].

Anagnostatos and co-workers [19; see also Anagnostatos` contribution in this book] proposed that there are aggregates of a small number of the original molecules surrounded by clathrate (cage-like) shells of the solvent; then, with subsequent serial dilutions and

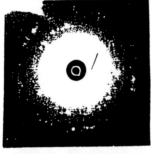

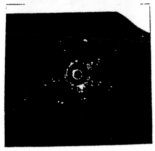

succussions, new mantle clathrates are formed around core clathrates which eventually become structured by the solvent. This concept is consistent with helical symmetry postulates.

The possibility of the formation of cavities in liquids during succussion is well recognised (see Auerbach's contribution in this book). The collapse of a cavity in a liquid under the pressure due to its surface tension would align the pentagonal platelet formations to produce the smallest closed structure possible in water, which in the limit must become a dodecahedron cluster bounded by pentagonal facets. Only two hydrogen atoms can be hydrogen bonded to each oxygen atom; any reversal would generate a temporary hydrogen ion - hydroxyl ion dipole and a strong electric field.

Benveniste [20] has commented that he found that serial dilutions by factors of 2, 4, 5, 6, 8 and 10 produce potencies extending beyond Avogadro's Number, whereas serial dilutions by factors of 3, 7 and 9 only give effects proportional to the degree of dilution. Those dilution sequences effective beyond Avogadro's Number represent the planes of symmetry which can be derived from a dodecahedron.

Evidence from the writer's laboratory based on the x-ray Laue crystallography of ice from pure water frozen in the presence of a 12mT steady magnetic field shows planes of symmetry consistent with these integers [21] as seen in Figure 2. Impure water shows a powder diffraction pattern, and pure water frozen in the geomagnetic field is intermediate.

Fig. 2: X-ray Laue crystallography of ice (from [22]). (above) is for tap water frozen in the geomagnetic field and shows powder rings due to impurities; (middle) is for double de-ionised and distilled water also frozen in the geomagnetic field; (below) is the same water as (middle) but was frozen in the presence of a 12 mT steady magnetic field.

Small dielectric changes have also been measured when water is potentized by a magnetic field as shown in Figure 3. The changes in capacitance and loss are both similar, suggesting that a hopping conduction process is being affected. The sharpness of the resonance, + /- 10 Hz in 50 kHz, was such that no further progress could be made without an oscillator giving frequency settings reproducible to ppm.

Dodecahedrons could join together to form pearl chains or larger clusters linked through their common pentagonal facts. Such chains would provide helical conduction paths through "hopping" of the hydrogen bonds [18].

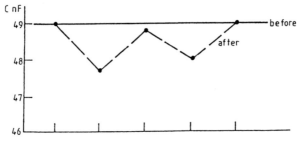

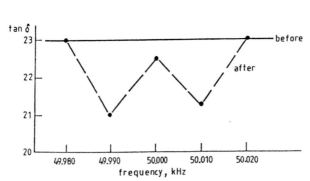

Fig. 3: Small dielectric changes measured when water in a capacitor is potentised by a magnetic field of approx 0.1 mT (rms). The changes in capacitance and loss are both similar suggesting that a hopping conduction process is being affected. The sharpness of the resonance is approx. + /- 10 Hz in 50 kHz.

The magnetic field produced by the current represented by the synchronous "hopping" of hydrogen atoms (protons) between adjacent oxygen atoms can induce a sufficient electric field to give self-sustaining oscillations at a frequency determined by the path length and velocity. This is, however, only a picture of classical physics.

The probably most important recent theoretical development is to be found in the work of Del Giudice and colleagues [22,23; see del Giudice's contribution in this book)], who, taking account of the previously neglected quantized radiation field interactions, have shown that water could behave as a two-fluid superconductor at ordinary biological temperatures.

Taking this theory into consideration and using only basic physical constants, the coherent water component should have coherence in its ground state and give domains of coherence about 100 μm in size. Within each domain, there will be molecules oscillating in-phase, like synchronous machines on a bus-bar. These domains are considered to be separated by incoherent regions of randomly fluctuating water molecules behaving like a gas and satisfying the requirements of thermodynamics for the water as a whole.

The coherent water domains should be able to communicate using the Josephson Effect, giving the possibility of a frequency to voltage inter-conversions at 500 MHz /μV (h/2e) and sensitivity to single quanta of magnetic flux (2.07 x 10^{-15} Wb).

Magnetic flux quantization is a fundamental property of the coherence of a magnetic field. In passive physical systems the necessary coherence and long-range order only occurs near the absolute zero of temperature. The laser and living systems achieve coherence by dynamical processes. However, water can have this coherence in the ground state, whereas a laser must be in an excited state. If a living system is able to sense magnetic flux quanta, then it will have the Josephson Effect available to it, since this has its basis in the quantization of magnetic flux. Evidence for the existence of such effects in biological systems has been presented by the writer [24].

Anomalous diamagnetic susceptibility measurements on lysozyme were interpreted in terms of the superconductivity theory, making the problem analogous to that of low-temperature superconducting colloidal mercury. This gave a critical temperature of about 100°C, the order of the denaturing temperature for proteins [25].

If water is able to react to a single quantum of magnetic flux, then a simple mechanism for the storage of frequencies of an applied magnetic field in water becomes available. A changing magnetic flux density will produce a change of one magnetic flux quantum within some (calculable) area, inducing an electromagnetic field around the perimeter. If this emf is able to stimulate the coherent hopping of about 4000 protons around a current loop in 10^{-11} s, then enough current will flow to regenerate a magnetic flux quantum through that area, thus continuing the process. If one takes a helix as the model, then only 20 hopping charges per turn are required. This is exactly the number of electrons associated with the five oxygen atoms in the water pentamer. The definitions of the magnetic flux quantum and the Josephson Effect mean that this mechanism is time-invariant so that any frequency can be stored.

Magnetic Vector Potential Phenomena

(Annotation by the editors: The concept of a vector potential field has been introduced in the theory of magneto- statics in order to solve the magneto-static equations in a general way, that is without requiring any special symmetry or intuitive guessing. Thus, the magnetic vector potential - in physics simply referred to as vector potential - has originally been a mathematical object. Since the beginning of quantum mechanics in 1926, however, it has been very obvious that the vector potential must have some more significance than just being a mathematical tool for convenience only. The vector potential is an inherent part of the quantum mechanical "equation of motion", called the Schrödinger equation. In classical mechanics, there is no vector potential. Thus, the vector potential is expected to interact or influence properties known in quantum mechanics such as the phase difference of two propagating waves or probability amplitudes. It was not until 1956, that Aharanov and Bohm suggested an experiment to prove that the vector potential was observable by physics. The experiment was extremely difficult to perform; however, Aharanov and Bohm were able to show that the vector potential acts on the phase difference of two - electron - waves, a phenomenon that is inherently part of the theory of quantum mechanics.

In the literature on ultra sensitive interaction of electromagnetic waves and living systems, the vector potential is frequently referred to as magnetic vector potential - MVP - for the magnetic field is the curl of the vector potential. For those not familiar with the fundamentals of electro-/magneto- statics, electrodynamics, and the principles of quantum mechanics, the term MVP might be misleading because it falsely implies that there is an electric vector potential. Experiments based on the influence of the vector potential on fields and living systems have to be performed in an extremely meticulous way because the phase changes of electromagnetic waves are extremely sensitive to any changes of external statical and dynamical electric and magnetic fields. - For a brief formal introduction into the concept of the vector potential field see Lectures on Physics, R.P. Feynman, R.B. Leighton, M. Sands, Addison-Wesley Publ., Reading, Massachusetts 1975; or any other standard book on electrostatics and electrodynamics -.)

The magnetic vector potential can interact with electron waves to produce a relative phase-shift even if the electron paths do not cross the magnetic flux lines. This is the

Ahranov-Bohm effect, an unusual but important effect in quantum physics not predicted by the classical Maxwell equations for electromagnetic fields. This effect has been most convincingly demonstrated at the Hitachi Research Laboratories, Tokyo [26].

At Wekroma AG, Switzerland, Werner Kropp and the writer showed that water and living systems appear to respond to the magnetic vector potential from a toroid [27; see also the contribution of Smith et al. in this book]. These effects are manifested through changes in the optical absorption and mass spectra of impurities, and through clinical effects. As with the early attempts to confirm the Ahranov-Bohm Effect, a rigorous exclusion of the leakage fields from the toroid was not possible. At Wekroma AG we demonstrated effects which were consistent and constant over magnetic fields from about 200 mT to 20 nT. In 1991, [28] the writer demonstrated the measurement of the frequencies of a series of potencies of candida, and showed that the optimum therapeutic potency could be selected for a candida patient by comparison with the body frequencies measured from a tube of water held in the hand.

METHODS AND RESULTS

Thresholds for Magnetic Effects in Water

Apparatus

A toroid contains the magnetic flux within the torus, but in the surrounding space it provides a dipole source of magnetic vector potential centred on its axis. For these experiments the toroid had originally been used in a magnetisation experiment from a teaching laboratory. The toroid had 333 turns, a mean diameter of 101 mm, a cross-section 25 x 3 mm, a core of relative permeability $\mu_r = 87$. The magnetic vector potential $\underline{A}$ was approximately = 4,3 µWb/m per amp, at a distance of 1 m. Any alternating magnetic field $\underline{B}$ leakage from the toroid and connecting wires could not be measured, being less than 1 µT.

For comparison, a solenoid of diameter 50 mm, length 200 mm, and wound 3000 turns/m was used giving $\underline{A} = 0,3$ µWb/m per amp at a distance of 1 metre. Inside the solenoid, $\underline{A} = 24$ µWb/m per amp and $\underline{B} = 3.8$ µT per amp.

Detection of Resonances

For the subjective detection of resonances in water (and homoeopathic potencies), the toroid or solenoid and the tube containing the water need to be orientated North-South with an accuracy of + /- 5° at the threshold, although a tolerance of + / - 45° suffices at higher values. This orientation may be required to avoid interference from the vector potential of the geo-magnetic field which is the East-West direction. The hands (or arms) of the experimenter need to be on the North-South axis, one on each side of the water being measured. One hand holds a pendulum which gives a positive indication when a resonance is reached as the other hand tunes the oscillator feeding the toroid or solenoid. The resonant frequency of the pendulum matches a natural 2 Hz resonance in the autonomic nervous system [29]. (For further information, see the contribution of Smith et al. in this book.)

The subjective detection of a 1 kHz resonance in water required the magnetic vector potential $\underline{A} \geq 17$ nWb/m (rms) at the glass tube containing the water. This threshold was found to be linear with distance from the toroid from at least 4 m to 0,3 m as seen in Figure 4, suggesting that a potential function rather than a field function is involved.

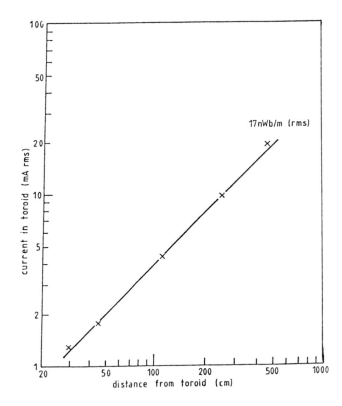

Fig. 4: The subjective detection of a 1 kHz resonance in water required a magnetic vector potential $\underline{A} \geq 17$ nWb/m (rms) in the glass tube containing the water. This threshold is proportional to the current in the toroid and the distance of the water from the toroid suggesting that a potential function rather than a field function is involved.

Potentising in a Solenoid

For potentising water by placing it in a glass tube inside the solenoid, the threshold conditions with the axis of the solenoid directed North-South were $\underline{A}$ = 47 nWb/m (rms), $\underline{B}$= 7,6 µT (rms). With the solenoid axis in the East-West direction, the water would potentise at half these values, which were independent of the frequency from at least 10 kHz to 10 Hz.

Since magnetic potentisation is a gentle and consistent process, it is possible to potentise water in a series of glass tubes of different internal diameters. As seen in Figure 5, there is a discontinuity at a tube diameter of 2,5 mm, below which the water potentises with the opposite sense of rotation of the hand-held pendulum which is probably responding to muscle tremor arising from changes in the automatic nervous system. This reversal may represent the transition from three-dimensional coherence to one-dimensional coherence.

These water potentisation thresholds have also been measured for a system of constant area (25 x 30 mm), as distinct from the above system of constant length (80 mm) in a series of capillary tubes.

The constant area cell containing the water was assembled from a pair of spaced microscope slides using a stack of pairs of cover-slips to provide separations between 200 micrometer and 3,8 mm. The potentisation threshold was measured with this cell inside a solenoid.

These results of water potentised in these cells also shown in Figure 5, give the discontinuity between pendulum rotations indicating "good" water (3-D coherence) and "bad" water (2-D coherence) between plate separations of 1,2 and 1,4 mm. For the water in the capillary tubes, this discontinuity is between tube diameters 2,0 and 2,5 mm.

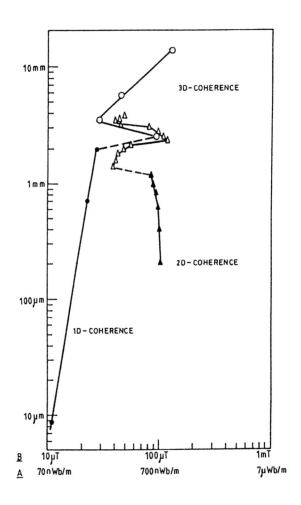

Fig. 5: Circles: Potentisation of water in a solenoid using tubes of different diameter, but a constant 8 cm length of water column. There is a discontinuity at a tube diameter of 2,5 mm, below which the water potentises with the opposite sense of rotation of the pendulum and which may be a transition from a three-dimensional coherence to a one-dimensional coherence.

Triangles: Potentisation of water in a solenoid using a cell of constant area (25 x 30 mm) and variable separation. There is a discontinuity at a separation of 1,25 mm, below which the water potentises with the opposite sense of rotation of the pendulum and which may be a transition from a three-dimensional coherence to a two-dimensional coherence.

There is a peak at a plate separation of 2,5 mm which disappears if well-boiled (air-free) water is used as seen in Figure 6. On the assumption that the the cell was behaving as a half-wave resonator, this would correspond to a wavelength of 5 mm or 60 GHz, the oxygen resonance. However, this distance would be that for the free-space velocity of propagation and not that appropriate for the dielectric constant of water at this frequency. This could mean that potentisation only involves *coherent water* and the observed resonance does not involve those electromagnetic interactions with individual water molecules which give rise to the dielectric constant and refractive index.

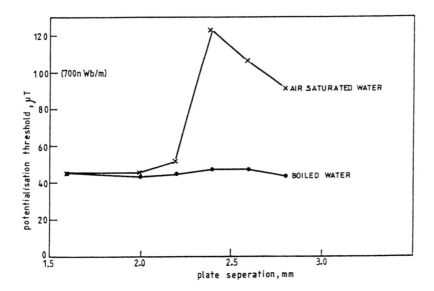

Fig. 6: Comparison of potentisation thresholds for air saturated water and boiled (de-aerated water) as a function of cell thickness (plate separation) showing that the peak in Figure 5 is dependent on dissolved air (oxygen).

By way of confirmation, water was potentised in a long measuring cylinder placed within the solenoid. To potentise an open column of water, 8,4 cm length required a current of 21,5 mA in the solenoid. At the maximum output of the oscillator (62,5 mA), lengths of 19,8 cm and 22,5 cm would just potentise, but it was not possible to get potentisation with the available facilities closer to the 21 cm (1,42 GHz) hydrogen resonance length. A small glass beaker was then floated on the water surface to make a "closed" water resonator. Now, for a water column of 21 cm, the water would potentise at 19 mA, but a column of 10,5 cm would not potentise at maximum oscillator output. This suggests that this effect at 2,5 mm (125 GHz) may be associated with the proton-hopping-frequency and which may in turn be involved in the process of potentisation. It certainly means that the dimensions of containers used in potentisation experiments are critical.

Many years ago, Herbert Fröhlich predicted that oxygen should have a stabilising influence on water structures and the writer made some molecular models for him to demonstrate this.

The straight line bond of the oxygen molecule in combination with water molecules can give ring structures without the need for bond-bending.

Potentising with Separated A & B.

Investigation of potentisation with the alternating magnetic vector potential separated from the magnetic field showed that there were two separate and independent critical conditions:

(a) an alternating magnetic vector potential of $A \geq 180$ nWb/m (rms) which is independent of frequency, and

(b) a steady magnetic field from a permanent magnet $\geq$ 1 mT. Raising the steady magnetic field to 60 mT did not alter the threshold of the alternating magnetic vector potential required for potentisation.

(c) Alternatively, an alternating magnetic field of 10-15 μT at a frequency less than that of the alternating magnetic vector potential as shown in Figure 7, in which the curves relate to magnetic vector potential frequencies of 100 Hz and 1000 Hz.

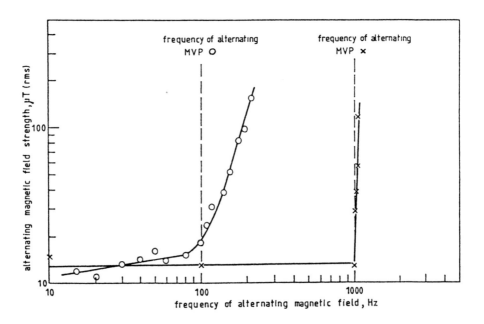

Fig. 7: Potentisation thresholds for a constant alternating magnetic vector potential (MVP) greater than the 185 nWb/m (rms) threshold and of frequencies 100 Hz (o) and 100 Hz (x) respectively, as a function of the alternating magnetic field. At frequencies less than that of the MVP the required magnetic field is 10-15 μT (rms). Above the MVP frequency, the required magnetic field rises by an order of magnitude in 100 Hz.

Potentisation by Succussion in an Alternating Magnetic Vector Potential.

To potentise water without the 1 mT steady field, succussion was necessary. A single impact of the glass tube against the wooden bench would provide potentisation in an alternating magnetic potential as small as $\geq$ 20 pWb/m (rms). There was no evidence of any potentisation even 5% below this threshold. Succussion before or after exposing the tube to the alternating magnetic potential produced no potentisation either.

This suggests that the function of succussion is to provide a coherent matrix of eddies of uniform size in which to write the information. From the Navier-Stokes equation for very small eddies, one can determine the eddy decay time as a function of its size [30]. Assuming that the eddy must not decay while the wavefront is crossing it, and assuming that a Mach 10 shock wave is produced in succussion, then the eddy size would be 26 nm.

Resonances can be detected in water dissolved in n-hexane. In this case, the threshold concentration of water corresponds to a mean molecular separation of about 20 nm. This may represent the maximum spacing between coherent water molecules in a Del Giudice 100 micrometer domain if it is like a sponge, containing both coherent and incoherent water. It is also of the order of the range of Van der Waal's forces.

An experiment was carried out at Wekroma AG whereby aliquots of water were exposed respectively on the North and South sides of a toroid producing a magnetic vector potential. The water was then placed in a 10 cm long cuvette (essential to find this effect) and the tracing shown in Figure 8 was produced in a Perkin Elmer differential spectrometer (Model Lambda 3,UV/VIS) using the unexposed water in the reference channel.

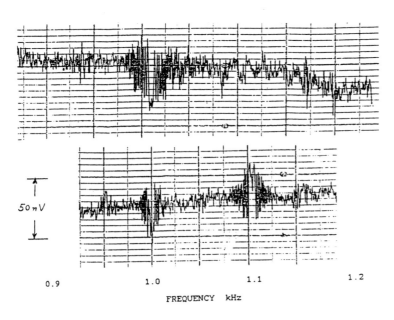

50 nV

0.9   1.0   1.1   1.2

FREQUENCY kHz

Fig. 8: The result of an experiment carried out at Wekroma AG whereby aliquots of water were exposed respectively on the North and South sides of a toroid producing a magnetic vector potential. These were then placed in a 10 cm path-length cuvette (essential to find this effect) in a Perkin Elmer differential spectrophometer (Model Lambda 3, UV/VIS) using unexposed water in the reference channel. This is shown in the centre tracing, the North and South side aliquots are shown in the upper and lower tracings.

DISCUSSION

Electromagnetic Fields in Coherent Media

One of the consequences of the Del Giudice two-fluid model for water is that electromagnetic fields propagate differently in a correlated medium from an uncorrelated one.

A weak external electromagnetic field which is not strong enough to disrupt the coherence can interact coherently with up to 3×10^8 molecules of the 10^{17} molecules in a 100 μm coherence domain in water without breaking the coherence. This reduces the velocity of propagation of electromagnetic radiation from 3×10^8 m/s to 1 m/s and gives it a longitudinal field component [23,24]. These slow waves may correspond to the waves discovered in the 1930's by Joseph Wüst [30].

Since the Poynting Vector is 'energy density times velocity', a sudden decrease in velocity of free-space radiation by a factor of 3×10^8 on interacting with a coherent medium will increase the energy density by the same factor. The internal electric and magnetic fields will increase by the square root of this factor, i.e. $1,7 \times 10^4$. When the incident power density is high enough to break the coherence, the usual Maxwell Equations for propagation in an incoherent medium will apply to all incident radiations. For example, an incident power flux density of 3 μW/m^2 on coherent water will give the same internal energy density as an incident power flux density of 100 W/m^2 on chaotic water under equilibrium conditions. This would impart to the volume of a typical biological cell an energy greater than thermal energy (kT).

Conclusions

The tentative conclusions from these threshold measurements for magnetic potentisation are that the magnetic vector potential carries the frequency information, and presumably the bio-information, while an additional magnetic field or a succussion is required to "format" the water in order to accept this information. Since threshold conditions appear to be independent of frequency over the very wide range of frequencies that can be imparted into water, from millihertz to gigahertz and up to X-ray, the mechanism of information storage must be frequency invariant. This rules out effects directly involving the electric field generated by $d\underline{A} / dt$, and favours Josephson coupling between coherent water domains.

ANNOTATION

A preliminary version of this paper was presented at the First International Conference on "Water Systems and Information" Kiev, Ukraine, May 12 - 17, 1992. The present paper was actualized and adjusted.

REFERENCES

1. Smith CW. Coherent electromagnetic fields and biocommunication. In: Electromagnetic Bio-Information, Popp F-A et al. eds., Munich: Urban & Schwarzenberg 1989, pp.1-17.
2. Smith CW. Electromagnetic effects in humans. In: Biological Coherence and Response to External Stimuli, Fröhlich H ed., Berlin: Springer-Verlag 1988, pp. 205-232.
3. Smith CW. Electromagnetic phenomena in living biomedical systems. Proc 6th Ann Conf IEEE Eng in Med & Biol Soc 15-17 Sept 1984; pp.176-180.
4. Smith CW, Jafary-Asl AH, Choy RYS and Monro JA. The emission of low intensity electromagnetic radiation from multiple allergy patients and other biological systems. In: Photon Emission from Biological Systems, Jezowska-Trebiatowska B et al eds., Singapore: World Scientific 1987, pp.110-126.
5. Del Giudice E, Doglia S, Milani M, Smith CW and Vitiello G. Magnetic flux quantization and Josephson behaviour in living systems. Physica Scripta 1989; 40: 786-791.
6. Smith CW, Choy RYS and Monro J. The diagnosis and therapy of electrical hypersensitivities. Clinical Ecology 1989; 6(4): 119-128.
7. Rea WJ, Pan Y, Fenyves EJ, Sujisawa I, Suyama H, Samadi N and Ross GH. Electromagnetic field sensitivity. J Bioelectricity 1991; 10 (1&2): 241-256.
8. Tsouris P and Smith CW. The detection of LF resonances in water (submitted for publication).
9. Jahn RG. The complementarity of consciousness. In: Tech Note PEAR 91006, Princeton NJ 1991: Princeton University.

10. Smith CW and Best S. Electromagnetic man: health and hazard in the electrical environment. London: JM Dent 1990; French & German edns. New York: St Martin's Press 1992.
11. Hahnemann S. Organon of Medicine: English Translation of Organon der rationellen Heilkunde. Los Angeles: Tarcher 1982.
12. Van Wijk R and Wiegant FAC. Homoeopathic remedies and pressure induced changes in the galvanic resistance of the skin. Alkmaar: VSM Geneesmeddelen 1989.
13. Barnard GO. Microdose paradox. A new concept. Am J Inst Hom 1965; 58: 205-212.
14. Fröhlich H and Kremser F eds. Coherent excitations in biological systems. Berlin, Heidelberg: Springer 1983.
15. Fröhlich H ed. Biological coherence and response to external stimuli. Berlin, Heidelberg: Springer 1988.
16.. Barrett TW and Pohl HA eds. Energy transfer dynamics: studies and essays in honour of Herbert Fröhlich. Berlin, Heidelberg: Springer 1987.
17. Clegg JS. Intracellular water, metabolism and cell architecture: Part 2. In: Coherent Excitations in Biological Systems, Fröhlich H and Kremer F eds., Heidelberg: Springer 1983.
18. Smith CW, Choy R and Monro JA. Water Friend or Foe? Laboratory Practice 1985; 34(10): 29-34.
19. Anagnostatos GS, Vithoulkas G, Garzonis P and Tavouxoglou C. Working Hypothesis on Homoeopathic Microdiluted Remedies. Proc 43rd Int Hom Congress, Athens May 1988.
20. Benveniste J. Molecular interactions with specific biological effects in the absence of molecular structures. Workshop Conference on Basic Issues in the Overlap and Union of Quantum Theory, Biology and the Philosophy of Cognition. Bermuda, April 1988.
21. Jaberansari M. Electric and magnetic phenomena in water and living systems. Ph D Thesis, Salford University 1989.
22. Del Giudice E, Doglia S, Milani M and Vitiello G. Electromagnetic field and spontaneous symmetry breaking in biological matter. Nuclear Physics 1986; B275 (FS17):185-199.
23. Del Giudice E et al. Magnetic flux quantization and Josephson behaviour in living systems. Physica Scripta 1989; 40: 786-791.
24. Del Giudice E et al. Magnetic flux quantization and Josephson behaviour in living systems. Physica Scripta 1989; 40: 786-791.
25. The living state II. In: World Scientific, Mishra RK ed., Singapore 1985.
26. Tonomura A et al. Evidence for Ahranov-Bohm effect with magnetic field completely shielded from electron wave. Phys Rev Lett 1986; 56(8): 792-795.
27. Wekroma AG. The use of magnetic vector potentials for materials. 19 Nov 1989: FRG Pat No 3938511.6.
28. Smith CW. Electromagnetic fields and diseases. 9th Ann Intl Symp. In: Man and His Environment in Health and Disease 1991, Dallas TX Feb - Mar 1991.
29. Smith CW. Electromagnetic fields and the endocrine system.10th Ann Int Symp. Man and His Environment in Health and Disease 1992, Dallas TX Feb - Mar 1991.
30. Cole GHA. Fluid dynamics. London: Methuen 1962, pp.188-191.
31. Wüst J and Wimmer J. Über neuartige Schwingungen der Wellenlänge 1 - 70 cm in der Umgebung anorganischer und organischer Substanzen sowie biologischer Objekte. Anatomische Anstalt der Univ München, Abt für Exp Biol, 1934; reprint: Wilhelm Krauth KG 1979.
32. 9th Int Conference on "Man and His Environment in Health and Disease", Dallas, Texas Feb - Mar 1991. Available on audio tape from INSTA TAPE Inc, PO Box 1729, Monrovia, CA 91017-5729, USA.

RESONANCE PHENOMENA OF AN ULTRA HIGH DILUTION OF THYROXINE - PRELIMINARY RESULTS

C.W. Smith, P.C. Endler

SUMMARY

Resonance phenomena between serially diluted agitated thyroxine and a coil fed by a laboratory oscillator were investigated. By testing different frequencies, a distinct pattern of frequencies that caused resonance leading to biological reactions (microtremor) in a human tester was found. Each succeeding step of dilution and agitation added two further and higher frequencies of resonance.

ZUSAMMENFASSUNG

Es wurden Resonanzphänomene zwischen „homöopathisch" zubereitetem verdünntem Thyroxin und der Spule eines Labor-Oszillators untersucht. Indem verschiedene Frequenzen getestet wurden, ergab sich ein bestimmtes Frequenzmuster, das Resonanzen, die zu biologischen Reaktionen (Mikrotremor) führten, hervorrief. Jeder der aufeinanderfolgenden Verdünnungsschnitte führte zu zwei weiteren (höheren) Resonanzfrequenzen.

INTRODUCTION

Our contribution on multicentered zoological experiments described the biological activity of an agitated high dilution of thyroxine on amphibia (see the contribution of Endler et al.). In this paper, a method standardized at the laboratory of C.W.Smith was applied to test the biological reaction (microtremor) of a test person exposed to what we think are resonance phenomena between the dilutions of thyroxine and the field from a coil fed by a laboratory oscillator. The range of frequencies from 0.01 Hz to 9.1 MHz for the agitated dilutions 10^{-5} to 10^{-30} was applied.

METHODS

Preparation of test solutions, precautions, transport

The dilutions of thyroxine (unsuccussed dilution log 4; succussed dilutions log 5 - log 30: D5 - D30) were prepared in Austria as described in the contribution of Endler et al. As a precaution, the water was pre-treated by heating it up to 75° C and then allowed to cool to room temperature in a site without electric equipment or wiring. The dilutions were prepared in hard-glass bottles. The 26 bottles containing the dilutions were each wrapped in aluminium foil and layers of paper to avoid external influences during transport through the mail.

Electrical device for exposure of test substances

In sequence, each of the dilutions (D5 to D30) was exposed to the field of a coil that was fed by an oscillator at a distance of 20 cm. The frequency generated by the oscillator was varied from 0.01 Hz to 10 MHz during the experiment. For investigation of frequencies up to 1 kHz, a toroid was used which prelimary generates a magnetic vector potential and, above this, a solenoid was used which generates both a magnetic field and a magnetic vector potential. These coils were chosen for experimental convenience. The oscillators were ordinary laboratory oscillators.

P.C. Endler and J. Schulte (eds.), Ultra High Dilution, 203–207.
© 1994 *Kluwer Academic Publishers. Printed in the Netherlands.*

Sensitive living measuring system

The coil, the dilution and the hands of the testing person are on a north-south axis. The trained test person, situated with his chest towards the dilution, placed his hand between the coil and the bottle containing the dilution. When the frequency was shifted from 0.01 Hz upwards, at certain distinct frequencies, the living organism showed a reaction in muscular microtremor (Smith 1991) at frequencies at which there obviously was a resonance interaction between the field emitted from the coil and the respective test substance. This reaction in muscular microtremor was amplified by a 2-Hz resonant hand pendulum held between the subject's thumb and forefinger. When the frequency was slowly varied, this microtremor-amplifier began to show pendulum oscillation when certain frequencies were reached. The frequencies were scanned for each of the 26 samples.

RESULTS

Control

When the test person was exposed to the frequency range from 0.01 Hz to 10 MHz as described in the methods section, with the test dilutions replaced by pure water heated to 75°C and cooled, no typical microtremor reactions occurred.

Exposure to the interaction between the field of the coil and the test dilutions

For all test substances from thyroxine D5 to D30, a reaction of the biological sensor occurred at 0.07 Hz. In addition, for the substances D6 to D30, a reaction was found at 0.107 Hz and 0.230 Hz. In addition, for D7 to D30, at 0.490 Hz and 0.700 Hz, for the remaining 23 test substances in addition at 1.3 Hz and 2.6 Hz and so on. Thus, the exposure of every test dilution to the field of the coil provoked all the reactions also occurring at the previous (lower) dilution, and two more (Fig. 1 and Tab. 1).

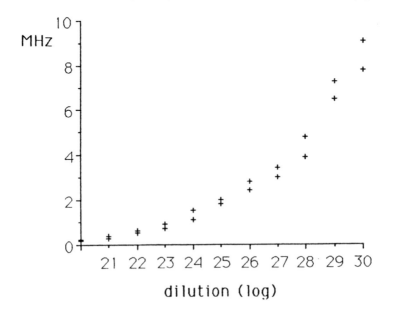

Fig. 1: Figure to Tab.1 from dilution D21 onwards. Ordinate: Frequency in MHz. Abscissa: Serial dilution of thyroxine in steps of 1 : 10. In the figure, only those frequencies that additionally occurred at each step of dilution are shown. For further explanation, see text.

Tab.1.:

Dilution	basic frequencies	+ Frequency 1	+ Frequency 2
log 4	(0.25 Hz), 0.04 Hz, 0.95 Hz		
D5		-0.07 Hz	
D6	see D5 (= 0.07 Hz) +	0.107 Hz +	0.23 Hz
D7	D6	0.49 Hz	0.70 Hz
D8	D7	1.30 Hz	2.6 Hz
D9	D8	4.8 Hz	6.3 Hz
D10	D9	8.4 Hz	10.0 Hz
D11	D10	18 Hz	42 Hz
D12	D11	67 Hz	89 Hz
D13	D12	300 Hz	670 Hz
D14	D13	1.2 kHz	2.4 kHz
D15	D14	4.3 kHz	5.7 kHz
D16	D15	8.6 kHz	13.5 kHz
D17	D16	24 kHz	ca. 60 kHz
D18			
D19			155 kHz
D20	D19	200 kHz	250 kHz
D21	D20	290 kHz	365 kHz
D22	D21	505 kHz	575 kHz
D23	D22	730 kHz	900 kHz
D24	D23	1.12 MHz	1.53 MHz
D25	D24	1.82 MHz	2.02 MHz
D26	D25	2.4 MHz	2.75 MHz
D27	D26	2.95 MHz	3.4 MHz
D28	D27	3.9 MHz	4.75 MHz
D29	D28	6.45 MHz	7.3 MHz
D30	D29	7.8 MHz	9.1 MHz

Tab. 1: Typical (resonance) frequencies that were attached to dilutions of thyroxine by measurement of microtremor in a human living system. For explanation, see text.

For example, thyroxine D30 differed from D29 in provoking additional resonances at 7.8 MHz and 9.1 MHz.

DISCUSSION

The results presented in this preliminary communication show that there is a biological reaction (change in microtremor) of a test person when exposed to the field of a coil fed with distinct frequencies of an oscillator, when a homoeopathically prepared dilution of thyroxine is brought into that field between the test person's hands. The frequencies where this reaction occurs are typical for each test dilution from D5 to D30. The result seems to be reproducable, as typical frequencies have been found with great regularity. However, further repetitions of the experiment, also under blind conditions, and also including other test persons, are, of course, necessary.

In order to explain the effect observed, our speculation is that, apart from the reaction of the living system, there are resonance phenomena between the magnetic vector potential or electromagnetic field produced by the coil fed by the oscillator and the homoeopathically prepared test dilution. This might be compared to the resonance effects of an undamped amplifier in an empty auditorium. The fact that these effects are also found when dilutions above Avogadro's limit (D25 - D30) are used points toward a physical effect in the solvent itself (see the following contributions in this book: Smith; Anagnostatos; Berezin; del Guidice; Endler et al.; Schulte). A possible specific sensitivity of a human tester to information from an agitated high dilution is also discussed in the contribution of van Wijk et al. in this book.

We wish to point out that these scientific approaches were undertaken by well-trained, skilled investigators, and that these data should not encourage any irresponsible diagnostic use of pendulum phenomena in human medicine.

Apart from the simple measuring method of microtremor of the human organism used in this study, better standardized methods to determine the effect of exposure could also be used. Further, it also seems worthwile to take other living systems as test objects, and to investigate e.g. the spontaneous climbing activity of juvenile highland frogs when exposed to these resonance phenomena of thyroxine when generated electromagnetically.

To investigate homoeopathic remedies apart from a living system, it will probably be necessary to simulate the electromagnetic frequency and coherence characteristics of a living system and measure the effects on the feedback path. This might be done by intercorporating the remedy in the feedback path of an electronic amplifier or a polarographic amplifier with a broad-band frequency response. (For details, see Smith´s contribution).

ACKNOWLEDGEMENTS

Special thanks are due to W. Pongratz for her careful preparation of the thyroxine dilutions.

REFERENCES AND FURTHER LITERATURE

Brillouin L. Fluctuations of current in a conductor. Helv Phys Acta 1934; 7, Suppl.2: 47-67.

Benveniste J. Molecular Interaction with Specific Biological Effects in the Absence of Molecular Structures. Workshop conference on Basic issues in the Overlap and Union of Quantum Theory, Biology and the Philosophy of Cognition, Bermuda, April 1988.

Clegg JS. Intracellular water, Metabolism and cell Architecture: Part 2. In: Coherent Excitations in Biological systems (eds. Fröhlich H, Kremer F). Heidelberg: Springer 1983.

Del Giudice E, Doglia S, Milani M, Vitiello G. Electromagnetic Field and Spontaneous Symmetry Breaking in Biological Matter. Nuclear Physics 1986; B275 (FS17): 185-199.

Frey AH. Electromagnetic field interactions with biological systems. FASEB J 1993; 7: 272.

Fröhlich H. Coherent electric vibrations in biological systems and the cancer problem. IEEE Trans MTT 26: 613-617, 1978.

Fröhlich H, Kremer F. Coherent Excitations in Biological Systems. Heidelberg: Springer 1983.

Fröhlich H. Coherence and the action of enzymes. In: Welch GR (ed.). The fluctuating enzyme. Chichester, Wiley 1986.

Fröhlich H. Coherent excitation in active biological systems. In: Gutman FH, Keyser (eds.). Modern Biochemistry. London: Plenum 1986.

Fürth R. On the theory of electrical fluctuations. Proc. Roy. Soc. 192A: 593-615, 1948.

Jaberansari M. Electric and Magnetic Phenomena in Water and Living Systems. Ph.D. Thesis, Salford University 1989.

Ludwig W. Eigenresonanzen verschiedener Wasserformen, Erfahrungsheilkunde 1987; 36: 952.

Popp FA, Warnke U, König HL, Pleschka W (eds.). Electromagnetic Bio-Information. München, Wien, Baltimore: Urban & Schwarzenberg 1989.

Ruth B. Experimental Investigations on Ultraweak Photon Emission. In: Popp FA, Warnke U, König HL, Pleschka W (eds.). Electromagnetic Bio-Information. München, Wien, Baltimore: Urban & Schwarzenberg 1989.

Smith CW, Best S. Electromagnetic Man. London: JM Dent 1989.

Smith CW, Choy R, Monroe JA. Water Friend or Foe? Laboratory Practise 1985; 34 (10): 29-34.

Smith CW, Jafary-Asl AH, Choy RYS, Monro JA. The Emission of Low Intensity Electromagnetic Radiation from Multiple Allergy Patients and other Biological Systems. In: Photon Emission from Biological Systems eds. Jezowska-Trzebiatowska B, Kochel B, Slawinski J, Strek W. Singapore: World Scientific 1987.

Smith CW. Coherent Electromagnetic Fields and Bio-Communication. In: Popp FA, Warnke U, König HL, Pleschka W (eds.). Electromagnetic Bio-Information. München, Wien, Baltimore: Urban & Schwarzenberg 1989.

C.W. Smith's contribution at the 9th. International Conference on "Man and His Enviroment in Health and Disease", Dallas, Texas, Feb. 27 - Mar 3, 1991. Available on audio tape from INSTA TAPE Inc., P.O. Box 1729, Monrovia, CA 91017-5729, USA;

For further information see the following contributions in this book: Anagnostatos, Berezin, Del Guidice, Endler et al., Popp, Schulte, Smith.

TRANSFER OF INFORMATION FROM MOLECULES BY MEANS OF ELECTRONIC AMPLIFICATION - PRELIMINARY RESULTS

M. Citro, C.W. Smith, A. Scott-Morley, W. Pongratz, P.C. Endler

SUMMARY

In the present study, it was attempted to transfer information from a molecular thyroxine suspension ($1:10^3$) to non-pretreated distilled water ("test liquid") by means of an electronic amplifier. Two transitions in the metamorphosis of Rana temporaria were investigated under the influence of this test liquid.

Interestingly, in these preliminary experiments, the test liquid, after an initial acceleration period, slowed down both the development from the two-legged to the four-legged tadpole as well as to the juvenile frog.

ZUSAMMENFASSUNG

In der vorliegenden Studie wurde versucht, mittels eines elektronischen Verstärkers Information von einer molekularen Thyroxin-Suspension ($1:10^3$) auf nicht vorbehandeltes destilliertes Wasser zu übertragen (= "Testflüssigkeit"). Unter dem Einfluß dieser Testflüssigkeit wurden zwei Übergänge in der Metamorphose von Rana temporaria untersucht.

Interessanterweise verlangsamte die Testflüssigkeit in diesen vorläufigen Experimenten nach einer anfänglichen Beschleunigungs-Phase sowohl die Entwicklung von der zweibeinigen zur vierbeinigen Kaulquappe als auch zum juvenilen Frosch.

INTRODUCTION

Biological effects of substances in ultra-high dilution (UHD) have been reported, where, theoretically, no original molecule can be present (see the respective contributions in this volume). Our experimental model is based on the fact that the hormone thyroxine plays an important steering role in the metamorphosis of amphibia (for details, see the contribution by Endler et al.). It was shown that the metamorphosis of these animals can be influenced by an UHD of thyroxine ($1:10^{30}$) in two typical ways: depending on the frequency of application, either an inhibitory or an accelerating effect was achieved. This led to the idea that information from the original thyroxine molecules was transduced to the diluent (water) during the dilution process (for theoretical - quantum mechanical - explanations of this information transfer see Berezin's, Del Giudice's, Schulte's and Smith's contributions).

In order to learn more about the nature of the information transduced from the molecule to the dilutent water, we performed experiments in which the UHD was not mixed with the water of the aquarium containing the animals. For this purpose it was applicated in a closed vial that was hung into the water basin and remained there during the course of the experiment. Here, too, comparable, statistically highly significant effects on the metamorphosis of the amphibia were found (see the contribution by Endler et al.).

The study described above and related studies (see the contribution by Pongratz et al. and the contribution by van Wijk et al.) led to the speculation that electromagnetic (or magnetic vector potential, see Smith's contribution) fields may play a decisive role in the information transfer from the UHD to the living system (see Popp's, Benveniste's and Smith's contributions and the Prospects on Elements of a Theory on UHDs).

P.C. Endler and J. Schulte (eds.), Ultra High Dilution, 209–214.

Our question was whether or not field phenomena also play a role in the information transfer from diluted molecules to the UHD, i.e., whether or not electromagnetic fields are linked to molecules (Fröhlich 1983, 1986a,b; see Benveniste's and Smith's contributions). In order to investigate this question, sealed vials of molecular thyroxine solution and of water for control, respectively, were placed in a coil connected with the input of a specially designed amplifier. Sealed water vials were placed in the output coil; the water of these vials was used to treat the amphibian larves with.

The method of using an electromagnetic device in order to transfer low energy information from UHDs was developed by Rasche (Morell 1990) with regard to information from UHDs (see also Smith's contribution). Moreover, the technique of transferring information directly from molecular substances was developed by Citro (Citro 1991). After preliminary experiments with an immunological model in the laboratory of J. Benveniste (Citro 1993), and a botanical model in the laboratory of F.-A. Popp (Citro 1993), the following experiments were performed.

METHODS

Animals: For the experiments, Austrian Rana temporaria from a site 400 m above sea level were taken. For this population, metamorphosis proceeds in May /June.

Staging: For the experiments we chose only those two-legged tadpoles which had just started to develop their hind legs, comparable to stage 31 according to Gosner's staging table (Gosner, 1960). The tadpoles were monitored until the animals had entered the four-legged stage as described in the contribution of Endler et al. (a). Furthermore, the animals were monitored until they had entered the stage with reduced tail (b).

Laboratories involved: The experiments were performed indoors at a site associated with the Research Site for Low Energy Bio-Information (Austria) by W. Pongratz.

Exposition to probes: 8 ml of the test dilution (thyroxine TFF; called "TFF" according to the original designation "pharmacological frequencial transfer" by Citro, for the method of preparation see below) or 8 ml of control (water TFF), respectively, were added blindly to the corresponding basins (each basin contained 8 l of water), followed by gentle stirring, every 8 hours. The corresponding amount of liquid (8 ml) was always pipetted out of the water basin. The transitions from the two-legged to the four-legged tadpoles and to the juveniles were examined in basins as described in the contribution of Endler et al. containing 8 l of water each. Having reached the stage with reduced tail, the animals were transferred into a natural biotope.

Further conditions: The positions of the basins were rotated in the course of the experiment. Indirect natural light was used. Temperature was kept above 18°C. The tadpoles were fed with cooked greens (lettuce). The experimental design was non-violent.

Preparation of testing solutions: Always 20 ml of an aqueous suspension of thyroxine sodium pentahydrate (Sigma) $1:10^3$ at 20°C were kept in a 30 ml bottle with an optical transmission spectrum starting from about 350 nm; the optical transmission spectrum is limited by the properties of the water to wavelengths less than about 2500 nm. This bottle was first agitated 30 times by pushing the partly filled bottle down at short regular intervals to arrive at an even suspension. The suspension was put in the center of a metal beaker that served as a coil which was connected with a specially designed amplifier (Mora III, Firma Rasche, FRG). A bottle containing 100 ml of water was put in the center of a metal beaker that served as the output coil. This coil (Rasche) included two weak permanent magnets. Further details on the electronic device are given in the respective annotation in the discussion section. Seven seconds of amplification 1:40 were always followed by an interval of 3 seconds during a time of 15 minutes. Then the liquid in the output coil was again submitted to the agitation process described above for the liquid in the input coil. It was called "thyroxine TFF". For the preparation of control, distilled water was prepared in an analogous way ("water TFF"). Several analogously prepared sets of liquids were used. Before adding the TFF-liquids to the basin water, they were agitated for several times once more. Always 8 ml of the basin water (8 l) were replaced by 8 ml of the test liquid thrice a day.

Independent solution coding: The sets for the experiment were coded at the University of Graz by A. Nograsek. All sets were applied blindly.

Data base: Six basins which contained 18 animals each were used. Three basins were used for treatment with liquid thyroxine TFF and 3 with water TFF, respectively.

Evaluation of the data: The cumulative frequencies of four-legged animals (F_a), compared to the cumulative frequencies of two- or three-legged animals were evaluated as a 4-field table by the chi-square test at intervals of eight hours (A). Analogously, the cumulative frequencies of animals with reduced tail (F_b) were evaluated. As the results are preliminary, the P-values are discussed, but not indicated in the Figures.

RESULTS

Experiments were performed with at total of 108 animals. In Fig. 1, the bias in the cumulative frequencies of the four-legged animals F_a between the groups treated with the test liquid and the control liquid, are documented. After an initial phase (I) where the F_a - values for thyroxine TFF - animals are above those for reference, the F_a - values for thyroxine TFF - animals are below those for reference at about 5 - 11% (phase II). Due to the small number of animals in the experiment, this difference could not be proven to be statistically significant. However, it can obviously be discussed as a clear trend.

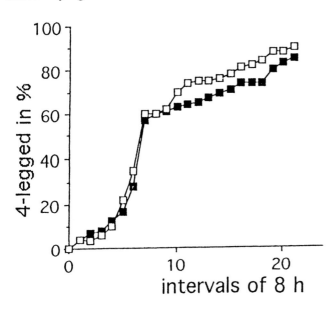

Fig. 1: The influence of the test liquid thyroxine TFF, when added at intervals of 8 hours, on the preclimax metamorphosis of lowland Rana temporaria. Ordinate: above: Cumulative frequencies of four-legged tadpoles in the control group. 100 % refers to 54 animals in the control group. Below: Bias in the cumulative frequencies of four-legged tadpoles between the groups treated with thyroxine TFF and with water TFF, respectively. Abscissa: time at intervals of 8 hours. For further details, see text.

In other words, after an initial increase of the number of animals that have entered the four-legged stage, the overall chance to enter the four-legged stage is generally smaller for the group treated with liquid thyroxine TFF than for the water TFF - group.

In Fig. 2, the bias in the cumulative frequencies of animals with reduced tail F_b is documented. After an initial phase (I) where the F_b - values for thyroxine TFF - animals are above those for reference , the F_b - values for thyroxine TFF - animals are below those for reference at about 5 - 28% (phase II). This difference could be proven to be statistically significant (P < 0.001) at the 10th interval.

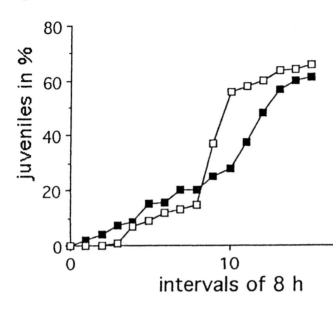

Fig. 2: The influence of the test liquid thyroxine TFF, when added at intervals of 8 hours, on the transition to the animal with reduced tail. Ordinate: above: Cumulative frequencies of animals with reduced tail in the control group. 100 % refers to 54 animals in the control group. Below: Bias in the cumulative frequencies of animals with reduced tail between the groups treated with thyroxine TFF and with water TFF, respectively. For further details, see Fig. 1 and text.

In other words, after an initial increase of the number of animals that have entered the four-legged stage, the overall chance to enter the four-legged stage is generally smaller for the group treated with liquid thyroxine TFF than for the water TFF - group.

DISCUSSION

In this study, it was attempted to transfer information from a molecular thyroxine dilution $(1:10^3)$ to non-pretreated distilled water ("test liquid") by means of an electronic amplifier. Two transitions in the metamorphosis of Rana temporaria were investigated under the influence of the test liquid.

Interestingly, in these preliminary experiments, the test liquid, after an initial acceleration period, slowed down both the development from the two-legged to the four-legged tadpoles as well as to the juvenile frogs.

The development of the idea

The finding that a molecular signal can be transferred via an electronic device is consistent with a study on information from histamine transferred and amplified in the same way as is described in our study on thyroxine. M. Citro and J. Benveniste performed experiments showing a significant variation of coronary flow of isolated hearts from immunized guinea-pigs (Citro 1993). This work has been verified by other scientists in Benveniste's working group (Aissa et al. 1993; Benveniste et al. 1993). Furthermore, in a study by Citro and F.A.Popp, it was shown that electrically transferred information from a cytotoxic poison (atracine) is able to influence unicellular organisms (Citro 1993).

Analogue effects in human medical therapy, observed e.g. in a study on drug-dependent patients, as well as in veterinary therapy have been reported by Citro and colleagues (Citro 1991, 1992a,b, 1993). These findings are able to support the idea that electromagnetic (or magnetic vector potential) fields may play a decisive role in the information transfer from biomolecules to the organism (Citro et al. 1993, see Prospects, p. 245ff. For further discussion, see Fröhlich 1983, 1986 a,b; Smith 1989; Smith's and Benveniste's contribution in this volume).

It goes without saying that the findings discussed here have to be repeated thoroughly and that further experiments have to be performed.

The electronic transfer and amplification

According to A. Scott-Morley, a specially designed amplifier (Mora III, Rasche, FRG), in this context, means that one of the "electroacupuncture" devices was used. Such instruments are, in principle, standard electronic amplifiers. If the instrument is operated whilst connected to the electrical mains there is some amplification and transmission of the 50 Hz frequency. However, the instrument is designed to be run from a re-chargeable battery. If operated only from the battery then the d.c. current only is used. It should be noted that even when the instrument is connected to the electrical mains this is only for purposes of re-charging the battery. The output signal still comes from the battery and hence is still d.c. The amplification of the 50 Hz is because of internal rectification of stray signals.

Work conducted at the laboratory of C.W. Smith (using a Bicom device, Brügemann, FRG) shows that the nature of this bio-information signal is a "propagating coherence" and not a "circulating current": hence, only a single lead is used to connect the sample to the amplifier. Such signal, surprisingly, is also amplified by ordinary transistors, the n-p-n bipolar or n-channel FET being effective; the p-n-p and p-channel do not propagate the signal. From this it is concluded that the signal is one of electron coherence, not hole coherence since holes do not propagate it.

The above mentioned antennae used for coupling the ampoules may be a solid beaker or a drilled block (matrix or honeycomb) of brass that may be gold-plated, or of aluminium. If a helical coil (spin-tester) is used to contain the ampoule, the direction of the winding is significant. The signals pass through glass from the internal liquid to the external metallic contact, but not through plastic (PVC) wire insulation. An ordinary banana plug and insulated lead can connect the antenna to the amplifier input.

The nature of the amplifier output viewed with an oscilloscope seems to be noise. The use of a signal analyzer or narrow-band filter can retrieve a coherent signal, but problems of repeatability remain. The device produces an effective coherent carrier with a wavelength of 6 cm on which the biologically effective signals are modulated.

The output antenna is equivalent to the input one, but with the signal passing in the reverse sense. Water slowly picks up a bio-information from a coil or antenna, but instantanously if a magnetic field is applied, as in the experiment discussed here, or if a shock wave is introduced (succussion).

REFERENCES

Aissa J, Litime MH, Attias E, Benveniste J. Molecular signalling at high dilution or by means of electronic circuitry. J Immunolog 1993; 150: 146A.

Benveniste J, Aissa J, Hjeiml M et al. Electromagnetic (EM) transfer of molecular signals. Poster presented at the meeting Americ Ass Adv Sci 1993.

Citro M. Vom Pharmakon zur Frequenz; Elektronischer Medikamententransfer. In: II. internat. Symp. Biokybernetische Medizin, Würzburg 1991.

Citro M. TFF, un'alchimia elettronica, basi teoriche e dati preliminari. Empedocle 2 - 3, 1992a.

Citro M. TFF dal farmaco alla frequenza. Vivibios, anno II, n.3, 1992b.

Citro M. Biologische Experimente mit Hilfe des TFF. In: IV. internat. Symp. Biokybernetische Medizin, Bad Homburg 1993.

Citro M, Pongratz W, Endler PC. Transmission of hormone signal by electronic circuitry. Poster presented at the meeting Americ Ass Adv Sci 1993.

Endler PC, Pongratz W, Van Wijk R et al. Effects of Highly Diluted Succussed Thyroxine on Metamorphosis of Highland Frogs. Berlin J Res Hom 1991a; 1: 151-160

Endler PC, Pongratz W, van Wijk R. Transmission of hormone signal by water dipoles. Poster presented at the meeting Americ Ass Adv Sci, 1993.

Endler PC, Pongratz W, Kastberger G et al. Climbing activity in frogs and the effect of highly diluted succussed thyroxine. Br Hom J 1991b; 80:194.

Endler PC, Pongratz W, Kastberger G et al. The effect of highly diluted agitated thyroxine on the climbing activity of highland frogs. Recommended for acceptance Vet & Hum Tox, 1993.

Fröhlich H, Kremer F. Coherent Excitations in Biological Systems. Heidelberg: Springer 1983.

Fröhlich H. Coherence and the action of enzymes. In: Welch GR (ed.). The fluctuating enzyme. Chichester: Wiley Publishers 1986.

Fröhlich H. Coherent excitation in active biological systems. In: Gutman FH, Keyser (eds.). M Biochemistry. London: Plenum Press 1986.

Morell F. The MORA Concept. Heidelberg: Haug Verlag 1990.

Smith CW. Homoeopathy, Structure and Coherence. In: ZDN (ed.). Homeopathy in Focus. Essen: Verlag für Ganzheitsmedizin 1990.

Smith CW. Radiesthesie: une technique scientifique. Proc Ann Congr Syndicat National des Radiesthesistes, Paris, 13-14 Nov 1993.

EFFECTS OF TYPICAL THYROXINE RELATED FREQUENCIES ON AMPHIBIA - PRELIMINARY RESULTS

K. Spoerk, W. Pongratz, P.C. Endler

SUMMARY

Amphibia were exposed to an electromagnetic field with a magnetic induction of 500 - 50 nT at 50, 100, 200 and 400 Hz as control frequencies (a) and at 42, 67, 89 and 300 Hz as frequencies that had been postulated to be typical resonance frequencies of thyroxine dilutions in previous experiments ("test frequencies thyroxine D30") (b). The rate of approach of the animals to the source of electromagnetic radiation ($500 - 50 \times 10^{-9}$ T) was monitored. In these experiments, the animals generally showed more avoidance of that radiation source in the experiments A, when the control frequencies were tested versus no field application at all. Interestingly, the animals generally showed more approach towards the radiation source when the frequencies B were tested versus no field application.

ZUSAMMENFASSUNG

Amphibien wurden elektromagnetischen Feldern unterschiedlicher Wellenlänge ausgesetzt. Die Tiere vermieden die Quelle verschiedener, willkürlich gewählter Kontrollfrequenzen, (50,100, 200 und 400 Hz), sie suchten aber die Quelle der Frequenzen 42, 67, 89 und 300 Hz auf. Diese letzteren Frequenzen wurden aufgrund vorangegangener Versuche zur Ermittlung spezifischer Resonanzfrequenzen von Thyroxinverdünnungen verwendet.

INTRODUCTION

A previous contribution (Endler et al., this volume) described the biological activity of an agitated high dilution of thyroxine on amphibia. Further, results from the laboratory of C.W. Smith were described that point towards specific resonance characteristics of agitated dilutions of thyroxine at the frequencies 42, 67, 89 and 300 Hz (see the contribution by Smith et al.). Moreover, experiments point towards the possibility of transferring information from thyroxine via electronic circuity and electronic amplification (see the contribution by Citro et al.). In the following chapter, amphibian experiments on the influence of an electromagnetic field at several frequencies, including the ones described ("test frequencies thyroxine D30"), are discussed.

METHODS

Electrical device

A 600 turn coil (20 by 5 cm) was fed by an ordinary laboratory oscillator at different frequencies (see results). The electromagnetic field measured in the direction of the axis 2 cm from the coil was ca. 500 nT; it decreases to ca. 50 nT at a distance of 19 cm. The magnetic field was inhomogeneous.

Animals

Rana temporaria from an Austrian highland pool as described in the contribution by Endler et al., that had entered the tailed, four-legged juvenile stage were used.

P.C. Endler and J. Schulte (eds.), Ultra High Dilution, 215-218.

Observation

The distribution of the animals based on their swimming activity was examined in one plastic basin (34 by 22 cm, filled with water up to a height of 5 cm). This basin was positioned with the smaller side 2 cm from the end of the coil, in order to cover the field 500 - 50 nT (see electrical device) with one half of their water-filled bottom (area I) and the area < 50 nT with the other half (area II). The coil and the basin were on the south-north axis. The basin was divided into two compartments by marks at its edges.

In each sub-experiment, one set of 14 non-pretreated animals was used. The number of animals in the area I with the magnetic induction 500 - 50 nT was counted 2.0, 2.5, 3.0, 3.5, 4.0, 4.5, 5.0 and 5.5 minutes after the start of the experiment. In each sub-experiment, one set of animals first was observed without influence of an experimentally created electromagnetic field, then it was observed under the influence of the electromagnetic field at the respective test frequency. In different experiments using different sets of animals, the frequencies a. 50, 100, 200 and 400 Hz and b. 42, 67, 89 and 300 Hz were applied. All experiments were performed between 6 p.m. and midnight.

Evaluation

The cumulative frequencies of animals in the two compartments of the basin were compared with regard to the sets of animals with and without treatment, respectively, at each of the 8 measuring points in time. Statistical evaluation was made by the chi-square test. As the results are preliminary, the P-values are discussed, but not indicated in the figures.

RESULTS

Control frequencies

Different experiments were performed on the influence of 50 Hz (42 animals), 100 Hz (126 animals), 200 Hz (42 animals) and 400 Hz (42 animals). In all these experiments, in general, during the application of the respective frequency, fewer animals approached the source of radiation (area I 500 - 50 nT). The differences were mostly small (< 10%), but were statistically significant in one or more of the 8 measuring points in time. Fig. 1 shows the difference between the control group (no frequency applied, normalized line above) and the 100 Hz test group.

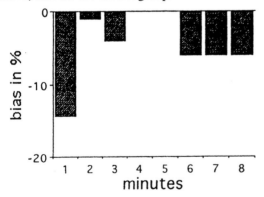

Fig. 1: Effect of an electromagnetic field (500 - 50 nT, 100 Hz) on the stay of highland Rana temp. in area I (maximal magnetic induction) of a water-filled test basin. Ordinate: the cumulative frequencies of animals under the influence of the electro-magnetic field compared to the lack of this influence, respectively. Abscissa: course of time in minutes. For further information, see text.

Test frequencies thyroxine D30 according to C.W. Smith

Different experiments were performed on the influence of 42 Hz (42 animals), 67 Hz (42 animals), 89 Hz (42 animals) and 300 Hz (126 animals). In all these experiments, in general, during application of the respective frequency, more animals approached the area I of 500 - 50 nT. The differences were mostly small, but were statistically significant at one or more of the 8 measuring points. Fig. 2 shows the difference between the control group (no frequency applied, normalized baseline) and the 300 Hz test group. This difference is significant at 7 measuring points.

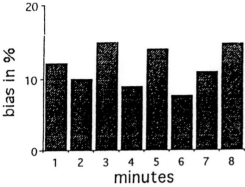

Fig. 2: Effect of an electromagnetic field (500 - 50 nT, 300 Hz). For further details, see Fig. 1 and text.

The results presented in this preliminary communication show that there are biological reactions of amphibia to an electromagnetic field (approach or avoidance of a test area with a magnetic induction of 500 - 50 nT, respectively) at different frequencies generated by a laboratory oscillator.

Among these reactions, the one towards a field with a frequency of 300 Hz was most marked (9-14% at 8 of 8 measuring points in time, 126 + 126 animals observed). This is consistent with our findings on the climbing behaviour of juvenile highland frogs with comparable conditions (magnetic induction 500 - 50 nT, frequency 300 Hz). In that study, among the animals that left the water, there were significantly more animals that left the water close to the source of radiation (500 - 50 nT) than in the control observations without radiation (a total of 220 + 220 cases were observed) (Fig. 3).

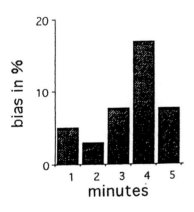

Fig. 3: Effect of an electromagnetic field (500 - 50 nT, 300 Hz) on the stay of juvenile highland Rana temp. in area I of a test basin. Ordinate: the cumulative frequencies of animals under the influence of the electromagnetic field compared to the lack of this influence, respectively. For further information, see Fig. 2 and text.

These - preliminary - results show the sensitivity of amphibia to electromagnetic fields in the range of a magnetic induction of 10^{-7} or even 10^{-8} Tesla (= 10^{-3} or 10^{-4} Gauß). This is consistent with biophysical statements on the dimensions of typical biological systems and the corresponding threshold fields capable of imposing on and ordering these systems (Smith in Popp - ed. - 1989, 3); an effect on tadpoles with a body length of ca. 1.5 cm should even be expected at weaker magnetic fields (e.g. 10^{-10} T).

The energy contained in a field of a given strength is proportional to the volume it occupies. The magnetic field can only be effective in establishing an order if it overcomes thermal fluctuations. With regard to this, an important concept which derives from the work of Brillouin (1934) and Fürth (1948) is that the magnetic interaction between the conduction electrons gives coherence to the whole circuit so that although the fluctuations in the charge carriers are the sum of random variations, the whole circuit behaves as if it has a single degree of freedom. In applying the concept of coherence to biological systems, Fröhlich (1978,1983, 1986a,b) considers them as non-linear systems. He concludes that they critically depend on the rate of the supply of energy, which, if exceeded, can lead to the coherent excitation of a single mode of oscillation, the excitation of a metastable highly polar state, and the possibility of limit cycles (Smith 1989).

As our preliminary study further showed, the reaction of the animals to the frequency 300 Hz is comparable to that to 42, 67 and 89 Hz. In a joint-venture study together with the laboratory of C.W.Smith, these frequencies have been described as particular resonance frequencies of homeopathically prepared dilutions of thyroxine. (For a complete list of frequencies up to radiofrequencies, see the contribution by Smith et al., Tab. 1.). When, in contrast, a field with identical magnetic induction, but with the control frequencies 50, 100, 200 and 400 Hz (according to Smith not significant for thyroxine dilutions) was applied, the animals significantly tended to avoid the electromagnetic field. Also when, in a further experiment, the influence of the frequency 300 Hz was directly compared to that of 100 Hz, more animals approached the source of radiation during the time intervals when 300 Hz were transmitted. This points towards a specific effect of distinct frequencies in the region of 500 - 50 nT.

More experiments will have to be performed a. with a larger number of cases, b. including other test and control frequencies, c. by other researchers and d. under blind conditions in order to investigate to which extent it is a specificity of the "thyroxine D30 resonance frequencies" (Smith) that causes a reaction of the test animals (approach of the source of radiation) different from the reaction to other (control) frequencies.

Further, the role of the specific susceptibility of amphibia in the state of metamorphosis to thyroxine in molecular solution and in agitated (high) dilution will have to be investigated with regard to the susceptability to the thyroxine D30 resonance frequencies.

ACKNOWLEDGEMENTS

Special thanks are due to G. Kastberger for his engineering expertise in the zoological laboratory.

For REFERENCES and further literature, see the contribution by Smith and Endler.

"Homoeo-pathy means to treat melancholy
with Goethe's Werther or songs by L. Cohen"

PART 4: THE HISTORICAL AND MEDICAL CONTEXT

"Stimulating immune and defense response, rather than controlling or manipulating symptoms, is a necessary and essential step forward in pharmacology. Such is the direction that pharmacology and medicine nees to go, and such is the direction that homoeopathy takes us."

D. Ullman, Meeting University Urbino, 1992

"Homoeopathic peculiarities
can be understood
as reflecting laws of nature"

F.A.C. Wiegant, personal communication

GLOSSARY ON HOMOEOPATHY

P. Andersch, P.C. Endler

In this book, the terms given below are used with the following meaning.

Homoeopathy: In a strict sense: medical treatment of or experimental influence on a living system according to the Law of Similarity (see there).

In a current (superficial) sense: use of specially prepared dilutions (see Preparation) in medical therapy or in scientific experiments.

Initial Aggravation of symptoms after application of a Homoeopathic Dilution is interpreted as a cybernetic reaction of the organism.

Individualisation: Individualized choice of a homoeopathic remedy with regard to the human patient's peculiar symptoms, with which not only the pathological clinical symptoms, but also further characteristics are meant. In research on lower animals or plants, the 'individual' reaction of different species is discussed.

Law of Similarity: Both the application of a molecular (crude) substance or a Homoeopathic dilution of that substance on healthy persons (see Remedy Proving) as well as an independent pathological process can induce similar, clearly defined symptoms. The very Homoeopathic Dilution that is able to induce the defined symptoms in healthy persons (Remedy Proving) is used in medical therapy of the analogous symptom (that has occurred due to the independent pathological process). Two kinds of examples are given:

(1) Cure after exposition to or protection from a toxin or allergen by a Homoeopathic Preparation of this toxin or allergen.

(2) Both a bee sting as well as ulraviolet rays can induce an inflammatory reaction. Homoeopathic Dilutions of bee venom are capable of inhibiting cutaneous erythema induced both by bee venom as well as by ultraviolet rays.

Potency Rule: Through the Preparation Process, information is transferred to the solvent. It is assumed that this information is more marked ("sharp"), the more often the process of agitation and dilution has taken place.

Preparation (potentisation, potenziation): of Homoeopathic Dilutions (remedies): Process of stepwise dilution of a substance e.g. in water or a water-alcohol mixture and input of exogene energy by agitation (e.g. succussing or vortexing) between the dilution steps. Non-watersoluble substances are first triturated with lactose. "D" - Dilutions are diluted in steps of 1 : 10, "C" - Dilutions in steps of 1 : 100.

Remedy Proving:, Drug Proving: Application of a crude (molecular) substance or a high or Ultra High Dilution (see there) to healthy volunteers with the aim of provoking symptoms ("empirical homoeopathic intoxication studies").

Sensitization: Homoeopathic Dilutions, like other (weak) stimuli, obviously exert their effects only in biological systems which are sensitive due to pathological, biological or experimental stress conditions.

Ultra High Dilution, UHD, High Dilution: Homoeopathically Prepared Dilution of infinitesimally small quantities of the original substance, in general diluted beyond 10^{-23}.

P.C. Endler and J. Schulte (eds.), Ultra High Dilution, 221.

CHARACTERISTICS AND SELECTED RESULTS OF RESEARCH ON HOMOEOPATHY

M. Righetti

SUMMARY

From its very beginnings, homoeopathy was in a sense based on exact empirical foundations of research. However, not much attention was payed to the specific aims and methods of research on homoeopathy by conventional medicine. This contribution discusses the foundations of homoeopathy itself and some selected older results from biochemistry, botanics, zoology and medicine.

ZUSAMMENFASSUNG

Seit ihren frühesten Anfängen basierte die Homöopathie in gewissem Sinne auf der Grundlage empirischer Forschung. Ihre spezifischen Forschungsziele und Methoden wurden von der konventionellen Medizin allerdings wenig beachtet. Der vorliegende Beitrag betrifft die Grundlagen der Homöopathie selbst, sowie einige ausgewählte Ergebnisse älterer Forschung mit Hochverdünnungen aus Biochemie, Botanik, Zoologie und Medizin.

INTRODUCTION

Scientific foundations

Homoeopathy

From its very beginnings, homoeopathy was in a sense based on exact empirical foundations of research. Not much attention was payed to the specific aims and methods of research on homoeopathy by conventional medicine. However, without a minimal knowledge of the fundamentals of homoeopathy, the correct judgement of its experimental - specially clinical - results is impossible. This is important as even today unreflecting prejudice predominates the discussion on homoeopathy. Homoeopathy is a medical science which - and this may astonish uninformed readers - is based on exact empirical facts (as is orthodox medicine or pharmacology). This means that observations are made according to clearly defined criteria. Nevertheless, the objective of experiments differs tremendously within the two medical disciplines.

The chemotherapeutics of orthodox medicine are based on extensive experimental testing of medicines in the laboratory and, thereafter, on patients. This is done mainly via so-called controlled randomized double-blind studies, i.e. according to the coincidental principle whereby neither the doctor nor the patient knows which patient receives the test substance or, as a control, the comparative substance.

Results from comparable studies with ultra-high dilutions using the orthodox research methods have little significance with respect to homoeopathic medical therapy, and they usually only serve to provide material for a dialogue with orthodox physicans.

For a better understanding of its research-related problems, a few basic facts of homoeopathy should be stressed. Homoeopathy, in a modern sense, was founded by Samuel Hahnemann (1755-1843). From the beginning, it was based on exact empirical foundations. Hahnemann was searching for a "rational" medical therapy and rejected the official medical educational viewpoint as speculative. Ironically, his science of homoeopathy was later regarded as "speculative", contrary to the facts. He undertook the first "Remedy

223

P.C. Endler and J. Schulte (eds.), Ultra High Dilution, 223–227.

Provings" on healthy persons (*Remedy Proving:* Application of a crude - molecular - substance or a high or Ultra High Dilution - see there - to healthy volunteers with the aim of provoking symptoms; "empirical homoeopathic intoxication studies", see Glossary, p. 221) and established the Law of Similarity in 1796: "Simila similibus curentur". (*Law of Similarity:* Both the application of a molecular - crude - substance or a Homoeopathic dilution of that substance on healthy persons in Remedy Provings as well as an independent pathological process can induce similar, clearly defined symptoms. The very Homoeopathic Dilution that is able to induce the defined symptoms in healthy persons is used in medical therapy of the analogous symptom that has occurred due to the independent pathological process, see Glossary). This means that the same remedy can be used to heal a sick person which, in Remedy Provings on healthy persons, had induced symptoms most similar to the patient's complaint. In the years that followed, an enormous amount of data has been accumulated by Remedy Provings on healthy persons, showing the specific symptoms and peculiarities of homoeopathic remedies (e.g. Kent 1989). The selection of remedies according to the Law of Similarity and the knowledge of homoeopathic drugs is therefore based on exact empircial evidence. Every sucussful application is, in the end, a human experimental confirmation of those specific foundations of homoeopathy. Each homoeopathic choice of remedy is the result of a comparison between the individual symptoms of the patient and the knowledge of the remedy's "intoxication" or "drug picture". Exact observation of the striking, individual and peculiar symptoms of a given patient is the decisive factor in the choice of remedy, while the allopathic diagnosis usually plays a subordinate role.

For chronic diseases, the choice of remedy is determined according to the patient, partly taken from his previous history. These symptoms are evaluated in a detailed examination which often lasts several hours. Various patients with the same medical diagnosis usually receive different remedies in homoeopathy, while a single, individually correctly chosen remedy can cure completely different and apparently incoherent "illness symptoms" in the same patient (Individualization Principle, see Glossary), i.e. migraine attacks, repeated tonsillitis and chronic eczema - because only one individual is ill!

In allopathy there is a more causal analytical, quantitative, reductional and linear approach to therapy on the grounds of Newton's mechanics (classical biomedical model). Homoeopathy is a systemic, holistic, quantitative, psychosomatic model which is closer to the new bio-psycho-social and systemic therapy models and to modern physics.

Ultra High Dilutions

Results from modern quantum physics, cluster research and solid-state chemistry also point to another special quality of homoeopathy which has been a stumbling block for some time: the Preparation of Homoeopathic Remedies (see Glossary), the so-called potentisation (or potenziation) process. Remedies are thereby diluted (potentized) in a water or water-alcohol solution by successive stepwise (ten- or hundredfold) dilution steps. After each step of dilution, the liquid is agitated (e.g. succussed) in a standardized way, often far beyond the limit where traces of the initial substance can be found. For solid substances, potentisation is achieved by trituration and dilution in milk sugar (lactose). From the point of view of biochemistry and pharmacology, any possible effect is no longer imaginable at such high 'dilutions'. This should not lead to the rejection of research on ultra-high dilutions and on homoeopathy, but rather to a detailed investigation of these unaccountable phenomena in which official science has, so far, not shown very much interest. (For information, see Righetti 1988 and the respective chapters of this book).

Research results

Research for its own purposes, i.e. Remedy Provings, is highly significant for homoeopathy, but not acknowledged very much by orthodox medicine (see Walach's contribution). The following studies exemplify the effectiveness of homoeopathy in different older research models which, however, can be criticised with regard to the standard of methods or evaluation, but usually are acceptable with respect to scientific criteria. (For further classical examples, see Poitevin's contribution.)

In vitro study: In 1954, Boyd published his meticulous study on the effects of mercury chloride ($HgCl_2$) D61 (= homoeopathically prepared dilution 10^{-61}. The 61st decimal dilution lies far beyond the molecular boundary of 6×10^{-23} (Loschmidt's or Avogadro's number) which corresponds with the 24th tenth (D24) or 12th hundredth dilution (C12). In a well-controlled and analyzed blind study with over 500 comparison examinations which lasted several years, Boyd significantly proved that $HgCl_2$ D61 accelerates the splitting of starch by the enzyme malt diastasis.

Plant studies: In this field, the untiring pioneer work of Lili Kolisko is to be mentioned, who, since 1923, has observed plants in connection with homoeopathic dilutions. It has become increasingly obvious that the use of remedies enhanced the growth of plant seeds. Using the same principles, an effect was also shown by Pelikan and Unger 1965 (silver nitrate D8-D19 on the growth of wheat germs). Jones and Jenkins (1981) were able to reproduce these results with similar silver nitrate dilutions (C6-C16 = 100^{-6} - 100^{-16}). The results of Kolisko's experiments with the potencies D24, D25 and D26 were recently reviewed by Pongratz and Endler (see the contribution of Pongratz et al.). (For a list of further older botanical studies, including important work by Boiron et al., see the contribution of Pongratz et al.)

Animal studies: An often used "standard model" are poisoning studies (see Roth 1991). These date back to Mrs. Wurmser and Lapp (1955 onward). The basic principle is similar in all studies: laboratory animals are poisoned with a toxic substance and protected or rather detoxicated by homoeopathically prepared dilutions of the same substance. The homoeopathic dilution increases the secretion of the poison in the urine and stools. The most elaborate and best controlled work comes from Cazin (1986, 1987), who in blind studies has carried out arsenic intoxication studies on 696 rats. Twelve hours after oral arsenic poisoning, rats were injected with a homoeopathically prepared arsenic solution. All arsenic dilutions tested (C5-C15 and D10-D30) showed a statistically highly significant increase in arsenic secretion via stools and urine and a lowering of the arsenic level in blood. However, the absence of influence of a dilution of lead on the excretion cinetics of lead has been observed by Fisher et al. (1987). Furthermore, the model has also been tested in artificial animal illnesses. One example is Diabetes mellitus induced by alloxane poisoning (toxic influence on the pancreas). The animals could be protected by potentised Alloxane (C7, C9) and did not or only rarely develop diabetes (see Righetti 1988).

In view of these results, the question of their therapeutical application arises. Is it possible to protect an organism against poisonings by giving the corresponding homoeopathic dilution? According to Souza Margo et al. (1986) an application of different homoeopathically prepared dilutions of gentamycine will diminish the damage to the kidneys known to be caused by that antibiotic. In the study by Gardes (1989) nalidixic acid C7 led to a significantly quicker excretion of this antibiotic. Finally, Labonia et al. (1986) were able to protect mice against the poison of the Bothrops snake by administering potencies of the very same poison.

Harisch and his team (1988) were successful in scientifically proving the biochemical effects of various homoeopathic remedies on the glutathione system of the liver of rats which had been poisoned with carbon tetrachloride in studies from 1984 onwards. A number of different effects could also be achieved by applying various homoeopathic remedies to rats with the aim of releasing histamine from the peritoneal mast cells of the rats.

Multicenter studies have been performed on amphibia during a developmental transition in which the animals show a comparatively high natural physiological thyroxine level. It is well known that in these stages, the metamorphosis of amphibia can be further accelerated by exogene molecular thyroxine. During the transition of non-pretreated animals, the team of Endler investigated the influence of thyroxine D30. Dependent on the way of application, both an inhibitory as well as an accelerating effect of the dilution on metamorphosis could be observed with sound statistical significance (see the contribution of Endler et al.). The studies on the inhibitory effect of thyroxine D30 on the metamorphosis could be discussed with regard to a "homoeopathic" effect of the dilution on the naturally thyroxine-stressed animals, whilst the studies on the accelerating effect of the same dilution, given under different conditions, could be discussed with regard to the principle that also underlies the "Remedy Provings" on comparatively healthy organisms.

The best-known *veterinary medical studies* are those on the birth process, performed by Wolter in 1966 (in. Gebhard 1980/1985). They might be criticised with respect to certain details, however. In blind studies Caulophyllum D30, under given circumstances, has proved to be very effective for the lack of labour pains of mother pigs.

In the field of *human medicine*, one of the best controlled double-blind studies goes back to World War II, when the Britons, expecting German air raids, examined the prophylaxis and therapy of mustard gas burns on the skin. Several homoeopathic remedies as well as a mustard gas dilution (C30) were statistically significant in comparison to the control group (Paterson 1944).

(For discussion of basic immunological studies, see Poitevin's contribution. The field of clinical studies is described in Haidvogl's contribution.)

ANNOTATION

This contribution is based on a comprehensive paper published in the Berlin J Res Hom 1991, 1(3): 195-203. It has been actualized and adjusted for this volume with the help of the author. Remarks referring to the Glossary are made by the biological editor. Thanks are due to L. Taylor-Spring, Homeopathic Links, Bern, for her help with the English.

REFERENCES

Boyd WE. Biochemical and Biological Evidence of the Activity of High Potencies. Br Hom J 1954; 44: 7-44.

Cazin J-C. Etude pharmacologique de dilutions Hahnemanniennes sur la retention et la mobilisation de l'arsenic chez le rat. In: Recherches en Homéopathie, Boiron J, Belon P and Hariveau E eds., Ste-Foy-les-Lyons 1986: Fondation Francaise pour la Recherche en Homéopathie, pp. 19-33.

Gardes E. Effect d'une dilution infinitésimale d'Acide nalidixique sur l'elimination de cette meme molécule chez l'homme sain. Diplome Pharmacie. Annales Homéopathiques Francaises 1989; 77 (5): 60.

Harisch G, Müller R, Meyer W. Beitrag zum Leberstoffwechsel der Ratte nach Tetrachlorkohlenstoff unter dem Einfluß von Nux vomica D6 bzw. Flor de Piedra D6 - Erste Befunde. Allg Hom Zeitung 1984; 229(5): 190-199.

Harisch G and Kretschmer M. Effekte homöopathischer Präparationen im Zellstoffwechsel. In: Stacher A ed., Wiener Dialog für Ganzheitsmedizin. Wien: J+V Edition 1988.

Harisch G, Kretschmer M, von Kries U. Beitrag zum Histaminrelease aus peritonealen Mastzellen von männlichen Wistar-Ratten. Dt tierärztl Wschr 1987; 94: 515-516.

Harisch G, Kretschmer M. Jenseits vom Milligramm-Die Biochemie auf den Spuren der Homöopathie. Berlin: Springer 1990.

Jones RL, Jenkins MD. Plant Responses to Homoeopathic Remedies. Br Hom J 1981; 70(3): 120-128.

Kent JT. Repertory of Homoeopathic Materia Medica. New Dehli: Jain Publishers 1989.

Kleijnen P, Knipschild P, Riet G. Br Med Journal 1991; 302: 316-323.

Kolisko L. Physiologischer Nachweis der Wirksamkeit kleinster Entitäten bei sieben Metallen - quoted from: Wurmser 1984, Dornach, Schweiz: Goetheanum Verlag 1926.

Labonia W, Prado MIA, Furlado MFD. Acao das dosis mínimas na protecao do enveneuamento oficido de animaes de laboratório. Proc Liga Med Hom Int, Rio de Janeiro 1986; 41.

Lapp C, Wurmser L, Ney J. Mobilisation de l'Arsenic fixé chez le cobaye sous l'influence de doses infinitésimales d'arseniate de sodium. Thérapie 1955; 10: 625-638.

Lapp C, Wurmser L, Ney J. Mobilisation de l'Arsenic fixé chez le cobaye sous l'influence de doses infinitésimales d'arseniate. Thérapie 1958; 13: 46-55.

Lapp C, Wurmser L, Krautele N. Mobilisation du Bismuth fixé chez le cobaye sous línfluence de doses infiniesimales d'un sel de bismuth. Thérapie 1958; 13: 438-450.

Paterson J - President of the Special Gas Research Committee. Report on Mustard Gas Experiments - Glasgow and London. J American Homeopathic Association 1944; 37: 47-50 and 88-89, also: Br Hom J 1943; 23: 131-142.

Pelikan W and Unger G. Die Wirkung potenzierter Substanzen - Pflanzenwachstums - Versuche mit statistischer Auswertung. Dornach 1965: Philosophisch - Anthroposophischer Verlag.

Righetti M. Forschung in der Homöopathie - Wissenschaftliche Grundlagen, Problematik und Ergebnisse. Göttingen 1988: Burgdorf Publisher.

Roth C. Literature Review and Critical Analysis on the topic of "In- and Detoxication Experiments in Homoeopathy". Berlin J Res Hom 1991: 1: 111-117.

Souza Magro IA et al. Reducao da nefrotoxidade induzida por aminoglucosideos. Proc Liga Med Hom Int, Rio de Janeiro 1986; 41.

Walach H. Die Wissenschaftliche Arzneimittelprüfung. Heidelberg: Haug Verlag 1992.

Wolter H. Wirksamkeitsnachweis von Caulophyllum D30 bei der Wehenschwäche des Schweins. In: Beweisbare Homöopathie, Gebhardt KH ed., 2 ed, Heidelberg: Haug Verlag 1985.

HOMOEOPATHY WITH SPECIAL REGARD TO IMMUNO-ALLERGOLOGICAL RESEARCH

B. Poitevin

SUMMARY

In this contribution, the history of two immunological models in research on UHDs is given. Furthermore, the special features of the ' Law of Similarity' and 'Sensitization' (see Glossary, p. 221) are discussed on the basis of these models.

ZUSAMMENFASSUNG

In diesem Beitrag wird die Geschichte zweier immunologischer Modelle zur Forschung an Hochverdünnungen dargestellt. Weiters werden das 'Simile-Prinzip' und die 'Sensitivierung' (siehe Glossar, S. 221) auf der Grundlage dieser Modelle diskutiert.

INTRODUCTION

Homoeopathy today has come to light as a complex discipline, situated at the intersection of medical, biological and physical sciences. It is for this reason that a diversified and pluridisciplinary approach to it is needed. In the light of this, we will discuss significant findings in immuno-allergological research we were engaged in for a long time. These findings can serve to illustrate peculiarities of homoeopathy.

METHODS AND RESULTS

Biological action

Immunoallergology is a privileged area of research for experimental studies on medicines for homoeopathic use, on the one hand due to the many concepts these two disciplines have in common, and on the other hand due to the large number of medicines for homoeopathic use which are employed in treating inflammatory and allergic syndromes. Useful results have been obtained in this area, results which have been published in international scientific journals (review in - 1 -).

Ultraviolet irradiation of mice

Experiments in this field were carried out for over more than one decade with homoeopathic preparations from bees and bee venom on erythema caused by ultraviolet radiation. These experiments are based on the reactional similarity (see 'Law of Similarity' in the Glossary) which exists between cutaneous phenomena due to bee venom and erythema due to UV radiation. Guinea pigs are UV irradiated before or after administration of the bee /bee venom dilution ("Apis mellifica", "Apium virus") or of the solvent, respectively and a blind rating is made.

The inhibitory action of Apis mellifica C7 has been proved in several studies directed by Aubin. In cooperation with the team of Bildet and Quilichini, we have studied the effect of Apis mellifica C5, C7, C9, C15 and Apium virus C5, C7, C9 and C15 (C15 = 100^{-15}) (2). The reference groups of 10 guinea pigs received a physiological salt solution, homoeopathically diluted to C15. Intraperitoneal injections of 1 ml of Apis mellifica, Apium virus or the saline solution were given 1 hour before radiation, immediately prior to radiation and four hours after. Erythema were measured 4 hours, 24 hours and 48 hours after radiation according to a 4 level grading system: absence of erythema, a barely visible spot, more marked erythema, clearly visible erythema. The effects of the different products were evaluated by percentage of erythema inhibition as compared to the reference groups.

P.C. Endler and J. Schulte (eds.), Ultra High Dilution, 229–232.
© 1994 *Kluwer Academic Publishers. Printed in the Netherlands.*

Four hours and twenty-four hours after radiation, there was no statistically significant effect for Apis mellifica and Apium virus. But, forty-eight hours after radiation, Apis mellifica C7 and C9 had a statistically significant inhibitory effect (using Mann /Whitney test) on erythema induced by UV radiation. This improvement was, respectively, 31,7 % and 41,5 %. At the same time, after radiation, Apium virus C5, C7 and C9 had a statistically significant inhibitory effect. The improvement was, respectively, 41,5 %, 39 %, and 34,1 % . The C5 and C15 dilutions of Apis mellifica and the C15 dilution of Apium virus did not lead to any significant improvement.

These experiments (although significant only in the range below Avogadro's value) show the interest of the Reactional Similarity which is the base of homoeopathy: just as a bee sting can induce an inflammatory reaction concommitant with cutaneous erythema, so also dilutions of bee and bee venom are capable of inhibiting cutaneous erythema induced by ultraviolet rays (for further information, see Glossary). They demonstrate also the possibility of obtaining substantial reproducibility over an interval of ten years, in the case of this experiment.

These results serve well to illustrate a characteristic which is specific to research on the pharmacological effects of high dilutions of homoeopathic medicines, namely the necessity to "sensitize" the biological material in order to demonstrate the effect of high dilutions (see Glossary).

Degranulation of human basophils

Another series of studies is based on the use of the in vitro degranulation test of human basophils. This test investigates the metachromatic capacities of these cells, using a method which involves counting the basophils by optic microscopy. This phenomenon, tied to modifications in ionic transportation to the surface of the granules, differs from the liberation of mediators, such as histamine, which are contained in basophil granules. The first studies covered the inhibition of the degranulation of basophils. In 1986 we pointed out the inhibitive effect of Apis mellifica C9 and C15 on degranulation induced by different allergens of hymenopter venom (3).

Davenas, Benveniste and myself have taken up these studies again for investigating the basophils of healthy subjects stimulated by an antiIgE serum, studying the effect of two products used in the homoeopathic treatment of allergic syndromes, Apis mellifica and Lung histamine (4). In the presence of the antiIgE we first observed a double degranulation curve whose peaks were centered around 10^{-9} and 10^{-17}M. On these two zones of degranulation we then studied the effect of successive dilutions of Apis mellifica and of Lung histamine.

The degranulation of basophils induced by the antiIgE antibody 1.66×10^{-9}M is significantly inhibited by the presence of Lung histamine C5 and by that of Lung histamine C15, by 28.8 % and 28.6 % respectively. It is inhibited by 65.8 % in the presence of Apis mellifica C9. The degranulation of the basophils induced by the antiIgE antibodies 1.66×10^{-16} to 1.66×10^{-18}M is also inhibited by high dilutions of Lung Histamine and Apis mellifica, with inhibition close to 100 % with Lung histamine C18 and Apis mellifica C10. An alternation in the zones of inhibition, of inactivity, and of stimulation is observed when the basophils are incubated in the presence of successive dilutions of Lung histamine and of Apis mellifica, giving curves which are "pseudosinusoidal" in appearance.

Apis mellifica and Lung histamine inhibit the in vitro human basophil degranulation, even at high dilutions wherein there are no longer any molecules present. Recently, also the inhibitory effect of Apis mellifica, from C15 to C20 has been studied by Davenas and

Benveniste. The inhibitive action of high dilutions of histamine on the degranulation of basophils was equally proved with inhibitory peaks centered around dilutions of C6/7 and C17/18 (5).

It is the same shape of a pseudosinusoidal curve which is found when the basophils of healthy subjects are stimulated by increasing dilutions of antiIgE up to dilutions of 1×10^{-120} (6). [Annotation by the editor: Controversial publicity and sensationalism surrounded this study on degranulation of basophiles enhanced by anti-IgE antibodies. One later attempt to repeat the study under independently controlled blind conditions was successful (7) and one failed (8). Furthermore, the effect of highly diluted antigen on IgM and IgG antibodies was studied and proved by an independent team (9). Backed up by all these studies, the hypothesis was formulated that water could transmit biological information (see also Benveniste's contribution and the "Prospects on Elements of a Theory on UHDs" in this book, pp 245ff.]

Taking into consideration the two immuno-allergological studies summarized here in order to get a historical/ medical survey, we note that high dilutions of medicines for homoeopathic use or of mediators seem to be active in vivo on biological systems (in animals), and in vitro on isolated cells. Further research is, of course, necessary (see the respective contributions in this book. (For further examples on immunological research, see Bastide's contribution.)

Research into the mechanism of action

This can be added directly as an extension to the research on the biological action. There is a form of biological information contained in high dilutions (10) and we need to work on its nature, its means of propagation and its targets.

On the nature of this information, we possess three types of indications :

- this information maintains a very narrow molecular specificity. This fact points out the importance of basic classical pharmacological and biological knowledge.

- it remains stable during successive dilutions, thus underlining the role of the solvent in receiving this information. The physicochemical characteristics of water and the ability of its dipoles to arrange themselves in a coherent and durable manner under the effect of a signal of molecular or electromagnetic nature are in favour of this hypothesis (11).

- lastly, the process of homoeopathic preparation by stepwise agitation and dilution occurs, imparting new physicochemical characteristics to the interactions between solute and solvent.

This information whose nature we ignore, propagates itself. Theoretically it follows the perlingual route. Nevertheless, the possibility of other networks of conduction should not be excluded, such as, for example, any resembling to those of acupuncture meridians.

This information then reaches targets. We know that high dilutions of medicines for homoeopathic use have a regulatory action on biological targets and work in an original manner, as is suggested by dose-response curves obtained with the high dilutions. Several theoretical arguments advanced by biologists and physicists tend towards the physical, electromagnetic nature of these targets (see the respective contributions and the Prospects Section).

DISCUSSION

It has been shown that homoeopathic medicines are active in biological systems.

The mechanism of action of this medicine remains to be investigated. Its regulatory action is original and its effect which extends beyond the threshold of theoretical molecular presence, most likely depends on physico-chemical mechanisms which remain to be investigated.

It is quite clear that many questions concerning homoeopathy have not been solved yet. Homoeopathy still cannot be explained in a coherent scientific manner. Nevertheless, none of the questions should be rejected without consideration. If homoeopathy "works", it is most likely in a totally paradoxical fashion in relation to contemporary scientific explanations. Thus, it would be preferable to wisely refrain from rejecting information which stems from traditional homoeopathy, since much of it could be indicative of new epistemological information.

ANNOTATION

The experiments carried out by the 200 Unit of INSERM have been carried out by E. Davenas under the scientific direction of J. Benveniste (see also - 11 -).

This contribution is based on a comprehensive paper by the author. It has been actualized and adjusted for this volume by the biological editor with the help of the author.

REFERENCES

1. Poitevin B. Recherche experimentale. In: Encyclopédie Médico-Chirurgicale. Homéopathie. Paris 1988; 3: 38-60, A30.
   Poitevin B. Relation générale entre homéopathie et immunoallergologie. Encyclopédie Médico Chirurgicale. Homéopathie, Paris 1988; 7: 38-55, A10.
   Poitevin B. Introduction à l'homéopathie. Bases expérimentales et scientifiques. Ed. CEDH, 1990.
2. Bildet J, Guyot M, Bonini F, Grignon MF, Poitevin B, Quilichini R. Mise en évidence des effets de dilutions d'Apis mellifica et d'Apium virus vis a vis de l'érythme provoque par un rayonnement U.V. chez le cobaye. Annales Pharmaceutiques Francaises 1989; 47: 24-32.
3. Poitevin B, Aubin M, Benveniste J. Approche quantitative de l'effet d'Apis mellifica sur la dégranulation des basophiles humains in vitro. Innovation et Technologie en Biologie et Médecine 1986; 7: 64-68.
4. Poitevin B, Davenas E, Benveniste J. In vitro immunological degranulation of human basophils is modulated by Lung histamine and Apis mellifica. British Journal of Clinical Pharmacology 1988; 25: 439-444.
5. Sainte Laudy J. Standardisation of basophil degranulation for pharmacological studies. Journal of Immunological Methods 1987; 98: 279-282.
6. Davenas E, Beauvais J, Amara J et al. Human basophil degranulation triggered by very dilute antiserum against IgE. Nature 1988; 333: 816-818.
7. Litime MH, Aissa J, Benveniste J. Antigen signalling at high dilution. FASEB J 1993; 7: A 602.
8. Ovelgonne JH, Bol AJWM, Hop WJC, van Wijk R. Mechanical agitation of very dilute antiserum against IgE has no effect on basophil staining properties. Experientia 1992; 48(5): 504-508.
9. Weisman Z, Topper R, Oberbaum M, Bentwich Z. Immunomodulation of specific immune response to KLH by high dilution of antigen. Abstracts GIRI Meeting, Paris 1992.
   Weisman Z, Topper R, Oberbaum M, Harpaz N, Bentwich Z. Extremely low doses of antigen can modulate the immune response. ProcVIII Intern Congress Immunol, Budapest, Hungary, 1992, 23-28 August: 532.
9. Poitevin B. Le devenir de l'Homéopathie. Lyon 1987: Ed Boiron; 285 pages.
10. Del Guidice E, Preparata G, Vitiello G. Water as a free Electric dipole laser. Physical Review Letters 1988; 61: 1085-1088.
11. Davenas E, Poitevin B, Benveniste J. Effect on mouse peritoneal macrophages of orally administered very high dilutions of Silicea. European Journal of Pharmacology 1987; 135: 313-319.

CLINICAL STUDIES ON HOMOEOPATHY.
THE PROBLEM OF A USEFUL DESIGN
By M.Haidvogl

SUMMARY

Numerous controlled clinical studies have proved the efficacy of homoeopathic remedies, but most of them could not meet sufficient methodological quality. However, especially homoeopathic studies have to show a maximum in scientific standard because they have to withstand much more criticism than pharmacological studies relating to the common paradigmas of clinical medicine. Clinical studies on homoeopathy should be planned in co-operation between homoeopathic practitioners and university clinics, because this allows for a maximum methodological standard and also for the possibility of publication in distinguished journals with a high impact factor, which usually are not available for homoeopathic practitioners. Controlled clinical human studies in the field of homoeopathy are only useful if they take into account the characteristics of homoeopathic drug prescription, i. e. the person-specific, individual choice of the remedy.

ZUSAMMENFASSUNG

Eine Vielzahl von kontrollierten klinischen Studien hat bisher die Wirksamkeit homöopathischer Medikamente gezeigt, die meisten davon allerdings, ohne die notwendige methodologische Qualität aufzuweisen. Indessen müssen spezifisch homöopathische Studien ein Maximum an wissenschaftlichem Standard aufweisen, da ihnen sehr viel mehr Kritik entgegengebracht wird als pharmakologischen Studien, die die allgemeinen Paradigmen der klinischen Medizin betreffen. Klinische Studien zur Homöopathie sollten in Zusammenarbeit von homöopathischen Praktikern und Universitätskliniken geplant werden, da auf diese Weise maximaler methodologischer Standard und auch die Möglichkeit der Publikation in anerkannten wissenschaftlichen Zeitschriften gewährleistet wird. Kontrollierte klinische Humanstudien auf dem Gebiet der Homöopathie sind nur dann von Nutzen, wenn sie die Charakteristiken homöopathischer Arzneimittelwahl, d.h. die personspezifische, individuelle Wahl des Arzneimittels, berücksichtigen.

INTRODUCTION

This article is written by a homoeopathic practitioner. Therefore it shall not deal with the delicate problems of advanced statistics, but give practical hints how clinical trials may be designed to satisfy the aim of a scientific proof without leaving the path of homoeopathic principles.

EVALUATION OF REFERENCES AND FURTHER METHODOLOGY
Clinical trials already performed

It is often claimed by the exponents of clinical medicine that homoeopaths have not yet tried to establish a scientific proof for the clincal efficiency of homoeopathic treatment. They are even accused of not being interested in such a task at all. Yet quite a lot of clinical studies on this subject have been performed.

In 1991 Kleijnen, Knipschild and ter Riet published an extensive metaanalysis of 107 clinical trials of homoeopathy in order to evaluate the methodological quality of the studies concerned. The authors developed a score divided among seven methodological criteria:

P.C. Endler and J. Schulte (eds.), Ultra High Dilution, 233–241.
© 1994 *Kluwer Academic Publishers. Printed in the Netherlands.*

1) Patient characteristics adequately described: Max 10 points

2) Number of patients analysed: Max 30 points

3) Randomisation: Max 20 points

4) Intervention well-described: Max 5 points

5) Double-blinding: Max 20 points

6) Effect measurement relevant and well-described: Max 10 points

7) Presentation of the results in such a manner that the analysis can be checked by the reader: Max 5 points.

Out of the 105 evaluated trials (2 trials compared the effect of several potencies of the same homoeopathic remedy), 81 showed a positive effect of homoeopathy, 24 trials failed to show an effectiveness of homoeopathy versus placebo or another clinical drug. Concerning only the 23 trials with a score over 55 (good methodological quality), 16 trials proved a positive effect of homoeopathy and 7 showed no effectiveness of the homoeopathic remedy compared to a placebo.

Does this mean that the effectiveness of homoeopathy has been proved?

Kleijnen et al. comment on the results of their study as follows: "The weight of the presented evidence will probably not be sufficient for most people to decide definitely one way or the other. (...) In our opinion, additional evidence must consist of a few well-performed controlled trials in humans with a large number of participants under rigorous double blind conditions. The results of the trials published so far and the large scale on which homoeopathy is brought into practice make such efforts legitimate."

The main problem of the studies hitherto published is the low methodological quality of a great number of these trials. This is not specific for homoeopathy but applies to many clinical trials in other fields as well (Gore, Jones et al., 1977; Pocock, Hughes, et al. 1987). But in such a controversial field of medical research, a high methodological standard is absolutely necessary.

Design of the studies as compared to the basic principles of homoeopathy

The usual design of a clinical trial in conventional medicine is a comparison between one drug and another or to a placebo in a group of patients with the same diagnosis. This concept does not apply to studies in the field of homoeopathy, because it does not take into account the basic principles of homoeopathy which do afford a highly individualised therapy according to the individual symptoms of the patient.

How did the studies mentioned above handle this problem: In only 14 trials the remedies were prescribed according to the rules of classical homoeopathy, in 58 trials the same single remedy was given to patients with comparable conventional diagnosis. In 26 trials complex remedies, that is a combination of several homoeopathic remedies, were tested, in 9 trials analogies to detoxification by the highly diluted toxin were investigated.

Concentrating only on the 23 trials with good methodological standard, this principal problem becomes even clearer. 9 of these trials evaluated the efficacy of complex remedies, that is remedies containing several homoeopathic preparations. Complex remedies can well be used in clinical medicine in a group of patients with the same disease. In our case they were tested with patients with common cold, upper respiratory tract infections, sinusitis and arthritis.

Complex remedies may be useful in the treatment of some acute diseases, but even in this case one may find patients whose symptoms do not fit into the drug picture of the remedies applied. Therefore most classical homoeopaths do not accept them at all. Thus, trials with complex remedies are easy to design, but do not clearly apply to the basic rules of homoeopathy.

In 10 trials the same single homoeopathic remedy was given to all patients with a certain diagnosis.The main indication was pollinosis, which can be treated with a remedy with a specific antiallergic action such as Galphimia glauca, and clinically proven remedies against postoperative ileus were a second indication.

Such a design seems only useful in a drug like Galphimia glauca which does not produce many individual symptoms in the remedy proving [for explanation, see Glossary, 219, Righetti's and Walach's contribution] but has a distinct, but unspecific antiallergic action. This probably does not apply to the remedies used in postoperative ileus and therefore these studies produced unequivocal results.

The well known study dealing with the action of Pollen C 30 (100^{-30}) on pollinosis by Reilly et al. 1986 is based on the treatment of a disease caused by a certain agent using a potency of the same agent. Nevertheless the trial is useful in showing the efficacy of a high potency.

The only trial performed according to the rules of classical homoeopathy was the study of Brigo (1987) on the treatment of migraine headache. In this randomised double-blind study Brigo included only patients whose symptoms corresponded with 8 preselected homoeopathic remedies in C30. After a careful selection of the appropriate remedy for the individual patient, randomisation was done and a double-blind controlled trial on the effect of the preselected individual homoeopathic remedy versus placebo was performed. The outcome of this study was excellent. After four months the number of attacks in the placebo group degreed from 9.9 to 7.9 attacks per month, while in the verum group it degreed from 10.0 to 1.8 attacks per month. The patients treated with homoeopathy also were able to cope much better with the severity of the remaining attacks. On a 10 cm visual analogue scale the severity changed from 9.1 to 2.9 in the homoeopathic group and from 8.4 to 7.8 in the control group. This means a statistically significant effect with high clincal relevance.

This study has only one major drawback: it was published in a congress proceedings and not in a journal with a high impact factor. And in the self-understanding of clinical scientists a paper is only of some value when published in such a journal.

Useful design for a homoeopathic trial

Basically a homoeopathic prescription is not done according to a clinical diagnosis but according to the individual and peculiar symptoms of a single patient. Therefore the method most suited to show the effectiveness of a homoepathic remedy should be a well-controlled case report, an individual case-analysis.

Individual case-analysis

In the case of acute illness, an individual case-analysis is not useful, because the success of a treatment can barely be differentiated from the natural history of the disease. In chronical illness a well-done case report can be conclusive.

There are several preconditions which have to be followed to make such a case report valuable (Gebhardt 1986):

1) The period of pre-observation has to be long enough in order to state the form and duration of the symptoms thoroughly.

2) A period of observation, which in most of the cases will correspond with the period of the administration of the remedy. This may, however, not be the case when high potencies are used, because in this case one dose will be sufficient for a longlasting curative action.

3) The period of post-observation has to be long enough to realize if the effect of the remedy persists, if the symptoms have been cured or if it is necessary to repeat the treatment.

This shall be exemplified in one case report (detailed further material has been accepted by the Br Hom J, 1993):

Christa S., age 15 years; Diagosis: Down-syndrom.

Pre-observation: the child has been in a school for handicapped children since the age of eight years. Within the last three years she has shown a profound lack of energy, most of the time she lies nearly motionless in a corner, has literally to be pushed and pulled to every slight action and responds with aggressive tantrums when disturbed. She is overweight, eats a lot and is chronically constipated.

Observation: after receiving the homoeopathic remedy Graphites D12 (10^{-12})she becomes a lot more active within two to three weeks, she now joins in activities of the group and is activateable without tantrums and is less constipated. Post-observation: after a treatment period of two months, Graphites is withdrawn, after another three to four weeks she slowly returns to inactivity, but responds to a second treatment period with Graphites of another two months. Then the effect lasted for another control period of one year.

Controlled clinical studies

A well-executed individual case analysis certainly comes closest to the methodology of homoeopathy and is certainly of scientific value.

Nevertheless we will have to do controlled clinical studies in order to evaluate the effectiveness of homoeopathy, as our clinical colleagues tend to see the presentation of mere case reports as some sort of an anecdotal approach to medicine, although impressive case reports have always been important for the progress of clinical medicine as well.

In designing a clinical study on the subject of homoeopathy, the basic principles in the selection of remedies must not be neglected. Therefore only two designs seem appropriate:

1) A randomised double-blind study without individual selection of remedies is only possible if it applies to a local symptom which is limited and exactly defined.

2) A study in a group of patients with a certain disease is only possible when the homoeopathic remedy is first individually selected for each patient, then the patients may be randomised to a verum and a placebo group in order to perform a double-blind study.

The problem of the useful design shall be explained by two rheumatological studies published in distinguished international journals with a high impact factor and a third one using a "hybrid" design between the two others:

1) Shipley et al., Lancet 1983; I: 97.

Randomised double-blind study, cross-over,

36 patients with gonarthrosis or coxarthrosis,

Not individualized homoeopathic remedy vs Fenoprofen or placebo,

Duration of treatment 2 weeks,

Fenoprofen effective, homoeopathic remedy and placebo ineffective

Gibson et al., Br J Clin Pharmacol 1980; 9: 453.

Randomised double-blind study,

46 patients with rheumatological symptoms,

Individualised homoeopathic remedy vs placebo

Duration of study 5 months,

Homoeopathy significantly effective (p < 0,01)

Fisher, Br Hom J 1986; 75: 142.

Double blind study, 24 patients,

Fibrositis defined in term of symptoms

3 homoeopathic remedies prescribed according to symptoms,

Duration of study 3 months,

Homoeopathy significantly effective on pain and sleep, when well-indicated acc. to symptoms.

In the study of Shipley et al. a single homoeopathic remedy in low potency was tested out on 36 patients with a degenerative joint disease of knee and hip versus an effective anti-inflammatory remedy and placebo. Considering the principles priorily stated it seems logical, that for such a chronical illness one single homoeopathic remedy could hardly be the effective remedy for each of the 36 patients.

A two weeks period of post-observation is hardly sufficient, because this is not an acute illness, in which one would expect sudden effectiveness. Therefore this study is unacceptable seen from the viewpoint of the basic principals of homoeopathy.

In the second study of Gibson et al., the homoeopathic remedy was first individually chosen to fit the symptoms of the patients, only afterwards the patients were randomised and the verum or placebo administered in a double-blind manner. This study was long enough and finally a significant effectiveness of the homoeopathic remedies was shown at least for a few symptoms.

In his study on fibrositis Fisher used a somewhat "hybrid" design. Fibrositis was defined not as a pathological entity, but purely in terms of its symptoms. He chose three remedies (Arnica, Rhus.tox and Bryonia) which covered most of the symptoms. The study was performed double blind with a duration of three months, with an assessment of symptoms every 4 weeks. When well indicated, the effect of the homoeopathic remedy was significantly better than placebo, whereas no significant difference was found when the remedy covered only one ore none of the symptoms.

Examples for useful designs of studies

The two following studies planned and carried out by the Ludwig Boltzmann Institut für Homöopathie show possible designs for useful studies:

1) *Clinical trial without individual selection of remedies*

"The effect of Lycopodium D12 in infantile colics culminating in the late afternoon"

Abdominal colics are one of the most frequent problems within the first three months of life and may well be very exhausting for mother and child. For the most frequent forms of these colics with a typical crying-hour between 5 p.m. and 8 p.m., a homoeopathic preparation of Lycopodium has proved to be very useful in the experience of homoeopathic practitioners. Often the discomfort disappeared the following day after the first administration of the Lycopodium dilution. The typical maximum of discomfort in the late afternoon is essential for the prescription of Lycopodium. For other kinds of infantile colics Lycopodium is not indicated.

2) *Randomised double-blind study with individual selection of remedies*

"Homoeopathic treatment of juvenile warts"

A randomized double-blind study was designed concerning the treatment of juvenile warts of the hands. From the experience of homoeopathic practitioners it is known that juvenile warts respond favourably to homoeopathic treatment. Included in this study are children from 6 to 14 years of age with juvenile warts of the hands whose constitutional picture (totality of symptoms) fits into the drug picture of 10 preselected homoeopathic remedies known to be effective in treating warts. The extension of the warts is stated by a clinical dermatologist by photography and a planimetric method. The individual remedy is then chosen by an experienced homoeopathic practitioner on the basis of the totality of symptoms of the patient. (Every study in classical homoeopathy does not only test the effectiveness of homoeopathy, but the wisdom of the homoeopath involved as well.) Then the patients are randomized and treated with verum or placebo in a strictly double-blind manner.

Other "specifically homoeopathic" designs

Overreaction as a proof of effectiveness

In a dissertation at the University of Freiburg i.B. G.Schwab tried to demonstrate the action of high potencies of Sulfur in dermatological patients. The most reliable distinction between the effect of Sulfur and the placebo was a sudden alteration of symptoms in the form of an overreaction to the homoeopathic drug, which was then in most cases followed by an amelioration of symptoms. This overreaction is a well-known event in patients with very reactive diseases, above all with patients with atopic dermatitis and could well be used as a basis for further studies.

Comparison between a homoeopathically potentised
drug with a dilution of the same remedy.

In a controlled randomized double-blind trial with 164 patients Wiesenauer and Gaus (1985) investigated the effectiveness of homoeopathically prepared Galphimia D6, a conventional dilution 10^{-6} and a placebo for the therapy of pollinosis. The average duration of treatment was about 5 weeks. Although no statistical significance was achieved, it is remarkable that there was a clear trend of the superiority of Galphimia D6 while the Galphimia dilution 10^{-6} was about equally effective compared with placebo.

Similar comparisons between high potencies and conventional dilutions of the same strength should prove to be interesting to show the importance of potentisation (that means dilution and succussion) on the effect of a homoeopathic preparation. It might well be that a mere dilution of a remedy will turn out to be the ideal placebo in clinical studies in the field of homoeopathy.

Comparison of the effect of different potencies of the same remedy

In another study with Galphimia, Wiesenauer compared the effect of different potencies of the remedy. This design is valuable to improve the knowledge of homoeopathic practitioners, but is of little value in the discussion with clinicians.

DISCUSSION

On the quality of clinical studies in homoeopathy

In the Berlin Journal on Research in Homoeopathy 1991 J.Hornung deals with the basic quality of a study meant to prove the effectiveness of a new form of treatment.

Under the title "Significant - and nothing else" - an article well-worth reading by everyone interested in the field of clinical trials - he states four major constituents of the value of a clinical study.

I) Formal correctness and methodological quality. Studies dealing with unproved medical aspects have to provide a perfect methodological quality as far as the design of the study, execution and statistical analysis are concerned. Therefore the participation of a well-versed biostatistician in the early phase of the planning of the study is essential.

II) Validity. Validity means that the chosen formulation of the question, the chosen way of treatment and the verified result of a study have to be meaningful and rational, relating to medical reality. We often tend to measure something only because we expect good results - but the well-known criterion to "measure something" should not be the basis of a good study.

III) Quality of publication. Quality of publication means that the patient's medication and the outcome of a study have to be presented in such a way that anyone is able to check the whole process of medication and outcome and to recalculate the statistical results in order to evaluate the correctness or incorrectness of the results presented and of the conclusions drawn by the author. The reproducibility of a study is actually the main problem in a scientifically not approved method like homoeopathy. A single study is of no value when not reproduced by other independent authorities.

Reproducibility is a particular problem in homoeopathy, because we do not operate with relatively high pharmacological doses, but with a minimal information which induces the self-regulation in an individual patient. This means that it is important to describe each patient properly, because it may well be that the effect only refers to a certain group of patients [this, obviously, seems to be different in studies with lower organisms, where individuals of one and the same species or at least biotope community can be compared] to a certain symptom within a given diagnosis. Similarly, the process of the production and administration of the remedy has to be exactly defined.

IV) Publication bias. Each of us, no matter whether searching in the clinical or homoeopathic field, tends to publish only those studies that have led to a positive result, i.e. a result corresponding with the ideas preconceived from the very beginning of the study. Many studies that have not led to the expected results therefore remain unpublished.

Especially in a new and controversial field it is essential to commit at the very beginning of a study that the result has to be published whether it leads to the expected result or not. This can be best obtained if the study is planned by both the supporters and sceptics of a method.

Conclusion

Numerous controlled studies have proved the effectiveness of homoeopathic remedies, but they are not sufficient to prove the effectiveness of homoeopathy because most of them could not meet sufficient methodological quality.

But especially homoeopathic studies have to show a maximum scientific standard because they have to withstand much more criticism than pharmacological studies relating to the common paradigmas of clinical medicine.

Homoeopathic studies should be planned in co-operation between homoeopathic practitioners and university clinics, because this allows for a maximum methodological standard and for the possibility of publication in distinguished journals with a high impact factor which usually are not available for homoeopathic practitioners.

Controlled clinical studies in the field of homoeopathy are only useful if they take into account the characteristics of homoeopathic prescription, i. e. the individual choice of the remedy.

And a last comment: We should always question if a double-blind study is really the only possible way to prove the effectiveness of any therapy. I am not sure if double-blinding is always ethically justified, even if the study has passed the ethic committee. To question this is not the sole responsibility of homoeopaths, but should be a concern for every conscientious physician.

ANNOTATION by the editor

The annotation in [] was made by the editor. In this context, also the human physiological studies of H. Walach (this volume), R. van Wijk et al. (this volume) and of G. Hildebrandt (Hildebrandt 1993) should be mentioned.

REFERENCES

Brigo B. Homoeopathic treatment of migraine. Proc Congress Liga Med Hom Int, Arlington 1987.
Fisher P. An experimental double-blind clinical Trial in Homoeopathy. Br Hom J 1992; 75: 142-147.
Gebhardt K-H. Beweisbare Homöopathie. 2nd ed. Heidelberg: Haug 1986.
Gibson R, Gibson Sheila LM et al. Salycilats and homoeopathy in rheumathoid arthritis; preliminary observations. Br J Clinical Pharmacol 1978; 6: 391-395.
Gibson R, Gibson Sheila LM et al. Homoeopathic therapy and rheumatoid arthritis: evaluation by double-blind clinical therapeutic trial. Br J Clinical Pharmacol 1980; 9: 453-459.
Gore SM, Jones IG et al. Misuse of Statistical Methods. Brit Med J 1977; 1: 85-87.
Hildebrandt G. Pharmakologie adaptiver Prozesse. In: Institut für strukturelle medizinische Forschung e.V. and Physiolog. Inst. Uni. Graz (eds.). Wasser und Information. Heidelberg: Haug 1993.
Haidvogl M. Naturwissenschaftliche Grundlagen der Homöopathie. Der Kinderarzt 1990; 21: 75-84.
Haidvogl M. Forschung in der Homöopathie. In: Stacher A ed., Ganzheitsmedizin, 2 Wiener Dialog. Wien: Facultas Verlag 1991, pp 349-354.
Haidvogl M. Klinische Studien zum Wirksamkeitsnachweis der Homöopathie. Der Kinderarzt 1992; 23: 1477-1484.
Hornung J. Significant-and nothing else? On formal correctness, validity, quality of publication, and meta-aspects of controlled clinical trials. Berlin J Res Hom 1991; 1: 206-214.
Kleijnen J, Knipschild P et al. Clinical trials of homoeopathy. Br Med J 1991; 302: 316-323.
Kleijnen J, Knipschild P et al. Clinical trials of homoeopathy. Berlin Jj215 Res Hom 1991; 1:175-194.

Pocock SJ, Hughes MD et al. Statistical problems in the reporting of clinical trials. New Engl J Med 1987; 317: 426-432.

Reilly D, Taylor M. Is homoeopathy a placebo response? Lancet 1986; II: 881.

Righetti M. Forschung in der Homöopathie; Grundlagen, Problematik und Ergebnisse. Burgdorf: Göttingen 1988.

Righetti M. Characteristis and selected results of research in homoeopathy. Berlin J Res Hom 1991; 1: 195-203.

Schwab G. Läßt sich eine Wirkung homöopathischer Hochpotenzen nachweisen? Thesis, Freiburg 1990. Nachdruck Deutsche Homöopathie, Union Karlsruhe 1990.

Shipley M, Berry H et al. Controlled trial of homoeopathic treatment of osteoarthritis. Lancet 1983; I: 97.

Wiesenauer M, Gaus W. Double-blind Trial Comparing the Effectiveness of Galphimia Potentisation D6 (Homoeopathic Preparation), Galphimia Dilution 10^{-6} and Placebo on Pollinosis. Arzneimitt Forsch /Drug Res 1985; 35: 1745 -1747.

Wiesenauer M, Gaus W. Wirksamkeitsvergleich verschiedener Potenzierungen des homöopathischen Arzneimittels Galphimia glauca beim Heuschnupfen-Syndrom. Dt Apotheker Zeitg 1986; 126: 2179-2185.

Puzzling to find a solid basis of water research!

PROSPECTS

"Anyone familiar with the epistemological foundations of natural science and its practical application is aware of the problems which occur when observations are introduced into the discussion which appear to have a new character and which cannot be satisfactorily explained in theoretical terms. In an advanced field of science a model usually precedes a goal-orientated experiment."

R. van Wijk, Pressure Induced [...], 1989

PRELIMINARY ELEMENTS OF A THEORY ON ULTRA HIGH DILUTIONS

J. Schulte, P.C. Endler

INTRODUCTION

The major question regarding homoeopathy is how the health bringing information is transferred to the carrier substance water, or how information can be stored in water permanently and how this information is transferred to the patient to be cured (plant, animal, human). These questions have not been answered within this book, but pieces of that puzzle have been turned around and elucidated. Furthermore, our matter of concern is to encourage others to further experiments in order to solve that puzzle.

Although the experimental basis can still be questioned in most cases, researchers can start from remarkable fundamental observations in order to explain the effects of UHDs. For some topics in question, it seems to be evident that the information to be transferred is stored in the specific properties of the diluted substance by means of long-range electromagnetic fields. Also, the interaction of the UHD and the organism seems to be based on the interaction of long-range electromagnetic interactions.

MAIN PART

Some unsolved questions

The different mechanisms **A.** of the interaction between the molecular mother substance and the solvent water and **B.** of the storage of molecule-specific information in the solvent, as well as **C.** the physiological basis of the sensitivity of the living organism towards the UHD and **D.** the mechanism of the interaction of the test dilution with the organism are all widely unknown. Several ideas, however, have been postulated.

Particle - field - vector potential

It is well known that biomolecules "broadcast" electromagnetic signals (Fröhlich, see the contribution by Citro et al.); Such information is even detected by radioastronomers.

Respective experiments were performed based on the fact that molecular thyroxine has a steering role in the development of amphibian larves (Endler et al., this volume, pp 39ff). A molecular suspension of thyroxine was sealed in quartzglass ampoules and hung into the water basin containing the animals. In contrast to the "direct" control treatment with molecular thyroxine, the speed of metamorphosis was not altered by this "indirect" treatment, but, surprisingly, in these preliminary experiments standard deviation was clearly higher in the group of animals treated with the biomolecule suspension shielded by quartzglass than in the control group, where pure water was sealed in the quartzglass ampoules.

It is a triviality to every physicist that atoms and molecules consist of quants, that cannot only be described as particles, but as well as waves, as energetic fields. *Dependent on the mathematical theoretical point of view, quantes are described either as corpuscular particles, electromagnetic fields, or vector fields.* However, this way of looking at things is not always taken into consideration when biological and medical concepts of life and health are concerned.

The charge of an electron is, course, electrostatic, but its behaviour is that of an electromagnetic wave. Thus, the information of atoms and molecules consists of the (smeared out) electromagnetic field of the electrons as well as of the nucleus. Due to the finite

245

P.C. Endler and J. Schulte (eds.), Ultra High Dilution, 245–251.
© 1994 *Kluwer Academic Publishers. Printed in the Netherlands.*

temperature of the molecular system, the molecules can be found in specific rotational and vibrational states (modes) that couple to other molecules, and external fields, in very complex ways. (These modes are subject of the NMR- and Raman-spectrocopy, see the introductory contribution by Schulte et al. in Part 2 of this volume.) The rotational states of the molecules are located in the IR-range, i.e. above the range of microwaves, and are followed by the thermal vibration of molecules and clusters at comparable or at much lower frequencies. (A remark on terminology: photones are the quants of the electromagnetic field - in all frequency ranges -; phonones are the equivalent for the vibrational oscillation.)

A. Transfer of information from the molecule to the UHD

Homoeopathic remedies are traditionally diluted in bipolar liquids such as water and alcohol or are diluted in lactose. An important role of hydrogen bonds (O-H) is what these substances have in common. Especially with regard to water, a theoretically well-based approach is formulated by Del Giudice (see Del Giudice's contribution): quantum electrodynamics allows an ensemble of molecules - beyond a density threshold - to move coherently and be kept in phase by an electromagnetic mode with 'coherence domains' the size of which is the wavelength of the mode (superradiance). Thus, with regard to the process of information storage in the solvent, the most common idea is that there is a coherent interaction between the electromagnetic or magnetic vector potential fields of molecules of the diluted mother substance and the dipoles of the solvent water, including the permanent polarisation of the water, which thus becomes coherent. The analogy to a laser is used, but in water the coherence is in the ground state. The speculation was raised that in the process of succussion, the water in the immediate environment of the biological molecule shows a certain capacity to act as an agent of transmission (see Benveniste's contribution). The radiation fields of the charged molecule might generate a permanent polarisation of thousands of water molecules in its environment (Fröhlich 1986, see the contribution by Citro et al.).

A number of experiments to transfer molecular information via (polarized coherent water dipoles in) UHDs are described in the respective chapters of the volume presented here. In parallel, experimental approaches were done towards the idea of perimolecular charges and the possibility of their transmission via a metal wire system (Endler, Ludwig et al., 1991, material at the Boltzmann Institute, Graz), via an electronic device including an amplifier equipped with a single input- and a single output-wire (Citro 1991, see the contribution by Citro et al.) and via an electronic circuitry including an amplifier (Benveniste 1993, see the contribution by Citro et al.). In our respective experiment on the amphibia bio-assay, an electronic device as used by Citro 1991 was also used to transmit information from thyroxine molecules to pure water, which was then tested versus water submitted to the same electronical process. A significant effect of information from thyroxine, transmitted via the electronic amplifier, was observed. The fact that an electronic device can transfer biological activity supports the electromagnetic nature of the molecular signal. These findings may radically change our concepts of molecular signalling.

In short, to our present knowledge derived from experiments, information from biomolecules obviously can be transferred by different methods as shown in Tab. 1.

1 - By means of "direct" contact of the partners of a chemical reaction, i.e. of contact of the respective fluid (water) layers surrounding them,

2 - By means of information transduction via the preparation process of UHDs and

3 - By means of information transduction via conduction in metals.

Tab. 1. For explication, see text.

Work conducted at the laboratory of C.W. Smith shows that the nature of the bio-information transmitted via electronic devices is a "propagating coherence" and not a "circulating current". Such signals, surprisingly, are also amplified by ordinary transistors, the n-p-n bipolar or n-channel FET being effective; the p-n-p and p-channel do not propagate the signal. From this it is concluded that the signal is one of electron coherence since holes do not propagate it. The nature of the amplifier output viewed with an oscilloscope seems to be noise. The use of a signal analyzer or narrow-band filter can retrieve a coherent signal, but problems in repeating remain. As far as is known, the frequencies of biomolecule - solutions have a harmonic series which is best described by a product equation.

The question of the profound understanding of biomolecular interaction was e.g. formulated by Popp (1979, 1989, see his contribution) and by Druker, Mamon and Roberts. These latter authors point out the importance of understanding "the means by which proteins communicate - that is, how signals are transferred from one protein to the next in the signal-transduction cascade." (New Engl J Med 1989; 321: 1383.)

In his poster presented at the meeting of the AAAS, Boston 1993, J. Benveniste sums up the following hypothesis:

"The electromagnetic field of a molecule suspended in water generates a permanent polarization of water dipoles (Del Giudice, Preparata, Vitiello, Phys Rev Lett 1988; 61: 1085), enabling water to transmit (amplify ?) radiated fields.

When the field matches the kinetic characteristics of a reaction, the latter becomes funtional at the optimum field strength (electro-conformational coupling - Tsong, Tr Biochem Sci 1989, 14: 89 -) as for a radio receiver (Frey, FASEB J 1993; 7: 272). This concept may provide the intimate mechanism of molecule recognition, shape change and signal transduction."

In the same meeting, confirmatory data were presented by the working group of the biological editor.

Along with the idea that energetic signals play a decisive role in the information transfer from biomolecules to the organism, the possible role of biomolecules as passive resonator systems has to be discussed. This way of looking at the phenomenon includes the recently well-established knowledge of coherent energetic excitations from living organisms, organs and cells (see below - C -) which then actively scan the biomolecules and cause resonances (- D -). The information from the biomolecules (or other particles) can then be transferred to different levels, reaching from the direct effect of the molecule and the electric charges which it carries to an indirect effect that specifically depends on the actual state of the organism (see below - D -).

During the succussion process, the perimolecular water then is separated from the molecule, but, continues to carry its bio-information, probably by means of specifically organized water dipoles. It is commonly known that water dipoles oscillate in a frequency range above the visible light; furthermore, oscillations in the range of THz (O-atom), 50 Hz (O-H-bindings) and MHz (H-atoms) have been described. The specific field effects of the soluted molecules influence /determine the dynamics of the water molecules (dipoles) surrounding the soluted molecules.

According to Del Giudice, the water molecules are polarized and oscillate in phase in a coherent manner. Little is known about the processes during agitation (see Auerbach's contribution), where exogen energy is brought into the dilution.

Experiments may be performed on a possible role of the magnetic field of the earth in the process of information of the carrier solution.

B. Storage of molecule-specific information in the solvent

Typical water structures such as clathrates, helical structures, different other types of clusters or typical water phases and network systems over the whole fluid are hypothetised and described, respectively (see Del Giudice's, Anagnostatos's, Schulte's, Smith's and Berezin's contributions; see Preface: Resch /Gutmann, Trincher). The editors have previously been engaged in a project giving a survey over respective current ideas (see Preface). In general and simple terms, one must keep in mind that the information of a carrier substance that is stable enough to provide long-term effects of UHDs must, to our present knowledge, be based on quantum physical processes that might support coherent structures.

The picture of information storage and transport that has been given in Schulte's comprehensive contribution may be a general picture for a variety of promising theories which have recently been developed. E.g. Berezin (see his contribution) has discussed the isotopic diversity of chemical elements as a physical foundation of homoeopathy. Homoeopathic remedies based on water or water-alcohol mixtures basically consist of molecules built of hydrogen (H), carbon (C) and oxygen (O) atoms. From nuclear physics it is known that, for almost all elements (atoms) there are atoms of the same type, but slightly different mass (caused by a different number of neutrons in the atomic nucleus). Some of those atoms with different mass (isotopes) are stable, most of them are unstable and suffer radioactive decay.

Thus, besides the different chemical and physical properties of atoms in a molecule, we get two more characteristics: the diversity in mass, and the diversity in abundance. With a different mass, the vibrational interaction among the molecules changes. From the natural abundance of the isotopes an average distance may be estimated, and vibrational modes and mode differences determined. Taking Berezin's theory on isotopic diversity into consideration together with a picture of cybernetic switches, we may imagine the isotopes as characteristic nodes in a pattern of information, and the diversity of vibrational modes as the dynamical carrier (amplifier, multiplier) of information. Thus, a drug molecule may change the pattern of nodes, which causes a different coupling of vibrational modes. The pattern of isotopic nodes may be stabilized (amplified, multiplied) by the coupling of vibrational modes, even when the drug molecule has been extracted from the system. This theory is asking for a new type of experiments in homoeopathy, the high resolution LASER spectroscopy.

The imprint theory developed by Del Guidice (see his contribution) is based on a coherence assumption similar to that of Berezin. Instead of isotopes keeping coherence long-time stable, Del Guidice concentrates on coherent states induced by electromagnetic polarization fields. It has been proven that polarization fields are extremely non-local, and may stabilize certain structural states. Although Guidice's model seemed, to a large extent , different from Berezin's isotope model, in a dynamical (thermal) theory both models work hand in hand through the electron-phonon coupling, i.e., the coupling of electron interaction and vibrational motion of the atom core. Thermal electrodynamics couples the amplitude of vibrational modes (phonons) and the polarizability of the atomic shell through a non-local integral over vibrational modes and frequency dependent polarizabilities (see Schulte's contribution).

Another promising theory on information storage and transfer in high-diluted solutions has been developed by Popp (see his contribution). In Popp's model the homoeopathic drug induces a resonance transfer of disregulatory energy from the patient's body (biological cell) to the absorbing homoeopathic dilution which then is excreted. Emission and absorption are described by means of coherent states of the "drug" field and the "disregulatory" field in the patient's body, respectively. The natural weak coupling of both fields keeps the states coherent. Popp also proposes that the long-time stability of homoeopathic remedies may be due to coherent phonon states induced by the succussion (agitation) during their preparation process. (For a discussion of the fluid dynamical part of the succussion process, see Auerbach's contribution.) With regard to Popp's model, it seems necessary to give specific model fields and Hamiltonian in order to develop it further.

In any case, with regard to models of energetic coupling between drug molecules and solvents, one should not be trapped by models that exclude further and more subtle interaction mechanisms (see Schulte's contribution).

An experimental evidence for the idea of a resonance capacity of UHDs was provided in studies on the exposure of a living system to the interaction phenomena between a coil fed by a laboratory oscillator and UHDs, e.g. serially diluted agitated thyroxine (see the contribution of Smith and Endler). Here, a distinct pattern of frequencies that obviously caused resonances in the thyroxine dilutions, which again led to biological reactions, has been found. Each succeeding step of dilution and agitation added two further and higher frequencies of resonance. In the range of the frequencies investigated (0.01 Hz to 10 MHz), the thyroxine dilutions gave regularly arranged resonances throughout this whole range. When, in return, in order to test the biological effect of such typical frequencies, with the help of a coil fed by a laboratory oscillator, their influence on highland Rana temporaria was compared to the influence of control frequencies, significant reactions of the animals were found (see the contribution of Spoerk et al.). It was further shown that the exposure of UHDs to external electromagnetic fields is able to inhibit their further biological effect (see Benveniste's contribution).

In studies on different living systems (see the contributions of van Wijk et al.; Endler et al.; Pongratz et al.) it was shown that UHDs can also exert their effects on living systems, when they are separated by a hardglass wall. In our respective study on amphibia, the UHDs of thyroxine and of water for control, respectively, were sealed in glass vials (ampoules) in order to avoid mixing with the water that contained the organisms.

It was shown that plexiglass can be used to block the informational interaction between the UHD and the organism. This may be the fact, because plexiglass has different electromagnetic properties. To verify this hypothesis, the experiment was carried out with different kinds of hardglass and with quartzglass, respectively. The biological effects were much more marked when the UHDs of thyroxine and of water were sealed in quartzglass vials rather than hardglass vials. The electromagntic (VIS) window that marked the difference in the transmission spectra between quartzglass and hardglass was situated between 210 and 310 nm (1.43 x 10e15 and 0.96 x 10e15 Hz). Further details are in preparation for print.

To conclude, the pathways of information transfer from biomolecules shown in Tab. 2 should be further discussed and experimentally investigated.

1 - a - Specific molecular modes causing specific electromagnetic radiation.

- b (?) - Polarized coherent water dipoles, electrodynamically coupled with isotopic diversity phenomena (specific vibrational modes of phonons).

2 - Polarized coherent water dipoles, electrodynamically coupled with isotopic diversity phenomena (specific vibrational modes).

3 - Propagating electron coherence via a metal wire (linked to an electronic amplifier).

Tab. 2. For explanation, see Tab. 1 and text.

Possible medical implications of this research work are further knowledge of the basis of homoeopathy (as it has therapeutically been used for more than 200 years) and of pharmacological frequency transfer (as it was developed by M. Citro in the eighties). A further example for a possible clinical implication of UHDs (or pharmacological frequency transfer-dilutions) of biomolecules is our research on the preparation of drugs from biomolecules. In human substitution therapy, the molecular application sometimes is difficult due to high costs or due to the danger of viral contamination.

C. The physiological basis of sensitivity towards UHDs

It obviously has to be the extremely high sensitivity of biological systems that provides the key to the understanding of the effects of succussed high dilutions. Living systems can be influenced by external coherent stimuli only to a small amount when the system is under good homoeostatic control, but to appreciable amounts when the system is under biological stress (see Popp's and Smith's contributions). This stress can be present under pathological conditions (in the case of disease) as well as under exceptional developmental conditions. It has been experimentally proven that life phenomena are linked to the emission of coherent electromagnetic waves - e.g. from the DNA lattice system - throughout the whole spectral range (see Popp's contribution). This emission sensitively depends on all biological processes, as cell cycle phase, growth or, obviously, metamorphosis. Living systems also have highly developed electromagnetic bio-communication systems involving transmitter oscillators, sensors and negative feedback control systems. The coherence of the signals provides a variety of extraordinary properties in living systems, e.g. the highest possible transparency for weak-intensity information transfer at the highest possible signal /noise-ratio. It seems that, in nature, such waves serve regulatory processes, or more generally, communication within living systems. Specific information can be transferred by electromagnetic frequencies being in phase, which makes them different from the variety of any other incoherent influences, and similar to a technical laser (see Del Giudice's, Popp's,Smith's and Schulte's contributions) . Very weak external influences are sufficient to change the intensity or the Fourier-pattern of the harmonics of electromagnetic emission significantly (Smith).

D. Interaction of the organism and the UHD

A concept has pointed out that the basic effect of succussed highly diluted substances is always a delocalisation of the energy in a resonance-like interaction between emitter (organism) and absorber (dilution) (see Popp's contribution). In the case of stress, the organism would work as a boson. Its typical - stress-linked - oscillations give rise to perturbations of homoeostasis. The agitated high dilution acts as a resonance absorber of stress-linked oscillations as soon as the adequate substance in the appropriate dilution is used. One may suppose that any drug behaves as a passive resonator causing the very gained amplifier of a bio-feed-back control system to oscillate. Both negative as well as positive resonance phenomena are to be expected. An analogy is the effect of a public address amplifier in an undamped empty auditorium. In our amphibia model (see the contribution of Endler et al.), the organism has completely ceased to oscillate with the frequency (frequencies) linked to the developmental stress after an initial superposition of frequencies out of phase, due to the constant presence of the passive resonator. In a further stage of phase shift, even a positive resonance occurred, which led to an enhancement of the reaction (development) that was initially slowed down. For a respective example in human studies on UHDs see Walach's contribution.

This is consistent with our preliminary experiments on electronically transferred information from molecules (see the contribution of Citro et al.). Here, we have observed both effects in the sense of those of the original molecular substance as well as effects opposite to those of the original molecular substance. This fact may be a hint concerning the importance of the actual state of the living system. In our respective amphibia experiment, an initial acceleration of development, expressed in terms   both of four-leggedness as well as of tail reduction was found. When the treatment with the electronically transferred information from the molecular thyroxine dilution was continued, a different situation was observed: the information from the test liquid caused an inhibition of reaching the observed parameters. Taking this into consideration, one may conclude that the original molecule in some way is the source of information that is able to interfere with the organism in various ways.

E. Homoeopathy between physics and psychology

As described above, speculations on the interaction between the electromagnetic /magnetic vector potential field of an organism and a homoeopathic remedy that works as a passive resonator are possible. Furthermore, it is evident that the interaction of organisms per se - as e.g. in the case of the relationship between a therapeut and his living test object - should be investigated with regard to the therapeutical effects of homoeopathic remedies. The idea may be productive for further research on therapeutic interaction where the high dilution can serve as a kind of specific amplifier in the conscious /inconscious interaction of a therapeut and his living test object. This point of view localizes homoeopathy in the field between physics, pharmacology, physiology and psychology, with possible unexpected implications in the relationship between an experimenter and a living test object in experiments on ultra high dilutions (see also the contribution by van Wijk et al.). A link between a quantum explication of information in an UHD and a quantum hypothesis of consciousness is subject to a speculation given in Berezin's contribution.

ACKNOWLEDGEMENT

The authors thank T.J. Milavec, W. Hohenau, R. van Wijk, C.W. Smith and F.A. Popp for reading the manuscript.

EPILOGUE

"In science, there is no place for dogmas. The scientific researcher is free to ask any question, question any proposition, search of any proof and to correct any fallacy. Whenever science was used in the past to develop a new dogmatism, such dogmatism in the end proved irreconcilable with the advancement of science; and ultimately the dogma was broken down, or else science and freedom passed away together."

J.R. Oppenheimer, The Open Mind, 1955

ALTERNATIVE RESEARCH OR
RESEARCH ON ALTERNATIVES?

F. Moser, M. Naradoslawsky, J. Schulte

INTRODUCTION

Recently problems of complementary or alternative research have gained the attention of "normal" sciences. For example: whereas until recently it has been mainly up to medicine and pharmacology to have contact with homoeopathy (although in a rather competitive atmosphere), today other sciences are also called upon to gain status for the results of the research on homoeopathy. This is especially true for chemistry and physics.

However, those who have followed the discussion about the research work presented by Davenas et al. [1] have probably become pensive about the heated discussion [2] as well as its subject. May the concrete study in question be reproducible or not, the question of openness and fairness in science remains. Throughout this controversial discussion it has become very clear how incompatible today's sciences have become. One origin for the controversy may be found in the historical development and methodology of the normal and the alternative research, respectively.

It was Kuhn [3] who has indicated that every science must have a paradigm in order to survive successfully. This paradigm merges into all dogmas that constitute the very frame of the corresponding science. Those dogmas are not questioned in the normal activities of science. To avoid depriving itself of a foundation upon which its future development and advancement lie, it is necessary for science to defend its paradigm.

The possibility of different results is inevitable if various sciences share research subjects based on different paradigms. In the event of the controversy between research in one science and the results of research in a different science, one can approximately expect the following reactions:

- The problem which leads to the controversial results becomes repressed.

- The results of the other research areas become rejected. Thus, not only will this individual result be drawn into effect, but the general consensus will also be to write off the entire research area as "non-scientific".

- The problem will undergo rigorous examination and will risk putting its own paradigm up for rejection.

Naturally the reaction stated above last will represent the obvious exception. How does this situation influence the relationship between homoeopathy and medicine? In this context it is important to determine that:

- Medicine and homoeopathy have the same research topic, namely human health,

- and that medicine and homoeopathy are based on completely different presuppositions (paradigms).

According to the opinion stated above, it seems unavoidable to conclude that medicine and homoeopathy are incommensurable, based on normal research. Assuming, however, that human health is a high value, and that it is not possible to refuse either science for the benefit of health, then this situation cannot be held for long. In such a case, one possible solution leading out of the dilemma is the non-ordinary alternative research. However, this alternative science of health should never be considered as merely a sum of medicine and

P.C. Endler and J. Schulte (eds.), Ultra High Dilution, 255–261.
© 1994 *Kluwer Academic Publishers. Printed in the Netherlands.*

homoeopathy, or even be considered coexistent or on the same level as medicine and homoeopathy. Above all it has to be a true alternative research. What is meant by the term "alternative research" shall be outlined in the following chapters.

THE RESEARCH PROCESS

In order to bring this topic closer, it is advantageous to primarily turn one's attention to the methodology of the procedure that we call "research". It is particularly interesting to understand which role, in this context, the paradigm plays in the methodology of research work.

The first influence that the paradigm has on the scientific development process is exerted in the choice of which problem to deal with. The fundamental motivation behind research activity is supplied by two sources:

- The researcher's environment serves the problems to deal with. This occurs through phenomena which are waiting for an explanation, as well as through problems which the researcher is called upon to solve.

- Based on prevailing paradigms, the Scientific Community decides whether a problem within the framework of various research directions is worth being scientifically investigated. This particular decision is especially important regarding the support of a paradigm.

If the problem or the phenomenon is interesting enough to the researcher as well as to the scientific community to be worth researching (i.e. concurring with the paradigm), then the development of the theory begins. According to K. Popper [4] there are three main sections in the development of a scientific theory:

1) The *Hypothesis Forming Phase*

2) The *Theory Building Phase*

3) The *Theory Testing Phase*

ad 1) The *Hypothesis Forming Phase* represents the creative portion of the research activity. A new hypothesis cannot logically be deduced out of the law or phenomena in question. It is up to the researcher's creativity to formulate it. This step is inherently il-logical. In order to arrive at a new hypothesis, the methodology is anarchistic, as P. Feyerabend [5] formulated: "Even a law and order science will only succeed where occasional anarchistic steps are permitted."

One could, like Feyerabend, formulate that this is the "anything goes" phase, meaning that, at this step of the research activity, any thinkable hypothesis is possible. It must however be mentioned that it is the researcher himself who is impressed by the paradigm of his science. As a result, principally his creativity is bound. Even in this phase he will favour some hypothesis which is in harmony with the paradigm.

ad 2) In order to build a theory out of a hypothesis, a systematic work process is necessary. A hypothesis formed out of creativity, which of course does not represent much more than a personal knowledge of the various contexts, must be transformed into an inter-subjective theory which can be proved and verified (e.g. through an experiment).

Unlike the first creative and anarchistic phase of scientific activity, the *Theory Building Phase* has to be systematic and logically strict. Through logical deduction, through analogies with existing theories, and through comparisons to other fundamental laws within the currently excepted paradigms, out of yet not clearly formulated hypothesis, a scientific theory will be formed. Simultaneously it must also be determined which methods and experiments are suitable to test and verify the new theory.

It is obvious that, in this phase, the work of the researcher is completely determined by the paradigm fundamental to his science. The endeavour of the scientist in this phase is of course to find a theory consistent with this paradigm.

ad 3) The last phase of the development of a theory begins after the theory is put into a scientifically adequate form and is not finished before the theory has been falsified. This is the *Theory Testing Phase*. The examination follows in succession with respect to the problem solving capacity of the new theory as well as the acceptability with respect to its paradigm. If an improved solution to the problem can be reached with this theory, compared to its predecessors, and if it is also consistent with the paradigm, then the new theory itself will become part of the frame of the particular branch of science. The accepted theory remains in this frame until it is replaced by a better theory.

Within the last phase, the theory will no longer be influenced by the paradigm if it is successful. It will become much more a self-integrated component of the paradigm's framework. As such it shall support the paradigm, and simultaneously, by the scientific community, it receives the same support as the paradigm itself.

The most prevalent meaning of the paradigm for the work of every single researcher very impressively follows from the above stated. It is in no way a pity-worthy progress-animosity when the scientific community seeks to defend the paradigm often under great sacrifices. More importantly, an undisputed paradigm is foremost the foundation upon which fruitful research activity is made possible.

RESEARCH ON ALTERNATIVES

Looking critically at the world's nature is a remarkable characteristic of the researching mind. This property naturally does not come to a halt even in its own science. The question about the research goals of their own science is often the result of researchers coming to terms with critical research.

In such a situation, research using alternatives begins. The worn-down paths of the research direction will become abandoned, and new goals for the researcher's own scientific activity are sought - goals that the majority of the colleagues do not consider important to the progress of science.

In this case, even with the goal conflicts, the science's own paradigm does not come into question. On the contrary, one could even state that alternative research promotes expansion of the paradigm's scope. After all, it is being attempted to reach new alternative goals with this paradigm in addition to currently acceptable conservative goals. Alternative research in no way departs from the paradigm; it expands its service.

It can be concluded that the scientific process presented in the last paragraphs is also applicable to research on alternatives. Probably the most important limitation is that the obstacle of research is quite often borrowed from other research directions. Intuition is also heavily required when dealing with topics foreign to the studied field.

Research on alternatives always supplies only local and confined knowledge. It is this type of research which Feyerabend sets in opposition to the worn-out paths of conservative research. Not only may the step of intuition be chaotic, but the choice of the research subject also may be considered anarchistic.

In spite of this anarchistic goal-finding, research-using alternatives must be totally conform with excepted scientific criteria. In addition it has to fulfill much more criteria, if it really wants to serve solving scientific problems. These additional criteria can best be expressed through the keywords:

- *meticulous experimental work,*

- *intellectual openness,*

- *human humbleness.*

What, after all, is meant by these guide terms? Let us discuss *meticulous experimental work* first. In most cases the dealing with phenomena outside the researcher's own field of science runs into experimental difficulties. The experimental framework of his /her own science is of course orientated to the phenomena of that science, and only needs to be successful there. As a result, phenomena that lie outside the defined scientific sphere, explained by the presently acceptable science paradigm, present a problem. The experimental single results are either insignificant, or they are very difficult to interpret, or both. However, this cannot be expected to differ when dealing with research on alternatives whose relevance is objected and often heavily opposed by the former of the presently acceptable paradigms. It would be a downright threat to paradigm if these phenomena led to unmistak-able experimental findings while examined with experimental methods of their field of science.

In this case, the experimental meticulousness becomes all the more important. The exertion in research using alternatives is clearly higher than in conventional research. Only truly first class experimenters will fulfill the demands required by this research. Only truly first class experimental data with respect to their quality as well as their quantity are able to survive the demands. Only open and honest presentations of these data can serve the basis of research on alternatives.

The second important point that concerns research using alternatives, is *intellectual openness.* This openness must be effective in two directions. First of all, because of the assumptions of various kinds of phenomena that are not solely explained by its own paradigm, it has to be a reputable foundation of the science. Second, and maybe much more important, it has to be open for criticism on the scientific work itself. Thus, it has to be worth this criticism.

Research on alternatives is clearly at a disadvantage when compared to normal research. Even though the research activity's foundation is the paradigm, the activity lies outside the paradigm's intellectual protection. Not only can criticism come from colleagues in their field, but also from anyone interested, whose paradigm may come from a different science. When dealing with research on alternatives, it must always be clear that one's own research represents only one possible alternative to the phenomenon's explanation.

This situation may be new to the researcher and therefore be a burden. He is ready to master this situation only if he has achieved the necessary *human humbleness. Human humbleness* is a necessity when it comes to discussing various scientific interpretations. Also, one has to be aware of the fact that no paradigm possesses sole validity in all decisive criteria. It is necessary to be respectful towards the work of others, and to be aware of the fact that one's own research basically illuminates only one aspect of the phenomena.

A whole series of dangers dwells in research on alternatives. If one violates the strict additional criteria described above, then unscientific results as well as the scientific community's rejection of the paradigm are caused. Consequently, the work accomplished will not lead to an expansion of the paradigm but rather to the damage of the field of knowledge. It will also lead to an interruption in the dialogues between the professionals of different fields.

If, on the other hand, the research on alternatives is taken seriously and is openly carried out, then it certainly will become part of a valuable scientific work. One example of such a scientific work worth mentioning here is that of R. Jahn [6], who investigated the effect of the human mind on material advancement. The scientist Jahn conducted a series of very interesting and scientifically indisputable experiments which tried to show the interaction between mind and matter. Although no final proof can be presented by these experiments, the results are so astonishing and interesting that these experiments deserve high ranking in future paranormal phenomena research.

ALTERNATIVE RESEARCH

Alternative to the research on alternatives, a completely different type of scientific activity is imaginable. We want to describe it as *alternative research*. Its main difference lies in its diversion from the mechanical view of the world as a foundation of various scientific paradigms. In alternative research, as opposed to normal research and research using alternatives, the fundamental paradigm of alternative research quite often comes into question. The basis of the research is no longer a mechanical interpretation of the world but rather a holistic view of it.

Totally different goals and criteria count for this type of scientific activity than for normal research. Whereas normal research (and research on alternatives) treats phenomena as detached from their embodiment into the wholeness of the phenomenon, alternative research focuses its activity onto the whole dynamic and complexity of the phenomenon. In the following paragraphs, the most important characteristics of alternative research will be presented.

The foundation of alternative research is the understanding that the world is an inseparable whole. An investigation of the whole network of connections, which portrays itself to us as reality, is impossible. It is also impossible to place oneself outside the field of research. Through his or her work itself the researcher (just as everyone else does) creates reality and is part of it (cf. H. Maturana [7]). Therefore the world is a whole, constantly in flow, whose character cannot be grasped completely by our senses. Nevertheless, man has to come to terms with it and gain understanding of it.

This view fundamentally changes the scientific process . It is no longer the goal of the research activity to explain the entire world (i.e. as a sum of its parts). It is far more important that the unit (the part) is contained in the system (the whole), and in no way that the system is part of the unit (compare K. Wilber [8]) The scientific process has become emancipated from an explanation of statistic states into a learning about dynamic processes. The rationalistic explanation of the phenomenon is no longer the center of the researcher's striving, but rather the understanding of the world through "the lingering on the phenomena."

The "lingering on the things" (also mentioned by Goethe), is after all a cognitive rather than an empirical process. This will radically change the work of the researchers. Before, the experiment was a necessary reflection of the static state to be described. For this reason, highest importance was assigned to its capability to be reproducible and to its intersubjectivity. In alternative research, however, the experiment is considered just a phenomenon in order to gain a broader perception. The experimenter's goal is no longer to screen all other influences which are not central to the subject of interest, but rather to compose these relations among them.

Hereby, the experiment also loses its characteristic of being the least valid scientific criteria. Also, the subject of investigation in alternative research itself causes to ask for experiments that no longer may be picked to define the quality of the research itself. Who could imagine an experiment in which the effect of the ozone layer's damage on the earth would be represented exactly? Who can propose an experiment in which the outcome of genetic interference in a person's hereditary storehouse would be made countable? These interesting systems are becoming more complex and the empires' ethical and scientific sacrifices more prohibitive.

Therefore, the researcher's main crutch in alternative research is the gedankenexperiment or the simulation. The importance of empirical experimenting loses its rank. It is no longer the foundation of the appraisal of scientific activity, but rather the basis of the researcher's intuition.

Along with the alternative criteria come new demands on the researcher's work. Let us begin with the demands on the experimental work. Even though it was stated that, in alternative research, the criteria of reproducibility is restricted, it does not mean that the experimenter may work less meticulously. It is quite the other way around. Alternative research especially, needs particularly exact experiments and empirical data. However, the relationship of the experimenter to the subject of the experiment has to be changed. It is no longer his creation or version of the world that is the subject of the experiment. It is the entire reality as co-created by the researcher which is subject of his observations. Therefore, the empirical researcher in alternative science must possess subtlety with re-spect to the nature and the dynamics of the studied processes. His work is no longer judged by how well he has changed nature to his wishes but rather by how well his observations have led to a deeper understanding of the world. The quantity of experiments is no longer the point of focus, but rather the quality of the understanding of nature that has led to the experiment is important.

A second problem is the supervision of scientific activity. An adequate standard must be found because experimenting is widely eliminated as a last desperation. This leads us to the problem of inter-subjectivity in the realization of a reality co-created by each person. In this situation inter-subjectivity can only be found by making the same impression on the subject. This requires a definition of the scientific process' participants. A short version of this definition might be "scientists are people who, on the basis of rational thinking and learning, try to attain an understanding of the reality they have experienced." Therefore, the work of each and every researcher must meet the criteria of rationality.

However, the criteria of the inter-subjectivity alone is in no way sufficient for an appraisal of scientific activity. It is a basic demand only. The appraisal of the scientific accomplishment has to be explained through the odds of the new goal. Alternative research is, as has been stated, a learning process. This learning process is not limited to the individual researcher, but is by its nature a collective process. This is also the basis of the appraisal of scientific work. The more the level of collective understanding of science is challenged through scientific accomplishments, the more valuable it becomes. The single dominant accomplishment which acts as an orientation for others is no longer the goal of the individual researcher. It is the service done to scientific understanding beyond the person himself which counts for much more.

A guideline to the question of if and how far the status of understanding of science has advanced can only be expected from the doctrines of wisdom. Although science, which by definition acts through rational means, would not come up with the same contents, one can infer from the comparison of the statements that the scientific thesis is valid.

COMPARISON BETWEEN RESEARCH ON ALTERNATIVES AND ALTERNATIVE RESEARCH

In the previous chapter we tried to show the difference between research on alternatives and alternative research as it has resulted from the above stated. How do these two categories of the scientific profession compare to each another?

It is abundantly clear that the mechanical view of the world is running into the limits of its possibilities. As a result, an expansion of the mechanical paradigm, which research on alternatives offers, appears to be irrelevant at first glance.

Nevertheless, research on alternatives fulfills an important function in developing a new encompassing paradigm. It sets, so to speak, an intellectual vanguard. The questions about the research goals of the old paradigm, and the work on paradigm-foreign problems, cause the new paradigm to be constantly evaluated. Meanwhile, new previously ignored phenomena may be pushed into the center of focus of scientific discussions.

Only through the open atmosphere of scientific discussion provided by research on alternatives, a new paradigm will be created and will succeed. Alternative research can only be possible (at least on a broad scale) following the existence of research on alternatives. Research on alternatives hereby fulfills the function of forming the new paradigm, and of making the alternative research possible, which is approximately comparable to John the Baptist's preparing humanity for Christian teaching.

Exactly from this view the importance of the quality of research on alternatives becomes clear. Great damage can be imposed on exactly the new paradigm's very important budding through unreliable and methodologically disputable work. Poor research on alternatives unavoidably leads to impenetrability of the frontier and to ignorance of the phenomenon outside the old paradigm, even though flexible settlements would be necessary. Besides pure temporary delay while forming acceptance of the new paradigm, unbearable dissipation of valuable scientific work also occurs. For this reason, all those researchers who work on the research on alternatives should be aware of their great responsibility to the future of science.

ANNOTATION (by the biological editor)

In addition to the qualities meticulous experimental work, intellectual openness and human humbleness mentioned by the authors, a more and more non-violent approach in experimental research on living systems might be necessary in order not only to describe, but also to explain the processes of life and health.

REFERENCES

1. Davenas E et al. Nature 1988; 333: 816.
2. Meddox J et al. Nature 1988; 334: 290
3. Kuhn TS. Die Struktur Wissenschaftlicher Revolution. Taschenbuch Wissenschaft. Frankfurt:: Suhrkamp Verlag 1976.
4. Popper K. Logik der Forschung. Tübingen: Mohr JCB 1976.
5. Feyerabend P. Wider den Methodenzwang, Skizzen einer anarchischen Erkenntnistheorie. Frankfurt: Suhrkamp Verlag 1977.
6. Jahn RG et al. Margins of Reality. The Role of Consciousness in the Physical World. New York: Harcourt Braces Jovanovic 1987.
7. Wilber K. Halbzeit der Evolution. Der Mensch auf dem Weg vom animalischen zum kosmischen Bewußtsein. Bern: Scherz Verlag 1984.
8. Maturana H. Erkennen der Organisation und Verkörperung von Wirklichkeit. Wiesbaden: Vieweg Verlag 1985.

SUMMARY

By the term Ultra-High Dilutions (UHDs), within the context of this book, standardized aqueous or aqueous-alcoholic solutions are meant where a substance has been diluted through a special dilution process in such a way that the concentration ratio of solute to solvent becomes of the order of Avogadro's number, or even far below it.

The objective of this book is to investigate whether effects and characteristics of UHDs can be quantified, and how they can eventually be explained in physiological, physical, biophysical and medical terms. We preferred to introduce the reader into the principles of UHDs and the therapeutic field related to such dilutions, namely homoeopathy, by using data collected in the biological laboratory as well as by presenting elaborated theories.

PART 1 of this book (PHYSIOLOGY) at first deals with hormesis effects: dose dependent reverse effects of low and very low doses (Menachem Oberbaum, Jean Cambar).

Then, emphasize is given on exemplary promising recent physiological experiments using UHDs (Madeleine Bastide, Waltraud Pongratz, Jacques Benveniste, Christian Endler, Harald Walach, Roeland van Wijk and their colleagues).

The first of these papers sketches impressive studies in the field of laboratory immunology and discusses parallels between immunological concepts and concepts in UHD research. The follow-up contribution critically re-investigates one of the oldest laboratory models in UHD research, a botanical model that has been quoted in homoeopathic literature for decades. Then, recent studies are discussed from the laboratory that opened high-level discussion on UHDs in 1988, backing up the epoche-making biophysical conclusions drawn at that time. Further, a paper on a multicenter zoological study opens the field of investigation of homoeopathy-related biophysical assumptions a. on (non-linear) biocybernetics and b. on bioinformation transfer. A human physiological study on a homoeopathic "intoxication" protocol that comes next illustrates one of these aspects, which is related to the gain of knowledge about remedies. The finding that homoeopathic dilutions can also be effective on living systems when sealed in glass vials, a finding which is also backed up by the previously mentioned zoological study, is next discussed with regard to another human model.

PART 2 (PHYSICS) (Jürgen Schulte, Emilio del Giudice, Georgos Anagnostatos, David Auerbach and coll.) is introduced by a critical outline of historical and current experimental methods to prove differences between UHDs and unprepared solvent. This is followed by an approach from the side of cluster physics that synthesizes main ideas from this part: One that emphasizes on quantum mechanically based coherence phenomena in UHDs, apparently due to polarization of solvent dipoles, and one that emphasizes on the diversity of isotopes of the solvent that is influenced by the preparation process of UHDs and is considered to influence living systems. The comprehensive view on coherence phenomena and on isotopicity includes long-range electromagnetic fields both in the information storage in the UHD as well as in the interaction with the organism. Furthermore, information on the preparation process of homoeopathic dilutions is included. Part 2 ends with suggestions for standards in further UHD research.

P.C. Endler and J. Schulte (eds.), Ultra High Dilution, 263–264.
© 1994 *Kluwer Academic Publishers. Printed in the Netherlands.*

PART 3, (BIOPHYSICS) in the first place comprises biophysical contributions (Fritz Popp, Cyril Smith). The pioneering concept of electromagnetic or magnetic vector potential radiation from living systems is sketched with regard to an explanation of the interaction of the UHD and the organism, including an active role of the living system and a passive role of the UHD as a resonator system.

Part 3 finally emphasizes joint-venture projects (together with Cyril Smith, Massimo Citro and coll.) starting from the zoological model on UHD research mentioned in Part 1. Here it is first attempted to find out specific resonance frequencies of the UHD used in that study; then to transfer signals from this UHD's mother substance via an electronic system rather than by homoeopathical diluting; and then to stimulate the organisms with electromagnetic frequencies in order to investigate whether or not they react to the respective frequencies when generated by a laboratory oscillator. The preliminary results of these joint-ventures are promising.

PART 4 on the HISTORICAL and MEDICAL CONTEXT (Peter Andersch, Marco Righetti, Bernard Poitevin, Max Haidvogl and coll.) gives research data on UHD experiments up to 1988, when the first UHD-paper was published in "Nature" and opened the public discussion in a somewhat hasty way. Furthermore, it is pointed out in this section that homoeopathy, from its very beginnings in the 18th century, bases the knowledge concerning its remedies on empirical homoeopathic intoxication studies.

A comprehensive PROSPECTS - section discusses Preliminary Elements of a Theory on UHD.

Finally, an EPILOGUE titled "Alternative Research or Research on Alternatives?" points out that, according to David Reilly-Taylor, there is no "homoeopathic" research, but only "good" or "bad" research.

This book has been written to encourage high standard research on complementary medicine, as well as to serve as an additional standard reference on current experiments and theories in research on UHDs and homoeopathy, respectively, and to encourage interdisciplinary research. In some contributions, we have tried to show that guidepost research on living systems can also well be performed on non-invasive, non-violent biological models. Furthermore, we think that, apart from their role in UHD research, such models will be helpful in the future in order not only to describe, but to understand life and healing processes better.

INDEX

electro-conformational coupling
37f,245ff
electrodynamics 113,156,194,245ff
electromagnetic hypersensitivity 187ff
electromagnetic interaction
1,2,25,35ff,100,112,118,133,135,158,160,
175,180,187ff,209ff,231,245ff
electron-phonon coupling 113,245ff
electroreception 180
energetic coupling - experiment
19ff,39ff,187ff,245ff
eppur ricorda 118
equilibrium energy 110
excitons 175

feedback loops 3
frequency of application of UHD 63ff
frequencies, UHD-related 203ff,215ff,245ff
flow, large scale vortex 129ff
-, saddle 129ff
-, shear 129ff
fluctuation 157,163

Hahnemann 69ff,129,171,189,223
health 6,113,181,258,263f
hidden parameter 114
historical context of UHD research
69ff,99ff,219ff
homoeopathy V,1ff,219ff,221
hormesis 5ff,167,179ff,263
hormoligosis 6
Hueppe's rule 6
human studies
63,69ff,81ff,177ff,203ff,223ff
hydrogen bond 108
hypersensitivity, electromagnetic 187ff

illness 181,219ff
immunological studies
27ff,35ff,210,229ff
imprint 100,108,112,245ff
individualisation
15, 221, 247ff
information from UHDs, non-molecular
19ff,39ff,81ff,245ff
- process, homoeopathy as an 245ff

- -, hormesis as an 15,247ff
inhibition of a process by energetic field
39ff, 209ff
- - - - by low dose or UHD
5ff,32,39ff,201ff,230f
initial aggravation 221
intermolecular communication
37f,245ff
intoxication with UHDs
64,69ff,221ff,263
isotopic attractor 166
- bifurcation 150
- composition 165f
- diversity 156
- freedom 140ff
- individuality 140ff
- informational dynamics 149f
- pattern 141,148,150
- rearrangement 162

Kolisko protocol 19ff

Lamor frequency 100
LASER 37,111,193,245ff
- spectroscopy 111
Laue crystallography 191
Law of Similarity
5,65,181,189,221ff
laws of nature, homoeopathic
peculiarities and 65,219
life phenomena 245ff,263f
ligands, seperation of 245ff
linear structure 107
longevity 5ff
long range electromagnetic interaction
1,245ff,263f
- - forces 110
magnetic field, inhibition of UHD-
effect by 35ff
- preparation of remedy 187ff,203ff
magnetic vector potential
187ff,193ff,203ff,245ff
medical context 65,219ff
- therapy 1,213ff,245ff
medicine test, Voll 81ff
memory 35,99,111,117ff,137ff,181f

Similia Rule see Law of Similarity
specificity of recognition 37f
spectroscopy 102,106
spin 100,140ff,181
spin-spin coupling 100
spontaneous symmetry breaking
  142,144,148
solitons 175
standard of research 171ff,255ff
stimulation of defense response 219
- of a process by energetic field
  39ff,209ff
- - - - by low concentration or UHD
  5ff,24ff,32,39ff,231,245ff
structure
  38,100ff,105ff,121ff,141ff,179,189ff
succussion
  19,35,37,99,108,113,125f,129f,172f,179ff,
  183,187,189,199f,215, 221, 231,239,245ff
superposition 147,157f,160f,245ff
surface tension 101
symmetry breaking 183
symptoms14,65,187ff,219ff
synergism 164

TFF, pharmacological frequentative transfer
209ff,245ff
theory on UHD, elements of 245ff
therapeutic V,39,69ff,187ff,263f
thermodynamics 163,178,193
thyroxine
  39ff,203ff,209ff,215ff,245ff
toxin 5ff,32,35,65,85,167,221, 225f
transition 144,157
transfer of information from molecules
  1,37ff,105ff,187ff,203ff,245ff
tunnelling factor 144

UHD (ultra-high dilution) 221
Ultra High Dilution 221
uncertainty principle 117
UV-absorption 102
UV-spectroscopy 102

vapor 118f
vibrational frequency 148,245ff

- mode 112ff,245ff
vertex 126
veterinary medical studies 226
vials, UHD experiments with closed
  19ff,39ff,81ff,187ff,203ff,245ff
viscosity 126,131ff,134,140
Voll medicine test 81ff
vortex 129ff

water dipoles 97ff,117ff, 245ff
- layer 113,122ff
-, liquid 105ff
- molecules, cluster of polarized 245ff
-,vapor 118
wave function 153,158ff,163

X-ray Laue crystallography 191
X-ray spectroscopy 102

zoological studies 39ff,215ff,245ff,263f